Regression Analysis in Medical Research

Ton J. Cleophas • Aeilko H. Zwinderman

Regression Analysis in Medical Research

for Starters and 2nd Levelers

Second Edition

 Springer

Ton J. Cleophas
Department Medicine
Albert Schweitzer Hospital
Sliedrecht, The Netherlands

Aeilko H. Zwinderman
Department of Epidemiology and Biostatistics
Academic Medical Center
Amsterdam, Noord-Holland, The Netherlands

This work contains media enhancements, which are displayed with a "play" icon. Material in the print book can be viewed on a mobile device by downloading the Springer Nature "More Media" app available in the major app stores. The media enhancements in the online version of the work can be accessed directly by authorized users.

ISBN 978-3-030-61396-9   ISBN 978-3-030-61394-5   (eBook)
https://doi.org/10.1007/978-3-030-61394-5

This Springer imprint is published by the registered company Springer Nature Switzerland AG
The registered company address is: Gewerbestrasse 11, 6330 Cham, Switzerland

Preface to the Second Edition

In 2017, the authors completed the edition *Regression Analysis in Medical Research for Starters and 2nd Levelers*. The edition covered in a non-mathematical way both traditional and novel approaches to regression analyses already applied in published clinical research. The edition was well received by medical/health workers and students with over 4000 downloads within the first month of publication.

It came to the authors' attention that, although the relevant methods of modern regression analysis were adequately accounted, information about history and background and purposes was sometimes somewhat lacking, and this called for a second edition.

In this second edition, each methodology has been extended with a two-to-three-page section of history and background, and information of purposes for medical practice.

Also textual and conceptual errors of the first edition have been corrected, and only original tables and graphs of the statistical software programs have now been used for improved usage and understanding of the methods. In addition, summaries in the form of condensed overviews of history and presence of various categories of analysis models, and of important contributors to the development of novel analysis models are given. Furthermore, attention has been given to regressions important to statistical testing of big data including canonical regressions and the implementation of optimal scalings in sparse canonical correlation analyses. Finally, quantile regression, a novel field of regression analysis, particularly suitable for analyzing skewed data like Covid-19 data, has been included. The 2020 version of SPSS statistical software has been applied for the purpose.

Preface to the First Edition

The authors, as professors in statistics and machine learning at European universities, are worried that their students find regression analyses harder than any other methodology in statistics. This is serious, because almost all of the novel methodologies in current data mining and data analysis include elements of regression analysis. It is the main incentive for writing a 26-chapter edition, consisting of:

- Over 26 major fields of regression analysis
- Their condensed maths
- Their applications in medical and health research as published so far
- Step-by-step analyses for self-assessment
- Conclusion and reference sections

The edition is a pretty complete textbook and tutorial for medical and healthcare students, as well as a recollection/update bench, and help desk for professionals.

Novel approaches to regression analyses already applied in published clinical research will be addressed: matrix analyses, alpha spending, gate keeping, kriging, interval censored regressions, causality regressions, canonical regressions, quasi-likelihood regressions, and novel non-parametric regressions. Each chapter can be studied as a standalone and covers one of the many fields in the fast-growing world of regression analyses. Step-by-step analyses of over 40 data files stored at extras. springer.com are included for self-assessment purposes.

Traditional regression analysis is adequate for epidemiology, but lacks the precision required for clinical investigations. However, in the past two decades, modern regression methods have proven to be much more precise. And so it is time that a book described regression analyses for clinicians. The current edition is the first to do so.

Sliedrecht, The Netherlands   Ton J. Cleophas

Amsterdam, Noord-Holland,   Aeilko H. Zwinderman
The Netherlands

Contents

Chapter 1
Continuous Outcome Regressions

General Principles Regression Analysis I

Abstract The current chapter reviews the general principles of the most popular regression models in a nonmathematical fashion. First, simple and multiple linear regressions are explained as methods for making predictions about outcome variables, otherwise called dependent variables, from exposure variables, otherwise called independent variables. Second, additional purposes of regression analyses are addressed, including

1. an exploratory purpose,
2. increasing precision,
3. adjusting confounding,
4. adjusting interaction.

Particular attention has been given to common sense rationing and more intuitive explanations of the pretty complex statistical methodologies, rather than bloodless algebraic proofs of the methods.

Keywords Linear regression · Multiple linear regression · Exploratory purpose · Precision · Confounding · Interaction

1 Introduction, History, and Background

The authors as teachers in statistics at universities in France and the Netherlands, have witnessed, that students find regression analysis harder than any other methodology in statistics. Particularly, medical and health care students rapidly get lost, because of dependent data and covariances, that must be accounted all the time. The problem is that high school algebra is familiar with equations like $y = a + b\,x$, but it never addresses equations like $y = a + b_1\,x_1 + b_2\,x_2$, let alone $y = a + b_1\,x_1 + b_2\,x_2 + b_1\,x_1 . b_2\,x_2$.

In the past 30 years the theoretical basis of regression analysis has changed little, but an important step, made by the European Medicine Agency (EMA) last year, was, that the EMA has decided to include directives regarding baseline characteristics in the statistical analysis of controlled clinical trials. And regression methods have started to obtain a reason for being applied here, while a short time ago their use was limited to hypothesis-generating rather than hypothesis-testing research.

The current chapter reviews the general principles of the most popular regression models in a nonmathematical fashion. First, simple and multiple linear regression is explained. Second, the main purposes of regression analyses are addressed. Just like with previous editions of the authors (References section), particular attention has been given to common sense rationing, and more intuitive explanations of the pretty complex statistical methodologies, instead of bloodless algebraic proofs of the methods. We will start with some history and background information.

1.1 Least Squares Method

The earliest form of regression, called the least squares method, was invented around 1800, by Legendre (1752–1833), a mathematician from Paris, and Gauss (1777–1855), a mathematician from Göttingen Germany.

About Legendre's early life little is known, except, that his family was wealthy enough to have him study physics and mathematics, beginning in 1770, at the Collège Mazarin (Collège des Quatre-Nations) in Paris, and that, at least until the French Revolution, he did not have to work. Nevertheless, Legendre taught mathematics at the École Militaire in Paris from 1775 to 1780. Legendre is also famous for his work on polynomials and Legendre transformations.

Gauss was the only child of poor parents. He was a child endowed with exceptional calculating abilities, and is generally recognized as the greatest mathematician of all times. His teachers and his devoted mother recommended him to the duke of Brunswick in 1791, who granted him financial assistance to continue his education locally, and, then, to study mathematics at the University of Göttingen. Gauss is famous for his contributions to many fields in mathematics and statistics.

1.2 Extension of Regression Theories and Terminologies

The above two mathematicians may have been the founders, but their theory was soon extended by many mathematicians very famous today. Quetelet (1796–1874) from Ghent Belgium was 7 years, when his father died. After the lyceum in Ghent where he showed himself to be very talented in mathematics, he felt he had to support his family, and became a schoolteacher, and later instructor in mathematics at the lyceum of Ghent. He worked and later published with Legendre. Galton (1822–1911), psychologist, from Cambridge UK, pioneered work on correlation coefficients. Pearson (1857–1936), statistician, from London UK, invented the Pearson correlation coefficient. Charles Spearman (1863–1945), psychologist, from London UK, believed, that intelligence is composed of a general ability he called "g", underlying all intellectual functions. He observed, that people who are bright in one area tend to be bright in other areas as well. He invented factor analysis, an unsupervised learning methodology, i.e., it has no dependent variable (Chap. 23). The term "regression" was coined by Francis Galton in the nineteenth century to describe a biological phenomenon. The phenomenon was, that the heights of descendants of tall ancestors tend to regress down towards a normal average (a phenomenon also known as regression toward the mean). For Galton, regression had only this biological meaning, but his work was later extended by Udny Yule (1871–1951 from Cambridge UK) and Karl Pearson from London UK to a more general statistical context. In the work of Yule and Pearson, the joint distribution of the response and explanatory variables is assumed to be normal, nowadays called Gaussian. This assumption was weakened by R.A. Fisher in his works of 1922 and 1925. Fisher (1890–1962 from London UK) assumed, that the conditional distribution of the response variable is Gaussian, but the joint distribution needed not be. In this respect, Fisher's assumption is closer to Gauss's formulation of 1821. The work of the above icons in regression theory will be more broadly addressed in many of the next Chaps. of this edition.

1.3   More Modern Times

In the 1950s and 1960s, economists used electromechanical desk "calculators" to calculate regressions. Before 1970, it sometimes took up to 24 h to receive the result from one regression. Regression methods continue to be an area of active research. In recent decades, new methods have been developed for robust regression, regression involving correlated responses, such as time series and growth curves, regression in which the predictor (independent variable) or response variables are curves, images, graphs, or other complex data objects, regression methods accommodating various types of missing data, nonparametric regression, Bayesian methods for regression, regression in which the predictor variables are measured with error, regression with more predictor variables than observations, and causal inference with regression. The current edition will also focus on a novel type of regression entitled quantile regression, included in the latest version of SPSS statistical software version 26 and released in April 2020. It is, particularly, suitable for non-normal data, data with many outliers, and big data sets. This is fortunate, because medical and health workers are pretty much at loss, when it comes to current analytical problems with skewed Covid-19 data due to excessive age-dependent morbidities and mortalities. More information of quantile regression will be given in the Chap. 27.

2   Data Example

An example is given. One of the authors is an internist. Constipation is a typical internists' diagnosis. This is a study of the effect of a new laxative as compared to that of a standard treatment. Thirty five patients are treated in a randomized cross-over study with a novel and a standard laxative, bisacodyl. The numbers of monthly stools is the outcome.

patient no.	new treatment	bisacodyl	patient no.	new treatment	bisacodyl
	(stools)				
1	24	8			
2	30	13	19	26	10
3	25	15	20	20	8
4	35	10	21	43	16
5	39	9	22	31	15
6	30	10	23	40	14
7	27	8	24	31	7
8	14	5	25	36	12
9	39	13	26	21	6
10	42	15	27	44	19
11	41	11	28	11	5
12	38	11	29	27	8
13	39	12	30	24	9
14	37	10	31	40	15
15	47	18	32	32	7
16	30	13	33	10	6
17	36	12	34	37	14
18	12	4	35	19	7

TRIAL

The above table shows, that the new treatment performed much better than the standard treatment did with stool frequencies rising from around 10 times per month to over 30 times per month. No further statistical analysis is needed. Statistical tests are only useful for small differences. In this example the effect is like with penicilline for infectious disease. It is simply fantastic. And so, the improvement is already obvious from the data table. No need for statistical testing. The analysis thus stops here. But the data can be used for another purpose shown in the next section.

3   Data Plot

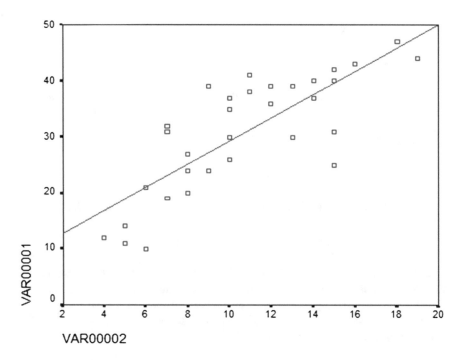

The above plot with the standard treatment results on the x-axis and the new treatment results on the y-axis shows something very interesting. There is a nice pattern in the data, the better the effect of the standard treatment, the better the effect of the new treatment. We can draw a line according to the equation y = a + bx. For every x-value this line provides the best predictable y-value in the population. The y-values measured are called the dependent variable, the x-values measured the independent variable. The b value in the equation is called the regression coefficient, in algebra, usually, called the direction coefficient, the a value is the intercept, the place where the line crosses the y-axis. The line is called the best fit line, there is no better line for describing the measured data. It has the shortest distance to all of the measured values.

4   Defining the Intercept "a" and the Regression Coefficient "b" from the Regression Equation y = a + bx

The a and b values from the equation y = a + bx can be found by the use of the least square method.

$$b = \text{direction coefficient} = \sum (x - \bar{x})(y - \bar{y}) / \sum (x - \bar{x})^2$$

Sum of products × times y/sum of squares×

SPxy/SSx

SPxy / SSx is an anova-like term and can be analyzed with anova (analysis of variance).

The enumerator of the above ratio, SPxy (sum of products x- and y-variable), is the covariance of the x- and y-variable. It is found by computing $(x - x_{mean})(y - y_{mean})$ for each patient, and then adding up the results. The denominator SSx (sum of squares x-variable) is equal to the squared standard deviation of the x-variable.

$$a = \text{intercept} = \bar{y} - b\,\bar{x}$$

Another important term in regression analysis is the R-value or r-value.

$$R = \text{correlation coefficient} = SPxy / \sqrt{(SSx.SSy)}$$

R looks a lot like b, and it is also an anova-like term that can be analyzed with analysis of variance. Why is it, just like b an important term in regression analysis? R is a measure for the strength of association between the measured x and y values. The stronger the association, the better the x predicts y. R varies between −1 and +1, −1 and 1 mean, that the association is 100%, 0 means it is 0%.

$$R^2 = SP^2xy / (SS^2 \times SS^2y) = \text{covariance xy} / (\text{variance} \times \text{times variance y})$$

R^2 varies from 0 to 1, it is nonnegative, and, therefore, often more convenient than R.

5   Correlation Coefficient (R) Varies Between −1 and + 1

R varies between −1 and + 1. The strongest association is either −1 or + 1. The weakest association is zero. The underneath graph gives an example of three crossover studies with different levels of association. Ten patients are treated with two treatments. The left study shows a very strong negative correlation: if one treatment performs well, the other does not at all. The study in the middle gives a zero correlation: the x-outcome can not predict the y-outcome. The right study shows a strong positive correlation: if one treatment performs well, the other does so too. In the left and right studies the x-data can be used as very good predictors of the y-data.

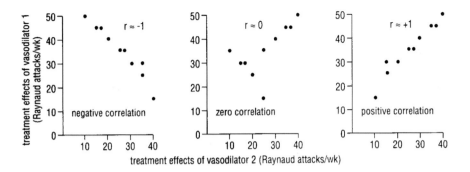

6 Computing R, Intercept "a" and Regression Coefficient "b": Ordinary Least Squares and Matrix Algebra

With big data files it is pretty laborious to compute R, "a", and "b". Regression analysis is also possible with just four outcome values, and the computations are similar but less laborious. A real data example of just 4 outcome values studies the effect of age of a magnetic resonance imager on its yearly repair costs. The relationship and best fit regression line between the age and the costs per year is in the underneath graph. We will use this small study to explain how computations must be done.

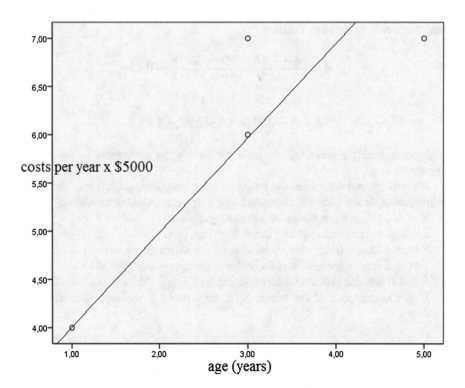

First, an ordinary least square computation will be performed.

age (years)	repair costs per year (x \$5000)			
x	y	xy	x^2	n
5	7	35	25	1
3	7	21	9	1
3	6	18	9	1
1	4	4	1	1
$\Sigma x = 12$	$\Sigma y = 24$	$\Sigma xy = 78$	$\Sigma x^2 = 44$	$\Sigma n = 4$

$Y = A + BX$ is the equation of a linear regression.

x, y, a, and b, and their upper case X, Y,....are often used pell-mell.

Y_m = mean of observed Y-values.

X_m = mean of observed X-values.

$$B = \frac{\sum (X - X_m)(Y - Y_m)}{\sum (X - X_m)^2} = \frac{\sum XY - n\,X_m Y_m}{\sum X^2 - n\,X_m^2}$$

$$A = Y_m - BX_m$$

We will now calculate B, the slope of the best fit regression line otherwise called regression or direction coefficient, and A, the intercept, the place where the best fit regression line crosses the Y-axis.

$$B = \frac{78 - 4.3.6}{44 - 4.3^2} = \frac{78 - 72}{44 - 36} = \frac{6}{8} = 0.75$$

$$A = 6 - 0.75.3 = 6 - 2.25 = 3.75$$

Second, matrix algebra will be used for finding the parameters of the best fit regression line

We will try and determine the intercept and regression coefficient of the linear regression equation with y as dependent and x as independent determinant.

$Y = A + BX$ is the equation of a linear regression.

It can be written and analyzed in the form matrices.

X is an n matrix (otherwise called column consistent of n x-values).

Y is an n matrix (otherwise called column consistent of n y-values).

A and B are also column vectors consistent of single values.

X' is the transpose of the vector X, if the column is replaced with the same value row.

Matrices can not only be transposed but also inversed, multiplied, added up.

$$X'X = 2 \times 2 \text{ matrix } X'X$$

$$X'Y = 2 \times 1 \text{ Column vector } X'Y$$

$(X'X)^{-1} =$ the inverse of the $X'X$ matrix
The $X'X$ and XY matrices can be easily calculated by hand.

$$X'X = \begin{bmatrix} 12 & 4 \\ 44 & 12 \end{bmatrix}$$

$$X'Y = \begin{bmatrix} 24 \\ 78 \end{bmatrix}$$

12	$= \Sigma x$
4	$= \Sigma n$
44	$= \Sigma x^2$
12	$= \Sigma x$
24	$= \Sigma y$
78	$= \Sigma xy$

It is, however, pretty complex to compute the inverse of the $X'X$ matrix by hand, and, therefore, the online matrix calculator "matrixcalc.org" was used. First enter the values of the $X'X$ matrix at the appropriate place, then press inverse. The result is given underneath.

$$(X'X)^{-1} = \begin{bmatrix} -3/8 & 1/8 \\ 11/8 & -3/8 \end{bmatrix}$$

Now the underneath three matrices can be written, and used for identifying the a and b values. The XY' and $(X'X)^{-1}$ must be multiplied as shown.

$$X'X = \begin{matrix} 12 & 4 \\ 44 & 12 \end{matrix}$$

$$X'Y = \begin{matrix} 24 \\ 78 \end{matrix}$$

$$(X'X)^{-1} = \begin{matrix} -3/8 & 1/8 \\ 11/8 & -3/8 \end{matrix}$$

$$(X'X)^{-1}.(X'Y) = \begin{matrix} -3/8 \times 24 & +1/8 \times 78 \\ 11/8 \times 24 & -3/8 \times 78 \end{matrix}$$

$$= \begin{matrix} -9 & +9.75 \\ 33 & -29.25 \end{matrix}$$

$$= 0.75 \quad +3.75$$

$$b = 0.75$$

$$a = 3.75$$

Or with traditional matrix brackets:

$$X'X = \begin{bmatrix} 12 & 4 \\ 44 & 12 \end{bmatrix}$$

$$X'Y = \begin{bmatrix} 24 \\ 78 \end{bmatrix}$$

$$(X'X)^{-1} = \begin{bmatrix} .3/8 & 1/8 \\ 11/8 & -3/8 \end{bmatrix}$$

$$(X'X)^{-1}.X'Y = \begin{bmatrix} -3/8 \times 24 & +1/8 \times 78 \\ 11/8 \times 24 & -3/8 \times 78 \end{bmatrix}$$

$$= \begin{bmatrix} -9 & +9.75 \\ 33 & -29.25 \end{bmatrix}$$

$$= [0.75 \quad +3.5]$$

$$b = 0.75$$

$$a = 3.75$$

The above computations, thus, give an overview of the procedure which provides the a and b values of the regression equation:

$$y = a + bx.$$

7   SPSS Statistical Software for Windows for Regression Analysis

SPSS statistical software for Windows is used for regression analysis. First enter the data, for example, from an Excel-file. The command is in the Menu.

Command:
Analyze....Regression....Linear....Dependent: stools after new laxative....Independent: stools after bisacodyl....click OK.

In the output sheets the underneath tables are given.

Model Summary

Model	R	R Square	Adjusted R Square	Std. Error of the Estimate
1	.794[a]	.630	.618	6.1590

a. Predictors: (Constant), VAR00002

ANOVA[b]

Model		Sum of Squares	df	Mean Square	F	Sig.
1	Regression	2128.393	1	2128.393	56.110	.000[a]
	Residual	1251.779	33	37.933		
	Total	3380.171	34			

a. Predictors: (Constant), VAR00002
b. Dependent Variable: VAR00001

Coefficients[a]

Model		Unstandardized Coefficients		Standardized Coefficients	t	Sig.
		B	Std. Error	Beta		
1	(Constant)	8.647	3.132		2.761	.009
	VAR00002	2.065	.276	.794	7.491	.000

a. Dependent Variable: VAR00001

We were mainly interested in the magnitudes of the b-values and the R or R square values. A lot of information is supplied, more than we asked for.

The upper table gives the R and R square values. If R square had been 0, then no correlation at all would have been in the data. If it had been 1, then correlation would have been 100%. We would be 100% certain about the y-value to be predicted from any x-value. However, the correlation was only 0.630. The level of correlation is 63%. We can be 63% certain about y knowing x. Generally R square values smaller than 0.25, means a very poor correlation, R squares between 0.25 and 0.50 means a reasonable level of correlation. A correlation level over 0.50 means a strong correlation.

In the middle is the anova table. The correlation is dependent not only on the magnitude of the R square value but also on the data sample size. Of course, with only 3 values on a line less certainty is provided than it is with 35 values like in our example. Anova (analysis of variance) is used to test, whether the r square value is significantly different from an R square value of 0. If so, then the line is not a chance finding, and it means, that it can be used for making predictions from future x-values about future y-values with 63% certainty. A typical anova-table is given. If the sum of squares regression, 2128.393, is divided by the sum of squares total, then you will find 0.630, which equals the R square value from the upper table. The sums of squares are adjusted for degrees of freedom, a measure for the sample size of the study, and a test statistic, the F value, is produced. It is very large and corresponds with a p-value < 0.0001. This means, that the correlation found is much larger than a correlation of 0.

At the bottom the coefficients table is given. In the B column the intercept and the regression coefficient are given, 8.647 and 2.065. The equation of the best fit regression line: $y = a + bx = 8.647 + 2.065$ times bisacodyl data. The b-term can just like the r square term be used for testing the level of correlation between x and y. If statistically significant, then x is a significant determinant, otherwise called independent or orthogonal determinant of y.

The above three tables are somewhat redundant. If you square the test statistic $t = 7.491$, then you will find the value 56.110, which is equal to the test statistic of the F-test of the middle table. Indeed, anova with two variables produces an F-value equal to the square root of the t-value from a t-test. Here an anova is equal to a t-test squared. Also redundant is the standardized beta coefficient of 0.794, because it is given already in the upper table as the R value of the data.

8   A Significantly Positive Correlation, X Significant Determinant of Y

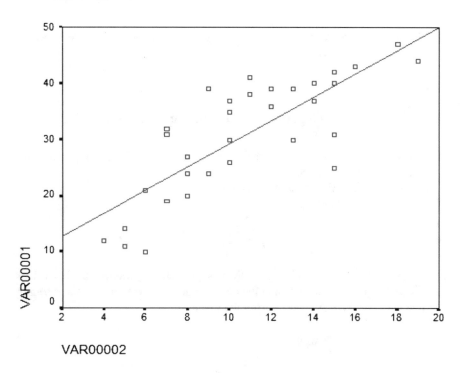

So far we have demonstrated a significantly positive correlation, x is a significant determinant of y. Maybe, also a positive correlation exists between the new laxative and the age of the patients in the study. For example, the new laxative may work better, the better the bisacodyl worked and the older the patients are. In this situation we have three observations in a single person:

1. efficacy datum of new laxative
2. efficacy datum of bisacodyl
3. age.

How do we test?
We first have to define the variables:

y variable presents the new laxative data
x_1 variable bisacodyl data
x_2 variable age data.

The regression equation for 3 variables is given

$$y = a + b_1x_1 + b_2x_2$$

The kind of equations have never been taught to you at high school. We will use here x_1 and x_2 to predict y.

9 Simple Linear Regression Uses the Equation y $=$ a+bx

With simple linear regression the equation y $=$ a+bx is used for making predictions from x about y. For example, if we fill out
x-value $= 0$, then the equation turns into y $=$ a

$$x = 1, y = a + b$$
$$x = 2, y = a + 2b$$

For each x-value the equation gives the best predictable y-value, and all y-values constitute a regression line which $=$ the best fit regression line for the data, meaning the line with the shortest distance from the y-values.

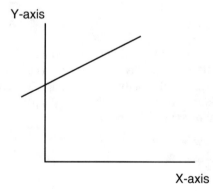

10   Multiple Regression with Three Variables Uses Another Equation

Multiple regression with 3 variables uses the equation $y = a + b_1x_1 + b_2x_2$.

The underneath orthogonal 3 axes graphical model can be used for visual modeling.

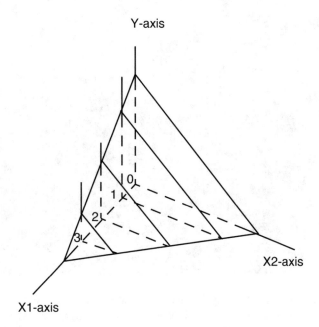

If we fill out:

$x_1 = 0$, then the equation turns into $y = a + b_2x_2$

$$x_1 = 1, a + b_1 + b_2x_2$$

$$x_1 = 2a + 2b_1 + b_2x_2$$

$$x_1 = 3a + 3b_1 + b_2x_2.$$

Each x_1 -value has its own regression line, and all of the regression lines constitute a regression plane, which is the best fit plane, i.e., the plane with the shortest distance from the x-values in a 3 dimensional space. Of course, this space does have uncertainty caused by (1) x_1- and x_2-uncertainty (explained) and by (2) residual uncertainty (unexplained).

11 Real Data Example

Real example							
patient no.	new tr y-var	bisacodyl x_1 -var	age x_2 -var	patient no.	new tr y-var	bisacodyl x_1 -var	age x_2 -var
1	24	8	25	19	26	10	27
2	30	13	30	20	20	8	20
3	25	15	25	21	43	16	35
4	35	10	31	22	31	15	29
5	39	9	36	23	40	14	32
6	30	10	33	24	31	7	30
7	27	8	22	25	36	12	40
8	14	5	18	26	21	6	31
9	39	13	14	27	44	19	41
10	42	15	30	28	11	5	26
11	41	11	36	29	27	8	24
12	38	11	30	30	24	9	30
13	39	12	27	31	40	15	20
14	37	10	38	32	32	7	31
15	47	18	40	33	10	6	29
16	30	13	31	34	37	14	43
17	36	12	25	35	19	7	30
18	12	4	24				

The 35 patient crossover study is used once more, with age (years) added as x_2 -variable. The regression equation is $y = a + b_1x_1 + b_2x_2$. In order to find the coefficients a, b_1 and b_2, just like with the simple linear regression, the least square method is used.

Essentially, three equations should be helpful to solve three unknown values a, b_1, and b_2. The underneath so-called normal equations are adequate for the purpose (n = sample size).

$$\sum y = na + b_1 \sum x_1 + b_2 \sum x_2$$

$$\sum x_1 y = a \sum x_1 + b_1 \sum x_1^2 + b_2 \sum x_1 x_2$$

$$\sum x_2 y = a \sum x_2 + b_1 \sum x_1 x_2 + b_2 \sum x_2^2$$

However, a lot of calculus is required, and we will, therefore, gratefully apply SPSS statistical software for computations.

12   SPSS Statistical Software for Windows for Regression Analysis

SPSS statistical software for Windows is used for regression analysis. First enter the data, for example from an Excel-file. Then command in the Menu:

Statistics....Regression....Linear....Dependent stools after new laxative....Independent stools after bisacodyl, age....click OK.

In the output sheets the underneath tables are given.

Model Summary

Model	R	R Square	Adjusted R Square	Std. Error of the Estimate
1	.848[a]	.719	.701	5.4498

a. Predictors: (Constant), VAR00003, VAR00002

ANOVA[b]

Model		Sum of Squares	df	Mean Square	F	Sig.
1	Regression	2429.764	2	1214.882	40.905	.000[a]
	Residual	950.407	32	29.700		
	Total	3380.171	34			

a. Predictors: (Constant), VAR00003, VAR00002

b. Dependent Variable: VAR00001

Coefficients [a]

Model		Unstandardized Coefficients		Standardized Coefficients	t	Sig.
		B	Std. Error	Beta		
1	(Constant)	-1.547	4.233		-.366	.717
	VAR00002	1.701	.269	.653	6.312	.000
	VAR00003	.426	.134	.330	3.185	.003

a. Dependent Variable: VAR00001

We were mainly interested in the magnitudes of the a- and b-values, and the R of R square values. A lot of information is supplied, more than we asked for. First, the above tables are the tables of a multiple regression rather than a simple linear regression. SPSS has named the y-variable VAR00001, the x_1 -variable VAR00002, and the x_2 -variable VAR00003.

The upper table gives the R and R square values. If R square had been 0, then no correlation at all would have been in the data. x_1 and x_2 would have determined y in no way. If it had been 1, then correlation would have been 100%. We would be 100% certain about the y-value to be predicted from any of the x_1 – and x_2 -values. However, the correlation was only $0.719 = 72\%$. The x_1 – and x_2 -values determine the y-values by 72%. The level of correlation is 72%. We can be 72% certain about y when knowing the x_1 – and x_2 -values. Generally R square values smaller than 0.25, means a very poor correlation, R squares between 0.25 and 0.50 means a reasonable level of correlation. A correlation level over 0.50 means a strong correlation. So, 72% is pretty good. As compared to the R square of the simple linear regression the level has increased from 63% to 72%. This is as expected: if you have additional predictors about y, then you will probably have greater certainty to make predictions.

In the middle is the anova table. The correlation is dependent not only on the magnitude of the R square value, but also on the data sample size. Of course, with 3 values on a line less certainty is provided than it is with 35 values like in our example. Anova calculates whether the r square value is significantly different from an R square value of 0. If so, then the line is not a chance finding, and it means that it can be used for making predictions from future x-values about future y-values with 72% certainty.

At the bottom the coefficients table is given. In the B column the intercept and the regression coefficient a are given, and 2.065. The equation of the best fit regression line: $y = a + bx = -1.547 + 1.701$ times bisacodyl data + 0.426 times age data. The b-terms can just like the r square term be used for testing the level of correlation between x-variables and y-variable. If statistically significant, then the x-variables are significant determinants, otherwise called independent or orthogonal determinants of y.

13   Summary of Multiple Regression Analysis of 3 Variables

A summary of the above multiple regression analysis with 3 variables is given.

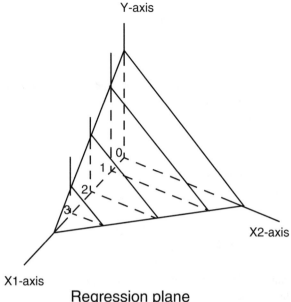

Regression plane

The regression equation of the above example is $y = -1.547 +$
$y = -1.547 + 1.701\ x_1 + 0.426\ x_2$

The overall correlation between the x-values and the y-values is $R^2 = 0.719$
($p < 0.0001$).

The independent determinants of y are x_1 and x_2.

The numbers of stools will be close to 0, if x_1 and x_2 are 0.

With every stool on bisacodyl, y will rise by 1.7

With every year of age, y will rise by 0.4.

Note: in case of >3 variables the multiple regression will get multi-dimensional,
and, visually, beyond imagination, but mathematically this is no problem at all.

14   Purposes of Multiple Linear Regression

The main purposes of multiple linear regression are four.

1. Exploratory purpose.

   When searching for significant predictors of a study outcome, here otherwise
called independent determinants of the outcome y, you may include multiple
x-variables, and, then, test, if they are statistically significant predictors of y. An
example of a linear regression model with 10 x-variables is given below.

   $y = a + b_1x_1 + b_2x_2 + \ldots\ldots\ldots b_{10}x_{10}$

   The b-values are used to test strength of correlation: b_1 to b_{10} values significantly larger or smaller than 0 means, that the corresponding x-variables are
significant predictors.

2. More precision.

    Twó x-variables give more precision for determining y than a single one does.

3. Assessing Confounding.

    In a two treatment study both treatments perform less well in a single subgroup.

4. Assessing Interaction (otherwise called heterogeneity of synergism).

    In a two treatment study one of the two treatments performs less well in a particular subgroup.

15 Multiple Regression with an Exploratory Purpose, First Purpose

In observational studies the question might be: what are the determinants of QOL (quality of life) in patients with angina pectoris?

$$
\begin{aligned}
y \text{ -data} &= \text{QOL score in patients with angina pectoris} \\
x_1 \text{ -data} &= \text{age} \\
x_2 \text{ ......} &= \text{gender} \\
x_3 \text{ ...} &= \text{rhythm disturbance} \\
x \text{ ...} &= \text{concomitant calcium antagonists} \\
x \text{ ..} &= \text{concomitant beta-blockers} \\
x \text{ ..} &= \text{NYHA (New York Heart Association) anginal class} \\
x \text{ ..} &= \text{smoking} \\
x \text{ ..} &= \text{obesity} \\
x \text{ ..} &= \text{cholesterol} \\
x \text{ ..} &= \text{hypertension} \\
x_{12} \text{ ..} &= \text{diabetes}
\end{aligned}
$$

The regression model would look like the one given underneath.

QOL-data $= a + b_1$ age-data+.......b_{12} diabetes-data

With 12 predictor variables multicollinearity must be tested. It means, that the x-variables should not correlate too strong with one another. It can be tested simply with linear regressions one by one. If the r-value is larger than 85%, then multicollinearity is in the data, and the model is not valid.

	age	gender	rhythm	vasc dis	ccb	bb	NYHA	smoking	bmi	chol	hypt
gender	0.19	1.00									
rhythm	0.12	ns	1.00								
vasc dis	0.14	ns	ns	1.00							
ccb	0.24	ns	0.07	ns	1.00						
bb	0.33	ns	ns	ns	0.07	1.00					
NYHA	0.22	ns	ns	0.07	0.07	ns	1.00				
smoking	-0.12	ns	0.09	0.07	0.08	ns	0.50	1.00			
bmi	0.13	ns	ns	ns	ns	0.10	-0.07	0.62	1.00		
chol	0.15	ns	ns	0.12	0.09	ns	0.08	0.09	ns	1.00	
hypt	0.09	ns	0.08	ns	0.10	0.09	0.09	0.09	0.07	0.41	
diabetes	0.12	ns	0.09	0.10	ns	0.08	ns	0.11	0.12	0.10	

The above table gives all of the significant R values from the example, ns means, that the R value was not significant here. Gender versus gender has a correlation coefficient of 1.

In addition to R-values, or as an alternative, often measures like tolerance or variance inflating factor are applied. Tolerance (t) = lack of certainty = (1−R square). It must be smaller than 20% in studies with 1 dependent and 3 exposure (independent) variables. Variance inflating factor (1/t) must likewise be larger than 5.

Unlike the R values, the B values are usually applied for testing the level and significance of the correlations. We will explore what number and what combination of B values produced the best fit model, the model with the best p-values. A step down method means multiple testing starting with a single B value and ending with a combined model with all of the B values. A step down procedure means the other way around. Ultimately the procedures will lead to the same result. We will use the step down method. The underneath table summarizes the B values of the predictors that are not statistically significant (No b is significantly different from 0).

x-variable	regression coëfficient (B)	standard error	test (T)	Significance level (P-value)
Age	-0.03	0.04	0.8	0.39
Gender	0.01	0.05	0.5	0.72
Rhythm disturbances	-0.04	0.04	1.0	0.28
Peripheral vascular disease	-0.00	0.01	0.1	0.97
Calcium channel blockers	0.00	0.01	0.1	0.99
beta blockers	0.03	0.04	0.7	0.43

No b is sig diff 0.

The underneath table is a model with all of the B values statistically significant, albeit not very significant. The results are not adjusted for multiple testing, but, then this is an explorative analysis, that waits for confirmation.

Covariate	regression coëfficient (B)	standard error	test stat (T)	Significance level p-value
NYHA-classification	-0.08	0.03	2.3	0.02
Smoking	-0.06	0.04	1.6	0.08
body mass index	-0.07	0.03	2.1	0.04
hypercholesterolemia	0.07	0.03	2.2	0.03
hypertension	-0.08	0.03	2.3	0.02
diabetes mellitus	0.06	0.03	2.0	0.05

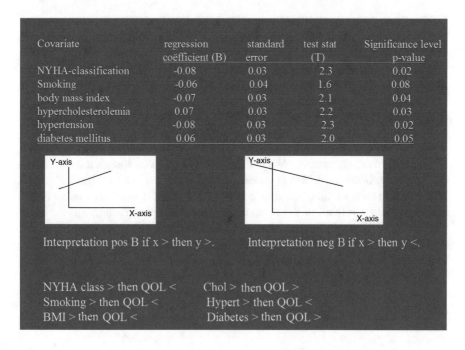

Interpretation pos B if x > then y >.   Interpretation neg B if x > then y <.

NYHA class > then QOL <   Chol > then QOL >
Smoking > then QOL <   Hypert > then QOL <
BMI > then QOL <   Diabetes > then QOL >

The test statistics (test stat) are T-values, and they are the ratios of b and its standard error. A positive B value means, that the larger the x value, the larger the corresponding y-value. A negative B value means, that the larger the x value, the smaller the corresponding y-value. The above inequalities obtained in this way allow for some interesting conclusions. The larger the anginal (NYHA) class, the smaller the quality of life (QOL). The more smoking, the less quality of life, the higher the body mass (BMI), the smaller the quality of life. It, however, also raises some puzzling results. Why should a higher cholesterol, as a major risk factor of heart disease, produce a better quality of life, and why should a more severe diabetes, another cardiac risk factor, produce a better quality of life? The answer is: linear regression does not preclude noncausal relationships, and multiple regressions are at risk of type I errors of finding differences where there is none.

With regressions it may, thus, be hard to define, what is determined by what, but, with clinical research, the search for causal factors is a principal goal, and x-values and y-values are often interpreted as respectively causal and outcome factors. With observational patient series it may be hard to define, what is determined by what, for example, a patient series may include the variables

1. Type operation
2. Type surgeon
3. Complications yes / no
4. Gender patients
5. Age patients
6. Required time for recovery

2 may determine 1, 3, and 6, but not 4 and 5, 4 and 5 maybe 1, 1 does not determine 4 and 5.

Also. for example, a patient series may include the variables

1. 2 types of anesthesia
2. Pain scores
3. Complications yes / no
4. Gender patients
5. Age patients
6. Comorbidity preoperatively
7. Qol after surgery.

1 determines 2 and maybe 3 and 7, but not 4, 5, and 6, 4, 5, and 6 may determine 1 and maybe also 2, 3, and 7. Regression can be nonsense, and still statistically significant , e.g., 1 determines 4, 5, and 6.

As an exercise of a multiple regression equation, and its interpretation, another example will be given.

Suppose in a multiple regression equation

y data = $24.4 + 5.6 x_1$ data + $6.8 x_2$ data,

y stands for body weight (pounds), x_1 for mobility(score),

and x_2 for age (years).

For each additional year of age, then, it can be expected that body weight will increase by 24.4 pounds.

1. Right   2. Wrong

The right answer is 2.

As another exercise of a multiple regression and its interpretations a subsequent example is given.

A study of independent determinants of longevity provides the results
model summary

s = standard error = 13.4 R^2 = 89.1%

Analysis of variance

	Sums of squares(SS)	df	mean square(MS)	f	sig.
Regression	7325.33	4	1831.33	10.19	0.013
Residual	898.28	5	179.66		
Total	8223.60	9			

	Coeff	St-error	t-ratio	sig.
Constant	82.237	81.738	1.01	0.361
School	-1.553	4.362	-0.36	0.736
Age	-1.685	1.253	-1.35	0.236
Psychological score	0.110	0.291	0.38	0.720
Social score	6.876	7.658	0.89	0.410

The regression equation for the given data is

a. $y = 82.2 - 1.55\,x_1 - 1.69\,x_2 + 0.11\,x_3 + 6.88\,x_4$

b. $y = 13.4 - 1.55\,x_1 - 1.69\,x_2 + 0.11\,x_3 + 6.88\,x_4$

c. $y = 81.74 - 4.36\,x_1 + 1.25\,x_2 + 0.29\,x_3 + 7.66\,x_4$

d. $y = 82.24 - 0.36\,x_1 - 1.35\,x_2 + 0.38\,x_3 + 0.90\,x_4$

The above tables are results from an observational study of predictors of longevity. The tables are similar to the SPSS tables shown in this chapter previously. The exercise question is: what is the appropriate regression equation of this study, a, b, c, or d. The answer is a. None of the predictors of longevity were statistically significant, which can be argued to imply, that an overall analysis is meaningless. However, all of the factors together produce a significant p-value of 0.013. In an exploratory fashion one could also argue, that, although none of the factors were significant, together they predict longevity with a $p = 0.013$. It would be consistent with the clinical belief of many professionals. Longevity is *not* determined by any of the predictors, but it is by their combination.

What percentage of y-values is predicted by the values from the x-variables ? Or in other words, how sure can we be about the magnitude of the corresponding y-values if we know the x-values?

a. 94 %
b. 82 %
c. 89 %
d. 13 %

The correct answer should be 89%.

(a) Is school an independent determinant of longevity?
(b) Is age an independent determinant of longevity?
(c) Is social score an independent determinant of longevity?
(d) Is longevity dependent on all of the x-variables?

The correct answer should be d.
R-square (R^2)is calculated from:

1. dividing SS reg by SS residual,
2. dividing SS reg by SS total,
3. dividing SS residual by SS total,
4. dividing MS reg by MS residual.

The correct answer should be (2).
The ... test is a statistic used to test the significance of a regression as a whole.
The correct answer should be F.
In a multiple linear regression of longevity a negative regression coefficient of determinant x indicates that

a. Longevity increases when the determinant increases,
b. Longevity decreases when the determinant increases,
c. None of these.

The correct answer should be b.

Signs of possible presence of multicollinearity in a multiple regression are

(a) Significant t values for the coefficients.
(b) Low standard errors for the coefficients.
(c) A sharp increase in a t value for the coefficient of an x-variable when another x-variable is removed from the model.
(d) All of the above.

The correct answer should be c.

16   Multiple Regression for the Purpose of Increasing Precision, Second Purpose

Multiple regression can also be applied for the purpose of increasing precision. In creasing precision means better p-values, and less uncertainty in your data. As a matter of fact the EMA (European Medicines Agency), has provided guidelines in its 2013 directives regarding the use of multiple regression in the analysis of confirmative clinical trials. Underneath the core results of a parallel group study of pravastatin versus placebo (Jukema, Circulation 1995; 91: 2528–40) is given.

Fall in ldl cholesterol after treatment (mmol/l)

patients		placebo	pravastatin	difference
n		434	438	
mean		-0.04	1.23	1.27
standard				
deviation		0.59	0.68	pooled standard error = 0.043

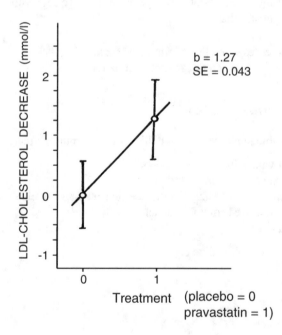

The above graph is a linear regression of treatment modalities versus ldl cholesterol decrease. The better the treatment, the better the ldl cholesterol decrease. The b value is equal to the difference between the mean values of the fall in ldl cholesterol on placebo and on pravastatin after treatment. the standard error of the b value is equal to the pooled standard error of the above mean differences. And, thus, simple linear regression is very much similar to unpaired t- tests. A major difference is, however, that, with regression analysis additional predictor variables can be included in the model, For example, add as 2nd variable the baseline ldl cholesterol values under the assumption that the higher the baseline the better the result of pravastatin treatment.

First we will draw a data plot.

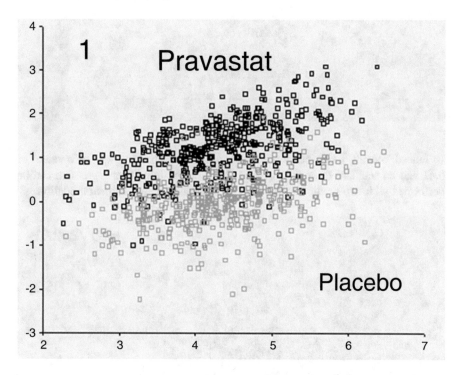

The above scatter plot with on the x-axis baseline ldl cholesterol and on the y-axis ldl cholesterol decrease may be helpful to choose the right model for analysis, a linear correlation between y and x values may be a reasonable concept from this scatter plot.

We will perform a simple regression analysis with all of the patient baseline ldl cholesterol values as x-variable, and all of the corresponding y-values as y-variable.

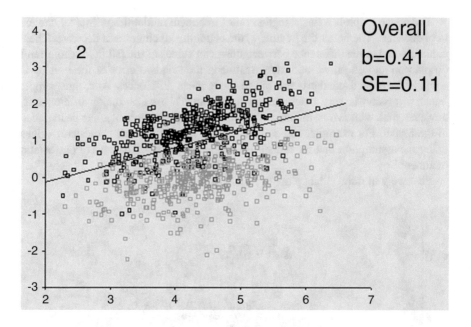

Indeed, overall, a significant positive correlation is observed with a "b value" of 0.41 and an standard error (SE) of 0.11. Next, a multiple linear regression can be performed with two predictors, baseline ldl cholesterol and treatment modality.

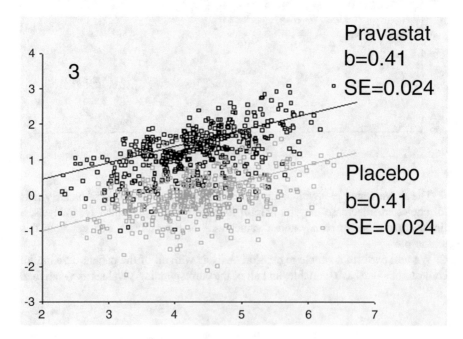

The multiple linear regression assesses, whether the data are closer to 2 parallel regression lines than will happen by chance. The underneath equation will be applied.

$$y = a + b_1 x_1 + b_2 x_2$$

y = ldl cholesterol decrease
x_1 = treatment modality (0 or 1)
x_2 = baseline ldl cholesterol

If $x_1 = 0$, then $y = a + b_2 x_2$ (equation of lower line).
If $x_1 = 1$, then $y = a + b_1 + b_2 x_2$ (equation of the upper line).

The data are closer to the two regression lines than could happen by chance (accidentally). The difference in level of the two regression lines is the difference in efficacy of placebo and pravastatin. Obviously, after correction for placebo pravastatin benefits patients with high baseline no better than patients with low baseline. However, why on earth should the two regression lines be perfectly parallel? When computing separately regression lines from the pravastatin and from the placebo data, no parallelism is observed anymore, as shown in the graph underneath. The separately calculated regression lines have significantly different directions with b values of 0.71 and 0.35. Obviously, the efficacy of pravastatin is very dependent on baseline: the higher the baseline ldl cholesterol the better the efficacy of treatment.

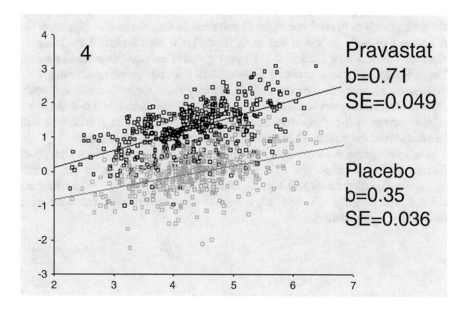

Multiple regression analyses can easily lead to erroneous conclusions. The above case is an example. The conclusion pravastatin works equally well in patients with high and in those low baseline cholesterol is not entirely correct. The problem is, that regression analyses may raise erroneous relationships, because interactions are often disregarded.

Confounding and interaction will soon be given full attention. More modern approaches to regression include standardized regression coefficients (Chap. 12), applied in DAGs (directed acyclic graphs) like principal components analysis (Chap. 23), Bayesian analyses (Chap. 12), and entropy methods working with the best cut-offs of impure internal nodes like classification and regression trees (Chap. 22). Additional methods for more unbiased regressions will be addressed in many of the next 26 chapters of the current edition.

By now we have come to realize limitations of linear regressions. A brief review will be given.

1. Risk of multicollinearity (x_1 is correlated with x_2).
2. Required homoscedasticity (every y-value has the same spread).
3. Spread is in the form of a Gaussian curves around y.
4. Linear correlation between x and y ($y=a + bx^2$ should not produce a better R than $y = a + bx$).
5. Risk of confounding (subgroups perform better).
6. Risk of interaction (subgroups perform better for 1 treatment).

We will now address the points 5 and 6 more broadly. In the underneath graphs the basic principles of confounding and interaction on the outcome are explained. On the x-axes of both graphs is treatment modality: 0 means control medicine or placebo, 1 means new medicine. The upper graph is an example of confounding. The subgroup of females performs less well than does the subgroup of males. This may have a peculiar effect on the analysis, if many males receive control, and many females receive the new treatment. The black dot of male controls and that of female new treatments will become very large, and the overall mean regression line will become almost horizontal. The treatment efficacy will, thus, be obscured.

The lower graph is an example of interaction on the outcome. The subgroup of females performs better, but only for the control medicine. Again the overall regression line of the data will become almost horizontal, and again the treatment efficacy will be obscured. And, so, a similar result is observed, but a different mechanism is responsible.

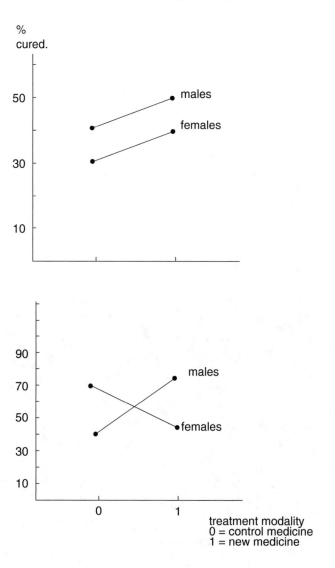

17 Multiple Regression for Adjusting Confounding, Third Purpose

Subgroups with better performance may cause confounding of parallel-group studies, if sample sizes are unequal. This problem can be adjusted several ways.

First, with a single <u>confounder</u> subclassification is possible. We will use the above example once more. The difference in efficacy-outcome can be calculated separately for the males, and, then, for the females. Subsequently, a weighted mean difference is computed with the help of the underneath weighting method

$$\frac{\text{difference}_{\text{makes}}/\text{variance}_{\text{males}} + \text{difference}_{\text{females}}/\text{variance}_{\text{females}}}{1/\text{variance}_{\text{males}} + 1/\text{variance}_{\text{females}}}$$

$$\text{variance}_{\text{males}} = \left[(\text{standard error mean}_{\text{treatment 0}})^2 + (\text{standard error mean}_{\text{treatment 1}})^2\right].$$

For the females the computation is similar. The weighted mean difference between the males and females should adjust for confounding bias.

Second, with more than a single confounder, regression-analysis with the treat modality and the confounders as x −variables is an appropriate method for adjustment.

Third, with many confounders we will rapidly have many differences in the data that will get smaller and smaller. The above method becomes powerless, and propensity scores will be needed for adjustment.

Propensity scores: for every subgroup calculate chance treat 1 vs chance treat 2 or their OR (used as relat chance).			
	treat 1	treat 2	chance treat 1/chance treat 2 or their OR
	n= 100	n = 100	p (vs OR 1)
1.Age>65	(chance)63 (%)	76	0.54 (63/37 / 76/24) 0.05
2.Age<65	37	24	1.85 (1/OR$_1$) 0.05
3.Dm	20	33	0.51 0.10
4.No DM	80	67	1.96 0.10
5.Smoker	50	80	0.25 0.10
6.No smoker	50	20	4.00 0.10
7.Hypertension	60	65	0.81 ns
8.No hypertension	40	35	1.23 ns
9.Cholesterol	75	78	0.85 ns
10.No cholesterol	25	22	1.18 ns
11.Renal failure	12	14	0.84 ns
12.No renal failure	88	86	1.31 ns

As shown above, for propensity scores calculate for every subgroup (age groups, diabetes (DM) groups, etc) the chance of having had treatment 1 versus treatment 2 or, rather, their odds ratios (ORs). Then multiply the statistically significant ORs for each patients. The multiplication terms are called propensity scores (prop scores).

Then multiply sig ORs for each patient.

	old y/n	dm y/n	smoker y/n	prop score = OR_1 x OR_2 x OR...		
Patient 1	y	y	n	0.54 x 0.51 x 4	=	1.10
2	n	n	n	1.85 x 1.96 x 4	=	14.5
3	y	n	n	0.54 x 1.96 x 4	=	3.14
4	y	y	y	0.54 x 0.51 x .025	=	0.06885
5	n	n	y			
6	y	y	y			
7	....					
8	....					

Multiplication terms have largely different sizes.

The patients are subsequently classified into subgroups according to the magnitudes of their propensity scores. An example with quartiles is given underneath. The mean treat difference per subgroup is calculated, and, then, a pooled treatment difference is computed similarly to the above subclassification procedure used with one confounder. The overall adjusted treatment difference is larger than the unadjusted treatment difference. The adjusted overall treatment result gives a better picture of the real treat effect, because it is adjusted for confounding.

Propensity scores are lovely. They can also be used different ways. For example, in a multiple regression analysis with treatment modality and propensity scores as x-variables. One or two caveats must be given. Propensity scores do not adjust for interactions. Propensity scores including irrelevant covariates reduce the power of the procedures and create spurious certainty in the data. Therefore insignificant (ns) ORs must be removed from the propensity scores.

Another relevant alternative for handling propensity scores is propensity score matching, sometimes applied with observational data. Only patients with identical magnitudes of their propensity score in the control group are maintained in the analysis. The rest is removed from the data analysis.

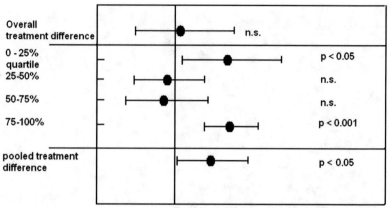

Treatment differences and their 95% confidence intervals

18   Multiple Regression for Adjusting Interaction, Fourth Purpose

With binary predictors, interaction of two x-variables on the outcome means, that one subgroup performs better on one treatment, but not on the other. With a significant interaction an overall data analysis is meaningless.

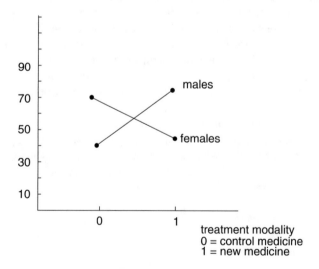

The above subgroup of females performs better, but only for the control medicine. The overall regression line of the data will become almost horizontal, and the treatment efficacy will be obscured.

	verapamil	metoprolol	
males	52	28	
	48	35	
	43	34	
	50	32	
	43	34	
	44	27	
	46	31	
	46	27	
	43	29	
	49	25	
	464	302	766
females	38	43	
	42	34	
	42	33	
	35	42	
	33	41	
	38	37	
	39	37	
	34	40	
	33	36	
	34	35	
	368	378	746
	832	680	

A 40 patient parallel group study of two treatments for number of episodes of paroxysmal atrial fibrillation (PAF) metoprolol performed overall better than verapamil with add-up counts of 680 versus 832. However, this was only true for one subgroup the males with add-up count of 464 versus 302 in the females. The interaction between gender and treatment modality on the outcome number of episodes of PAFs can be adjusted several ways.

First, the difference in counts between verapamil and metoprolol in the males is assessed. Then in the females. The mean results of the males and females are computed, and then subtracted from one another (SD = standard deviation, SE = standard error)).

	Males	Females
Mean verapamil (SD)	46.4 (3.23866)	36.8 (3.489667)
mean metoprolol (SD)	30.2 (3.48966)	37.8 (3.489667) -
Difference means (SE)	16.2 (1.50554)	-1.0 (1.5606)

Difference between males and females = 17.2 (2.166)
A t-test is performed, showing, that a very significant interaction is in the data.

$$t = 17.2 / 2.166 = 8....$$
$$p < 0.0001.$$

A t-test is performed, showing, that a very significant interaction is in the data.

$$t = 17.2/2.166 = 8....$$

$$p < 0.0001.$$

Second, instead of a t-test analysis of variance (anova) can be used.

Second method ANOVA		verapamil	metoprolol	
	Males	52	28	
		48	35	
		43	.	
		50	.	
Assess whether differences in the data due		+	+	
to interaction large comp to those that are		464	302	766
due to chance (residual).				
	Females	38	.	
		42	.	
		.	35	
		+	+	
		368 +	378+	746+
		832	680	1512

SS total = $52^2 + 48^2 + ...35^2$ - $\dfrac{(52+ 48+ +...35)^2}{40}$ = 1750.4

SS treat by gender = $\dfrac{464^2 +...378^2}{10}$ - $\dfrac{(52+ 48+ +...35)^2}{40}$ = 1327.2

SS residual = SS total − SS treat by gender = 423.2

SS rows = $\dfrac{766^2 + 746^2}{20}$ - $\dfrac{(52+ 48+ +...35)^2}{40}$ = 10.0 (= SS gender)

SS columns = $\dfrac{832^2 + 680^2}{20}$ - $\dfrac{(52+ 48+ +...35)^2}{40}$ = 577.6 (= SS treatment)

SS interaction = SS treat by gender − SS rows − SS columns = 1327.2 − 10.0 − 577.6 = 739.6

Analysis of variance assesses, whether the differences in the data due to differences between genders is large as compared to those, that are due to chance (residual
effect, SS (sum of squares) random). SS magnitudes determine whether the factors
are responsible for the differences in the data.

ANOVA-table					
	SS	dfs	MS	F	P
Rows (gender)	10.0	1	10	0.851	ns
Columns (treat)	577.6	1	577.6	49.1	<0.0001
Interaction	*739.6*	*1*	*739.6*	*62.9*	*<0.0001*
Residual	*423.2*	*36*	*11.76*		
Total					

Anova uses F-tests to test, whether a significant treatment and a significant interaction effect are in the data. With normal anova testing MS (mean square) terms are divided by MS residual. However, with large interaction effects it is better to use a random effects instead of fixed effects model. MS treatment is, then, divided by MS interaction (577.6 / 739.6 = 0.781). The resulting F value of only 0.781 corresponds with a p-value > 0.05, and the treatment effect is, thus, no longer statistically significant. This random effect analysis is, of course, more in agreement with the observed results. Interaction obscured the treatment efficacy.

Third, instead of a t-test or anova, multiple regression can be used for adjustment. The regression model consists of

$$y = \text{count of episodes of PAF}$$

$$x_1 = \text{treatment modality}$$

$$x_2 = \text{gender.}$$

Note, instead of the x-values 1 and 2 for the treatments 1 and 2, 0 and 1 must be applied, otherwise it is impossible to construct an interaction variable.

$$y = a + b_1.\text{treatment modality} + b_2.\text{gender}$$

An interaction variable, x_3, must be added

$$x_3 = x_1.x_2.$$

How to determine individ values interaction-variable			
4 possibilities	treat modal(0,1)	gender(0,1)	interaction(0,1)
1	0(no metop)	0(no female)	0x0 = 0
2	1	1	1x1 = 1
3	0	1	0x1 = 0
4	1	0	1x0 = 0

The above table shows how individual (individ) values of the interaction variable can be calculated. Only in patient no 2 interaction is equal to 1, in the patients 1, 3, and 4 it is equal to 0. A multiple regression analysis with three x-variables shows, that all of the x-variables are statistically significant predictors of y. The t-value of 7.932 is equal to the square root F-value of 62.916. The F-value from the above anova MS interaction of 62.9 is equal to the t from the regression analysis squared. The methods look different but, mathematically, they work largely the same. We should add, that statistical software programs have generally limited possibilities for random effects regressions.

Sometimes, interaction is not a disaster.

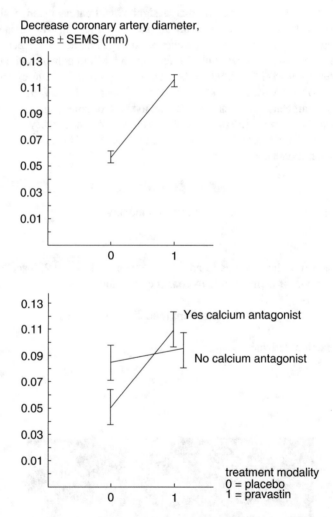

In the above pravastatin study of 884 patients an important endpoint was coronary diameter (Circulation 1995; 91: 2528–40). With calcium antagonist 0.095 mm improvement was observed, without only 0.010 mm improvement. The interaction p-value was 0.011. It was concluded, that calcium antagonists have added benefits, and a combination pill was taken into production.

19 Conclusion

Students find regression analysis harder than any other methodology in statistics. Particularly, medical and health care students rapidly get lost, because of dependent data and covariances, that must be accounted all the time. The problem is that high school algebra is familiar with equations like $y = a + b\,x$, but it never addresses equations like $y = a + b_1 x_1 + b_2 x_2$, let alone $y = a + b_1 x_1 + b_2 x_2 + b_1 x_1 \cdot b_2 x_2$. Particularly, with multiple predictor variables and with interaction variables computations will rapidly get complex, and without the help of a computer will take days instead of a few seconds.

In the past 30 years the theoretical basis of regression analysis has changed little, but an important step made by the European Medicine Agency (EMA) last year, was that the EMA has decided to include directives regarding baseline characteristics in the statistical analysis of controlled clinical trials. And regression methods have started to obtain a reason for being, while, a short time ago, their use was limited to hypothesis-generating, rather than hypothesis-testing research.

The current chapter reviews the general principles of the most popular regression models in a nonmathematical fashion. First, simple and multiple linear regression were explained. Second, the main purposes of regression analyses were addressed, including

1. an exploratory purpose,
2. increasing precision,
3. adjusting confounding,
4. adjusting interaction.

Particular attention has been given to common sense rationing, and more intuitive explanations of the pretty complex statistical methodologies, rather than bloodless algebraic proofs of the methods. We should add, that multiple regressions produce a global F-value and multiple t-values. These t-values are not usually adjusted for multiple testing, because they stem from a family of null hypotheses with many interactions and within a single experiment.

Reference

To readers requesting more background, theoretical and mathematical information of computations given, several textbooks complementary to the current production and written by the same authors are available: Statistics applied to clinical studies 5th edition, 2012, Machine learning in medicine a complete overview, 2015, SPSS for starters and 2nd levelers 2nd edition, 2015, Clinical data analysis on a pocket calculator 2nd edition, 2016, Understanding clinical data analysis, 2017, all of them edited by Springer Heidelberg Germany.

Chapter 2
Dichotomous Outcome Regressions

General Principles Regression Analysis II

Abstract Linear regression was reviewed in the Chap. 1. Plenty of regression models other than linear do exist. This chapter reviews (2) logistic regression as a model that, instead of a continuous outcome variable has a binary outcome variable, and (3) Cox regression as an exponential model where per time unit the same % of patients has an event.

Keywords Dichotomous regression · Logistic regression · Cox regression

1 Introduction, History and Background

In the Chap.1 only linear regression was reviewed, but plenty of other regression models exist. A few examples are given (ln = natural logarithm):

1 $y=a+b_1 x_1 + b_2 x_2 +\ldots\ldots\ldots b_{10} x_{10}$	linear
2 $y=a+bx+cx^2+dx^3+\ldots$	polynomial
3 $y=a + \text{sinus } x + \text{cosinus } x +\ldots$	Fourier
4 $\ln \text{ odds} =a+b_1 x_1 + b_2 x_2 +\ldots\ldots\ldots b_{10} x_{10}$	logistic
5 $\ln \text{ hazard} = a+b_1 x_1 + b_2 x_2 +\ldots\ldots\ldots b_{10} x_{10}$	Cox
6 $\ln \text{ rate} = a+b_1 x_1 + b_2 x_2 +\ldots\ldots\ldots b_{10} x_{10}$	Poisson.

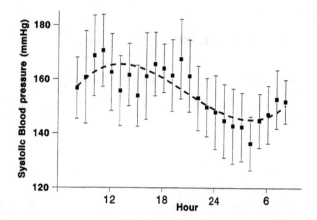

Above an example is given of a polynomial regression. The hourly means and the standard deviations of 10 mildly hypertensive patients are drawn. The best fit regression curve is given by a polynome of the 7th order.

$$y = a + bx + cx^2 + dx^3 + \ldots \ldots hx^7$$

standard deviations from the overall mean is 17 mm Hg, from the polynome it is only 7 mm Hg, and, so, the polynome provides a much better fit for these data than the overall mean does. Polynomial regression is fine for the purpose of studying circadian blood pressure rhythms. The general principles of any regression analysis can be summarized as follows.

1. Regression-analysis calculates the best fit "linear /exponential / curvilinear etc curve", which has the shortest distance from the data, and, then, it tests how far distant from the curve the data are.
2. A significant correlation between y- and x-data means that the y data are closer to the model than will happen with random sampling.
3. Statistical testing is usually done with t-tests or anova (analysis of variance).
4. The model-principle of regression analysis is, at the same time, its largest limitation, because it is often no use forcing nature into a model.

1.1 Logistic Regression

In statistics, the logistic model is used to model the probability of a certain class or event to occur, such as passing / failing to pass, winning / losing, being alive / dead or healthy / sick. This can be extended to model several classes of events such as determining whether an image contains a cat, dog, lion, etc. Like all regression analyses, the logistic regression is a predictive analysis. Logistic regression is used to describe data, and to explain the relationship between one dependent binary variable and one or more nominal, ordinal, interval or ratio-level independent variables. A detailed history of the logistic regression is given in "The origins of logistic regression", Discussion paper by Cramer from the Tinbergen Institute University of Amsterdam (2002). The logistic function was developed as a model of population growth and named "logistic" by Pierre François Verhulst (1804–1849), a Belgian student, and, later, a close colleague of Quetelet (Chap. 1) in the 1830s and 1840s. He was well aware, that the logistic model was based on exponential relationships, and lead often to impossible values, and that some kind of adjustment was required. Like Quetelet, he approached the problem of population growth by adding extra terms to the logistic model-equation. The logistic function was independently developed in chemistry as a model of autocatalysis (Wilhelm Ostwald, 1832–1932 from Riga in Latvia). An autocatalytic reaction is one in which one of the products is itself a catalyst for the same reaction, while the supply of one of the reactants is fixed. This naturally gives rise to the logistic equation for the same reason as population growth: the reaction is self-reinforcing but constrained. The logistic function was independently rediscovered as a model of population growth in 1920 by Raymond Pearl 1879–1940 and Lowell Reed 1886–1966, biologists and biochemists from Hopkins University Baltimore, published by them in Proc Nat Academ Sci (1920), which led to its use in modern statistics. They were initially unaware of Verhulst's work, and, presumably, learned about it from Gustave du Pasquier (mathematician 1876–1957 from Switzerland, but they gave him little credit, and did not adopt his terminology. Verhulst was acknowledged, and the term "logistic" revived through Udny Yule (Chap. 1) in 1925, and has been followed since. Pearl and Reed also first applied the model to the population of the United States, and, initially, fitted logistic curves by making them pass through three points; as with Verhulst, this, again, yielded pretty poorly fitted results. In the 1930s, the probit model was developed and systematized by Chester Ittner Bliss (1899–1979 from Columbia Ohio), who coined the term "probit" in his Science article (1934), and by John Gaddum (1900–1965 from Cambridge UK) in a special report series article of the Medical Research Council London UK of 1933, and the model was fitted to maximum likelihood estimation by Ronald A. Fisher in Science (1935), as addenda to Bliss's work. The probit model is much like the logistic model, but the two models have slightly different dependent variables, namely respectively log-probability of responding and log-odds of responding. The former was principally used in bioassay, and had been preceded by earlier work dating back to 1860. More information of this subject will be provided in the Chap. 18 of this edition. The logit model was initially dismissed as inferior to the probit model, but "gradually achieved an equal footing", particularly between 1960 and 1970. By 1970, the logit model achieved parity with the probit model in use in statistics journals and thereafter surpassed it. This relative popularity was due to the adoption of the logit outside of the bioassay, rather than displacing the probit within the bioassay, and its informal use in practice; the logit's popularity is credited to the logit model's computational simplicity, mathematical properties, and generality, allowing its use in varied fields. Various refinements occurred during that time, notably by papers by David Cox (Research papers in statistics, ed. by Wiley

London UK in 1966). The multinomial logit model was introduced independently by Cox in the same paper of 1966, and by Theil (in International Economic Review 1969; 10: 251), which greatly increased the scope of application and the popularity of the logit model. In 1973 Daniel McFadden, the Nobel prize laureate, linked the multinomial logit to the theory of discrete choice, showing that the multinomial logit followed from the assumption of independence of irrelevant alternatives and from the interpretation of odds of alternatives as relative preferences. This gave a theoretical foundation for the logistic regression models.

1.2   Cox Regression

David Cox, 1928–2016) was born in Birmingham UK. His father was a die sinker in a jewellery shop. He attended Handsworth Grammar School. Cox studied mathematics at St John's College, Cambridge, and obtained his PhD from the University of Leeds in 1949. He viewed survival models as consistent of two parts: first the underlying baseline hazard function, often described as the lambda, describing, how the risk of event per time unit changed over time, and, second, the effect parameters, describing, how the hazard varied in response to explanatory covariates. A typical medical example would include covariates, like treatment assignment, as well as patient characteristics, like patient age at the start of a study, gender, and the presence of other diseases at the start of a study, in order to reduce variability and /or control for confounding. The proportional hazards condition states, that covariates are multiplicatively related to the hazard. In the simplest case of stationary coefficients, for example, a treatment with a drug may, say, halve a subject's hazard at any given time, while the baseline hazard may vary. Note, however, that this does not double the lifetime of the subject; the precise effect of the covariates on the lifetime depends on the type of the covariate, and is not restricted to binary predictors; in the case of a continuous covariates it is, typically, assumed, that the hazard responds exponentially; with each unit of increase in results being proportionally scaled to the hazard. The Cox partial likelihood, is obtained by using Breslow's estimate of the baseline hazard function, plugging it into the full likelihood, whereby the result is a product of two factors. The first factor is the partial likelihood, in which the baseline hazard has "canceled out". The second factor is free of the regression coefficients, and depends on the data only through the censoring pattern. The effect of covariates estimated by any proportional hazards model can, thus, be reported as hazard ratios. Sir David Cox observed, that, if the proportional hazards assumption holds (or, is assumed to hold), it will be possible for estimating the effect parameter(s) without any consideration of the hazard function. This approach to survival data is called application of the Cox proportional hazards model, sometimes abbreviated to Cox model or to proportional hazards model. However, as Cox also noted, biological interpretation of the proportional hazards assumption can be quite tricky. Proportional hazards models are a class of survival models in statistics. Survival models relate the time that passes, before some event occurred, to one or more covariates, that may be associated with that quantity of time. Cox regression (or proportional hazards regression) is, thus, a method for investigating the effect of several variables upon the time that a specified event will happen. In the context of an outcome such as death, this is known as Cox regression for survival analysis.

2 Logistic Regression

Logistic regression is immensely popular in clinical research. Instead of a continuous outcome (dependent) variable, it has, in its simplest version, a binary outcome variable, which could be having had an event or not or any other yes / no variable. For understanding, how logistic regression works, we need to rehearse some high school logarithms.

$$^{10}\log 1000 = 3, \text{ because } 10^3 = 1000$$
$$^{10}\log 100 = 2$$
$$^{10}\log 10 = 1$$
$$^{10}\log 1 = 0, \text{ because } 10^0 = 1$$

$$^{e}\log e^3 = 3$$
$$^{e}\log e^2 = 2$$
$$^{e}\log e = 1$$
$$^{e}\log 1 = 0$$

Anti $^{10}\log 3 = 1000$
Anti $^{e}\log 3 = e^3$
Anti $^{e}\log 1 =$
Anti $^{e}\log 0 = 1, \text{ because } e^0 = 1$

Logistic regression works with odds, or, rather, log odds, or, to be more precise, ln odds, where ln = natural logarithm (naperian logarithm, the logarithm with an e-value (Euler value) to the base, e = 2.71828). As an example in a population:

$$\text{the odds of an infarction} = \frac{\text{number of patients with infarct}}{\text{number of patients without}}.$$

The odds of infarct is correlated with age.

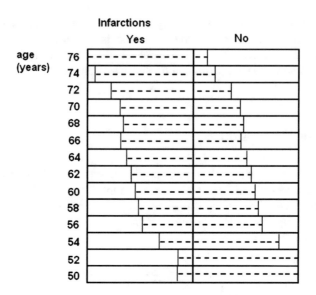

The above graph shows equally sized dotted intervals of patients at high risk of infarct in population of increasing ages. The vertical cut-off line shows that at the age of 50 few of them will have an infarction in 1 year of observation, while at the age of 76 many of them will have an infarction. And so, the odds of infarct is obviously correlated with age, but this correlation can be demonstrated not to be linear at all. The underneath graph shows, that the correlation is curvilinear, but, surprisingly, if you replace the odds values with ln odds values, then, all of a sudden, the relationships between x and y values have started to be linear. This has nothing to do with mathematical manipulations, but, rather, with a gift from nature, like other helpful gifts to statistics, like the Gaussian and other frequency distributions.

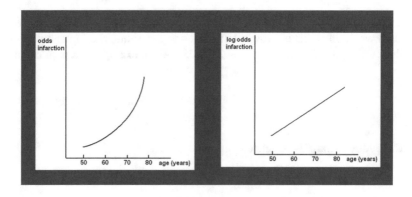

If we transform linear model.

$$y = a + b \, x$$

into a loglinear model.

ln odds $= a + b \, x$ ($x =$ age at the time of entry in the underneath study),

then the left graph changes into the right graph. This "loglinear" model can be used for making prediction about chances of infarction in patients with different ages. Take 1000 patients of different ages and follow them for 10 years. Then see who had an infarction or not. For analysis the data have to be entered in a statistical software program. We will use SPSS statistical software.

Enter the data and command:

binary logistic regression....dependent variable infarct yes / no (0 / 1)....independent variable age.

The statistical software program has used the underneath equation:

$$\text{ln odds} = \ln = \frac{\text{number of patients with infarct}}{\text{number of patients without}} = a + b \, x,$$

and it has produced the best fit a- and b-values just like with the linear regressions from the previous sections of this chapter.

$a = -9.2$.
$b = 0.1$ (standard error $= 0.04$, $p < 0.05$).

Age is, thus, a significant determinant of the odds of infarct (where odds of infarct can be interpreted as $\approx$ risk of infarct). We can now use the ln odds equation to predict the odds of infarct from a patient's age.

$$\text{Ln odds}_{55 \text{ years}} = -9.2 + 0.1.55 = -4.82265$$

Turn the ln odds value into an odds value to find the odds of infarction for a patient aged 55 years (for anti ln values on a scientific pocket calculator enter the anti ln value and click: invert (or 2ndf) button....ln button.

$$\text{odds} = \text{anti ln odds} = 0.0080 = 8/1000.$$

At the age of 75 the ln odds has changed dramatically.

$$\text{Ln odds}_{75 \text{ years}} = -9.2 + 0.1.75 = -1.3635$$

$$\text{odds} = \text{anti ln odds} = 0.256 = 256/1000.$$

The odds of infarct at 75 has risen over 30 times. The odds of an infarct can be more accurately predicted from multiple x-variables than from a single one. An example is given.

10,000 pts. are followed for 10 years, and the numbers of infarcts in those 10 years as well as the patients' baseline-characteristics are registered. The logistic regression model is given underneath:

dependent variable	infarct after 10 years "yes / no"
independent variables	gender
	age
	Bmi (body mass index)
	systolic blood pressure
	cholesterol
	heart rate
	diabetes
	antihypertensives
	previous heart infarct
	smoker.

SPSS produces b-values (predictors of infarct)

	b-values	p-value
1.Gender	0.6583	<0.05
2.Age	0.1044	"
3.BMI	-0.0405	"
4.Syst blood pressure	0.0070	
5.Cholesterol	0.0008	
6.Heart rate	0.0053	
7.Diabetes	1.2509	
8.Antihypertensives	0.3175	
9.Previous heart inf	0.8659	
10.Smoker	0.0234	
a-value	-9.1935	

The underneath regression equation is an example of a multiple logistic regression: the best fit a and b values have been computed by the computer.

Regression equation "ln odds infarct = a + b$_1$ x$_1$ + b$_2$ x$_2$ + b$_3$ x$_3$ +...." used to calculate best predictable y-value from every single combination of x-values.

-Male (x$_1$)
-55 years of age (x$_2$)
-BMI 28.7 (x$_3$)
-syst blood pressure 165 mmHg (x$_4$)
-cholesterol 6.4 mmol/l (x$_5$)
-heart rate 85 beats / min (x$_6$)
-dm (x$_7$)
-antihypertensives (x$_8$)
-previous infarct (x$_9$)
-15 cigarettes / day (x$_{10}$)

We can now apply the best fit a and b values for calculation of the odds of infarct for making predictions about persons with particular characteristics, their x-values.

Calculation odds infarct:

	b-values	x-values	
Gender	0.6583 .	1 (0 or 1) =	0.6583
Age	0.1044 .	55 =	5.742
BMI	-0.0405 .	28.7 =	..
Blood pressure	0.0070 .	165 =	
Cholesterol	0.0008 .	6.4 =	
Heart rate	0.0053 .	85 =	
Diabetes	1.2509 .	1 =	
Antihypertensives	0.3175 .	1 =	
Previous heart inf	0.8659 .	0 =	
Smoker	0.0234 .	15 =	
a-value		=	-9.1935 +

Ln odds infarct = -0.5522
odds infarct = 0.58 = 58/100

For a person with the above characteristics the odds of infarct is 0.58 (58%).

Odds is often interpreted as risk. However, the true risk is a bit smaller, and can be found by the equation.

$$\text{risk of event} = 1/(1 + 1/\text{odds})$$

If the odds of infarct = 0.58, then the true risk should be 0.37.

   The above logistic regression of longitudinal observational data is currently an important method for determining, with limited health care sources, treatment choices of patients with different chances of fatal events. and other major health decisions.

   Logistic regression models like the above one are increasingly used to decide who will be:

operated,
given expensive medications,
given the right to treatment or not,
given the do not resuscitate order,
etc.

   In addition to logistic models for risk profiling, logistic models are being used for efficacy analysis of clinical trials. An example is given. A parallel group study assesses two treatment groups for numbers of responders and non-responders (gr = group).

	responders	non-responders
new treatment (gr 1)	17 (E)	4 (F)
control treatment (gr 2)	19 (G)	28 (H)

$$\text{odds of responding} = E/F \text{ and } G/H,$$
$$\text{odds ratio (OR)} = E/F/G/H$$
$$\approx \frac{\text{chance responding gr 1}}{\text{chance responding gr 2}}$$

   No linear relation exists between treatment modality and odds responding, but a close-to-linear relation exists between treatment and logodds of responding to treatment, ln odds (natural log of odds) even better fits.

linear model: $y = a + b\,x$
transformed into: ln odds $= a + b\,x$

Ln odds $=$ dependent variable
x $=$ independent variable (treatment modality: 1 if
                                    new treatment, 0 ...)

Instead of   lnodds $= a+bx$,
describe equation as   odds $= e^{a+bx}$,
if new treatment, x=1   odds $= e^{a+b}$,
if control treatment, x= 0   odds $= e^{a}$
the ratio of two treatments
                          odds ratio $= e^{a+b} / e^{a}$
                                      $= e^{b}$,
OR $\approx$ chance of responding in group 1 / group 2
                                      $= e^{b}$.
The software calculates the best fit b for the given data,
if b $= 0$ $e^{b} = $ OR $= 1$,
if b sig > 0, then the OR significantly > 1, and a significant difference exists between
new and control treatment efficacy.

The computer computes the best fit a and b values for the data.

	coefficients	standard error	t	p
a	−1.95	0.53	.....	.....
b	1.83	0.63	2.9..	0.004

b $=$ significantly different from 0 at p $= 0.004$,
   Thus, a significant difference exists between the new and control treatment
efficacy.
   OR $= e^{b} = 2.718^{1.83} = 6.23$.
   In conclusion, the new treatment performs significantly better than the control
treatment at p 0.004. A similar result can be obtained with the traditional chi-square
test, but logistic also provides ORs: one treatment is about 6 time better than the
other. Also, logistic regression can adjust for subgroup effects. For example, the
effect of age groups.

	responders	non-responders	responders	non-responders
	\> 50 years		<50 years	
Group 1 New treatment	4	2	13	2
Group 2 Control treatment	9	16	10	12

The data are entered in, e.g., SPSS statistical software, and the computer will use the model [ln odds $= a + b_1 x_1 + b_2 x_2$].

The SPSS outcome table is given:

	values	standard error	t (z)	p-value	OR
a	−2.37	0.65			
b_1	1.83	0.67	2.7..	0.007	$e^{1.83} = 6.23$
b_2	0.83	0.50	1.6..	0.10	$e^{0.83} = 2.29$

The new treatment is 6.23 times better than the old treatment. The younger perform 2.29 times better than older patients. If you have new treatment and are young, then you will perform $6.23 \times 2.29 = 14.27$ times better than you would, being in the old treatment, and being older.

Graphs are given to explain, how the above multiple logistic model works.

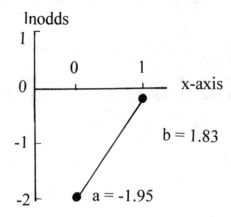

In the linear regression with log odds of responding on the y-axis and treatment modality on the x-axis, the log odds with control treatment is close to -2, with new treatment it is significantly larger.

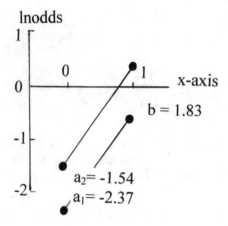

The subgroup analysis with two age groups shows two regression lines one for the older one for the younger patients ($a_2 = a_1 + b_2$). However, why should the best fit regression lines be parallel?

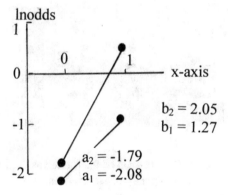

Indeed, the separately calculated regression lines are not parallel. The younger line starts higher and, in addition, runs steeper. We have to conclude that the multiple logistic model is not entirely correct. It concludes, that the younger start higher, and respond similarly to the older, while, in reality, the younger start higher, and, in addition, respond better. Multiple logistic regression is, obviously, sometimes a major simplification of the reality. It is, therefore, better suitable for explorative non-confirmative research. An example is given of logistic regression for exploratory purpose. In an observational study the determinants of endometrial cancer in postmenopausal women is searched for.

y-variabele = ln odds endometrial cancer

x_1 = estrogene consumption short term
x_2 = estrogene consumption long term
x_3 = low fertility index
x_4 = obesity
x_5 = hypertension
x_6 = early menopause

Logistic model
ln odds endometrial cancer = a + b_1 estrogene data +...............
+.......b_6 early menopause data

The data are entered in the statistical software program and the commands are: Regression Analysis....Binary Logistic Regression....Dependent: cancer yes or no (1 or 0)....Independent: 6 risk factors.....click OK.

The underneath table is in the output sheets.

risk factors	regression coefficient(b)	standard error	p-value	odds ratio (e^b)
1.estrogenes short	1.37	0.24	<0.0001	3.9
2.estrogenes long	2.60	0.25	<0.0001	13.5
3.low fertility	0.81	0.21	0.0001	2.2
4.obesity	0.50	0.25	0.04	1.6
5.hypertension	0.42	0.21	0.05	1.5
6.early menop	0.53	0.53	ns	

The odds ratios in the right column can be interpreted as the chance of cancer with / without the risk factor. For example, the chance of cancer with estrogenes shortly is 3.9, with low fertility it is 2.2. The chance of cancer with the factors 2–5 present (answers 1) are equal to

$$e^{b2+b3+b4+b5} = e^{b2}.e^{b3}.e^{b4}.e^{b5} = 75.9 \text{ times as large.}$$

However, this conclusion implying a 76 fold chance of cancer, seems clinically unrealistic! This result may be somewhat biased due to effects of interaction between the x variables on the outcome variable.

A nice thing about logarithmic transformation, because that is what we are doing, is that a complex exponential function becomes a simple linear function:

$$\text{odds of responding} = e^{b1} . e^{b2} . e^{b3}$$

$$\text{ln odds of responding} = b_1 + b_2 + b_3$$

The problems with multiple logistic regression is, that it is a major simplification of reality. The logistic model massages the data. An unadjusted odds ratio of 76 is clinically unrealistic. The model does not adjust interaction, and clinicians can tell you, that there must be interaction, for example, between obesity, hypertension, and low fertility. We should add, that, for the rest, logistic regression has the same caveats as linear regression has (risks of multicollinearity, heteroscedasticity, nongaussian curves around the outcome values, lack of linearity, etc).

Multiple Binary logistic regressions including a categorical predictor without a stepping function should not be analyzed without adjustment of the categorical variable into multiple binary variables using the command Category. An example is given.

In 60 patients of four races the effect of the race category, age, and gender on the physical strength class was tested. The effect of race, gender, and age on physical strength was assessed. A binary outcome (physical strength < or ≥ 70 points) was applied.

race	age	gender	strengthbinary
1,00	35,00	1,00	1,00
1,00	55,00	,00	1,00
1,00	70,00	1,00	,00
1,00	55,00	,00	,00
1,00	45,00	1,00	1,00
1,00	47,00	1,00	1,00
1,00	75,00	,00	,00
1,00	83,00	1,00	1,00
1,00	35,00	1,00	1,00
1,00	49,00	1,00	1,00

race 1 = hispanic, 2 = black, 3 = asian, 4 = white.
age = years of age.
gender 0 = female, 1 = male.
strength score 1 = ≥ 70 points, 0 = < 70 points.
The entire data file is in "chapter39categoricalpredictors", and is in Extras. springer.com. We will start by opening the data file in SPSS.

Command:
Analyze....Regression....Binary Logistic Regression....Dependent: strengthbinary.... Covariates: race, gender, age....click OK.

The underneath table is in the output.

Variables in the Equation

		B	S.E.	Wald	df	Sig.	Exp(B)
Step 1[a]	race	-,040	,292	,018	1	,892	,961
	age	-,047	,023	4,065	1	,044	,954
	gender	1,692	,629	7,238	1	,007	5,429
	Constant	1,899	1,689	1,265	1	,261	6,682

a. Variable(s) entered on step 1: race, age, gender.

Race is not a significant predictor of strength. Next a categorical analysis will be performed.

Command:
Analyze....Regression....Binary Logistic Regression....Dependent: strengthbinary....
Covariates: race, gender, age....click Categorical....Categorical Covariates: enter race....Reference Category: mark Last....click Continue....click OK.

Variables in the Equation

		B	S.E.	Wald	df	Sig.	Exp(B)
Step 1[a]	race			13,140	3	,004	
	race(1)	-1,423	1,066	1,782	1	,182	,241
	race(2)	1,229	1,324	,861	1	,353	3,417
	race(3)	-4,210	1,405	8,973	1	,003	,015
	age	-,043	,029	2,199	1	,138	,958
	gender	1,991	,910	4,791	1	,029	7,323
	Constant	2,527	1,636	2,385	1	,122	12,520

a. Variable(s) entered on step 1: race, age, gender.

The above table shows the results of the analysis. As compared to the whites (used as reference category),
hispanics are not significantly less strengthy (at $p = 0.182$),
blacks are not significantly more strengthy (at $p = 0.353$),
asians are, however, significantly less strengthy than whites (at $p = 0.003$).
Age is not a significant predictor of the presence of strength.
Gender is a significant predictor of the presence of strength.

3 Cox Regression

Another regression model is just like logistic regression immensely popular in clinical research. That is Cox regression. It is based on an exponential model: per time unit the same % of patients has an event (which is a pretty strong assumption for complex creatures like man. Exponential models may be adequate for mosquitos, but less so for human creatures. Yet Cox regressions are widely used for comparisons of Kaplan-Meier curves in humans.

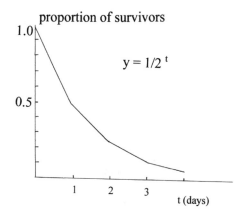

Cox uses an exponential model. Above an example is given of mosquitos in a room with concrete walls. They only die when colliding against the concrete wall. Unlike with complex creatures like humans there are no other reasons for dying. In this setting a survival half life can be pretty easily computed. In the above example, after 1 day 50% of the mosquitos are alive, after the 2nd day 25% etc. A mathematical equation for proportion survivors $= (1/2)^t = 2^{-t}$. In true biology the e value 2.71828 better fits the data than does 2, and k is applied as a constant for the species. In this way the proportion survivors can be expressed as.

$$\text{proportion survivors} = e^{-kt}$$

Underneath an example is given of Kaplan-Meier curves in humans, and their exponential best fit Cox models, the dotted lines.

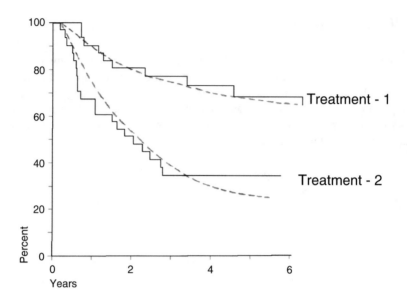

The equations of the Cox models run like:

$$\text{proportion survivors} = e^{-kt-bx}$$

x = a binary variable (only 0 or 1, 0 means treatment 1, 1 means treatment 2), b = the regression coefficient. If x = 0, then the equation turns into.

$$\text{proportion survivors} = e^{-kt}$$

If x = 1, then the equation turns into.

$$\text{proportion survivors} = e^{-kt-b}$$

Now, the relative chance of survival is given.

$$\text{relative chance of survival} = e^{-kt-b}/e^{-kt}$$
$$= e^{-b}$$

and, likewise, the relative chance of death, otherwise called the proportional hazard ratio.

$$\text{hazard ratio} = e^b.$$

Just like with the odds ratios used in the logistic regressions reviewed earlier in this chapter the hazard ratios are analyzed with natural logarithmic (ln) transformations, because the ln hazards are linear.

$$\ln \text{ hazard ratio } (HR) = b$$

The computer can calculate the best fit b for a data file given. If the b value is significantly >0, then the HR (= antiln b) is significantly >1, and a significant difference between the hazard of death is between treatment 2 and 1. The hazard is interpreted as the risk similar to the odds of an event interpreted as the risk of an event.

The data from the above graph will be used for a statistical analysis. We will apply SPSS statistical software. Enter the data file, and command:

Analyze....Survival....Cox Regression....Time: follow months....Status: event.... Define Event....Covariates: treatment....click Categorical Variables....Continue.... click Plots....Survival....Treatment....Separate Lines....click Hazard....click continue....click OK.

In the results sheets the underneath table is given:

b	standard error	t	p
1.10	0.41	2.68	0.01

The hazard ratio of treatment 2 versus treatment 1 equals $e^b = e^{1.10} = 3.00$.

This hazard ratio is significantly larger than 1 at $p = 0.01$. Thus, the treatments are significantly different from one another. For statistical testing the significance of difference between the risks of death between two treatments also a chi-square test can be applied. It would produce an even better p-value of 0.002. However, Cox regression can adjust for subgroups, like relevant prognostic factors, and chi-square tests cannot do so. For example, the hazard of death in the above example could be influenced by the disease stage, and by the presence of B-symptoms. If these data are in the data file, then the analysis can be extended according to the underneath model.

$$HR = e^{b1x1 + b2x2 + b3x3}$$

All of the x-variables are binary.

$x_1 = 0$ means treatment 1, $x_1 = 1$ means treatment 2.
$x_2 = 0$ means disease stage I-III, $x_2 = 1$ means disease stage IV.
$x_3 = 0$ means A symptoms, $x_3 = 1$ means B symptoms.

The data file is entered in the SPSS software program, and commands similar to the above ones are given.

The underneath table is in the output sheets.

Covariate	b	standard error	p-value
treatment modality(x_1)	1.10	0.45	< 0.05
disease stage(x_2)	1.38	0.55	< 0.05
symptoms(x_3)	1.74	0.69	< 0.05

HR of treatment 2 versus 1 $= e^{1.10} = 3.00$
HR of disease stage IV versus I-III $= e^{1.38} = 3.97$
HR of B-symptoms versus A-symptoms $= e^{1.74} = 5.70$
The HR adjusted for treatment, disease stage and B-symptoms $= e^{1.10+1.38+1.74}$
                                                                 $= 68.00.$

We can conclude that the treatment-2, after adjustment for disease stage IV and b-symptoms, raises a 68 higher mortality than does the treatment-1 without adjustments.

We need to account some of the problems with Cox regressions. Cox regression is a major simplification of biological processes. It is less sensitive than the chi-square test, it massages data (the above modeled Kaplan-Meier curves showed that few died within first 8 months, and that they continued to die after 2 ½ years ??!!). Also a HR of 68 seems clinically unrealistically large. Clinicians know darn well that the disease stage and the presence of B symptoms must interact with one another on the outcome. In addition, Cox regression produces exponential types of relationships only.

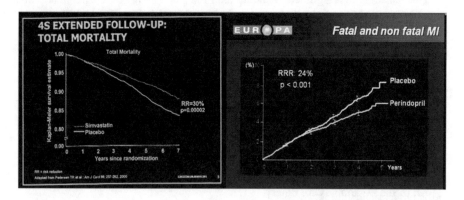

The above graphs from famous international mortality studies show that Cox regression analysis has sometimes been inadequately applied. It is inadequate, if a treatment effect only starts after 1–2 years, or if it starts immediately like with coronary interventions, or if unexpected effects start to interfere like with graph versus host effects.

4 Conclusion

Regression analysis is harder to make students understand than any other methodology in statistics. Particularly, medical and health care students rapidly get lost, because of dependent data and covariances, that must be accounted all the time. The problem is, that high school algebra is familiar with equations like $y = a + b\ x$, but it never addresses equations like $y = a + b_1\ x_1 + b_2\ x_2$, let alone $y = a + b_1\ x_1 + b_2\ x_2 + b_1\ x_1.\ b_2\ x_2$. In the past 30 years the theoretical basis of regression analysis has changed little, but the EMA (European Medicines Agency) has just decided to include directives regarding the baseline characteristics in the statistical analysis of controlled clinical trials. Regression methods have, thus, obtained a reason for existence in this field, while a short time ago their use was limited to hypothesis-generating rather than hypothesis-testing research.

The Chaps. 1 and 2 review the general principles of the most popular regression models in a nonmathematical fashion, including simple and multiple linear regression, the main purposes of regression analyses, the methods of logistic regression for event analysis, and Cox regression for hazard analysis. Just like with previous editions of the authors, e.g., "Understanding clinical data analysis, learning statistical principles from published clinical research, Springer Heidelberg Germany, 2017", particular attention has been given to common sense rationing and more intuitive explanations of the pretty complex statistical methodologies, rather than bloodless algebraic proofs of the methods.

Finally, the explorative nature of regression analysis must be emphasized. In our edition 2012 edition Statistics applied to clinical studies 5th edition we stated as follows. There is always an air of uncertainty with regression analysis. Regression in the context of clinical trials is exploratory mostly. Regression massages the data, and, thus, massages reality. It is incredible, that regression, nonetheless, took so much possession of clinical trials. Generally, you should interpret regression analysis as interesting, but an analysis that proves nothing. Today, only 5 years later, regression is getting more mature, and is increasingly accepted for the analysis of controlled clinical trials. However, limitations must be mentioned. In controlled clinical trials only a single covariate like a baseline characteristic is recommended by drug administrations like the EMA. The covariate should be well founded on prior clinical and biological arguments. Multiple testing must be accounted, e.g., with the help of gate keeping (Chap. 3). A regression analysis as main outcome procedure for a clinical trial should have a sound prior hypothesis and should be assessed in the form of null hypothesis testing (Chap. 3).

References

Five textbooks complementary to the current production and written by the same authors are
(1) Statistics applied to clinical studies 5th edition, 2012,
(2) Machine learning in medicine a complete overview, 2015,
(3) SPSS for starters and 2nd levelers 2nd edition, 2016,
(4) Clinical data analysis on a pocket calculator 2nd edition, 2016,
(5) Modern Meta-analysis, 2017,
all of them edited by Springer Heidelberg Germany.

Chapter 3
Confirmative Regressions

Baseline Characteristics and Gate Keeping Upgrade Regressions to an Adult Methodology

Abstract Traditional regression analysis is adequate for epidemiology, but lacks the precision required for clinical investigations. However, in the past two decades modern regression methods have proven to be much more precise. And so it is time, that a book described regression analyses for clinicians. The current edition is the first to do so. Three examples are given of regression analyses as primary analysis from controlled trials.

Keywords Confirmative regressions · Baseline adjustments · Gate keeping

1 Introduction, History, and Background

The first rule of all kinds of scientific research is commonly called a clearly defined prior hypothesis. It is otherwise often called the start of the scientific method. The condensed version of "the scientific method" is as follows: reformulate your scientific question into a hypothesis, and try and test this hypothesis against control observations. The scientific method is routinely used in randomized controlled trials, but, otherwise, it is not the basis of all kinds of scientific research. The scientific method is not usually applied with observational research. The scientific method is believed to be the form of scientific research that is least biased of all forms of scientific research. The daily life of clinical professionals largely consists of routine, with little need for discussion. However, there are questions, that they, simply, do not know the answer to. Some will look for the opinions of their colleagues or the experts in the field. Others will try and find a way out by guessing, what might be the best solution. The benefit of the doubt doctrine (Ordronaux, The jurisprudence of medicine in relation to the law of contracts, and evidence, Lawbook Exchange, 1869) is, often, used as a justification for unproven treatment decisions, and, if things

Electronic Supplementary Material The online version of this chapter (https://doi.org/10.1007/978-3-030-61394-5_3) contains supplementary material, which is available to authorized users. The videos can be accessed by scanning the related images with the SN More Media App.

T. J. Cleophas, A. H. Zwinderman, *Regression Analysis in Medical Research*,
https://doi.org/10.1007/978-3-030-61394-5_3

went wrong, another justification is the expression: clinical medicine is an error-ridden activity (Paget, Unity of mistakes, a phenomenological interpretation of medical work, Comtemp Sociol 1990; 19: 118–9). So far, few physicians routinely follow a different approach, the scientific method. In clinical settings, this approach is not impossible, but, rarely, applied by physicians, despite their lengthy education in evidence based medicine, which is almost entirely based on the scientific method. One thousand years ago Ibn Alhazam (965–1040) from Iraq argued about the methods of formulating hypotheses, and, subsequently, testing them. He had been influenced by Aristotle and Euclides, from Greece, 300 years BC). Ibn Alhazam on his turn influenced many of his successors, like Isaac Newton (1643–1727), at the beginning of the seventeenth century, from Cambridge UK, a mathematician, famously reluctant to publish his scientific work. His rules of the scientific method were published in a postmortem publication entitled "Study of Natural Philosophy". They are now entitled the Newton's rules of the scientific method, and listed in the Oxford English Dictionary, and today routinely used. They are defined, as a method, or, rather, a set of methods, consisting of:

1. a systematic and thorough observation, including measurements,
2. the formulation of a hypothesis regarding the observation,
3. a prospective experiment, and test of the data obtained from the experiment,
4. and, finally, a formal conclusion, and, sometimes, modification of the above hypothesis.

Clinical intervention trials of new treatments have been defined as the only form of clinical research that is evidence based, particularly so, if they are placebo-controlled and double-blinded. Other lower quality types of research are observational data, the type of data you will observe in epidemiological research. Traditional regression analysis is adequate for epidemiology, but lacks the precision required for clinical investigations. However, in the past two decades modern regression methods have proven to be much more precise, particularly regressions that use standardized regression coefficients (Chap. 12). In clinical efficacy studies the outcome is often influenced by multiple causal factors, like drug – noncompliance, frequency of counseling, and many more factors. Structural equation modeling (SEM) was only recently formally defined by Pearl (In: Causality, reason, and inference, Cambridge University Press, Cambridge UK 2000). This statistical methodology includes:

path analysis (Chap. 12),
factor analysis (see also Chap. 23), and
linear regression (see Chap. 1).

An SEM model looks like a complex regression model, but it is more. It extends the prior hypothesis of correlation to that of causality, and this is accomplished by a network of variables tested versus one another with standardized rather than unstandardized regression coefficients. The network computes the magnitudes of the standardized covariances in a multifactorial data file, their p-values of differences versus zero, and their correlation coefficients. It is also used to construct a DAG (directed acyclic graph), which is a probabilistic graphical model of nodes (the variables) and connecting arrows presenting the conditional dependencies of the nodes. Multistep path statistics, as with SEM models, brings the search for causalities one step further. If you have multiple arrows like the ones in the graphs of the Chaps. 12 and 23, and if they point directly or indirectly to the same direction, and, if your add-up regressions coefficients are increasingly large, then causal relationships will get increasingly probable.

And so, it is time, that a book described regression analyses for clinical investigators. The current edition is the first to do so. When talking of regression analysis, we are actually talking about correlations. Correlations are almost always used, when you measure at least two variables, and they are considered pretty inappropriate, when one variable is something you experimentally manipulate like the primary outcome of a controlled clinical trial. This chapter will show, that, today, we do have ample arguments for changing that traditional point of view.

2 High Performance Regression Analysis

Much of regression analysis is based on the least squares (Chap. 1), while other methods like analysis of variance and t-statistics is based on the partitions of variances. The least squares will work well, if your data have small numbers of groups with many members. The partition approach will also work well, if your data have large numbers of small groups. Indeed, epidemiological research tends to involve small numbers of large groups, while experimental research is more costly and, therefore, tends to be modest in data size. Ironically, the performance of regression analysis methods in practice may be outstanding, if it is used in interventional data. Examples will be given in the next section. We should add, that regression analysis, unlike partition methods like chi-square, logrank, t-tests, analysis of variance, can easily adjust for subgroup effects. These two benefits were the main reason for a recent update of guidelines from the European Medicines Agency's (EMA's) Committee for Medicinal Products for Human USE April 2013. New directives emphasized, that the scientific method of a prior null hypothesis be taken into account, and, that experimental data rather than observational data are needed in the field, but, at the same time, that the adjustment for one or two baseline characteristics is appropriate, although, only if noted in the study protocol prior to its scientific committee's approval. The protocol must also state eventual multiple comparisons / outcomes adjustments.

3   Example of a Multiple Linear Regression Analysis as Primary Analysis from a Controlled Trial

As a first example, the data of the REGRESS (Regression growth evaluation statin study, Jukema et al. Circulation 1995; 91: 2528–40) will be used. A summary of the study's main characteristics is given:

- patients with proven CAD (coronary artery disease),
- patients randomized between placebo (n = 434) and pravastatin (n = 438),
- main outcome low density lipoprotein (LDL) cholesterol decrease.

The primary data analysis produced the following results after 2 year treatment: an average LDL-decrease on

pravastatin:	average 1.23 mmol/l (SD (standard deviation 0.68), se (standard error) = 0.68/√438,
placebo:	average -0.04 mmol/l (SD 0.59), se = 0.59/√434
efficacy	$1.23 - - 0.04 = 1.27$ mmol/l $se_{pooled} = \sqrt{[(0.68/\sqrt{438})^2 + (0.59/\sqrt{434})^2]} = 0.043$ mmol/l.

Usually, we will apply the unpaired Student's t-test here:
t = difference of the two means / se (of this difference)
t = (1.23 - -0.04) / 0.043 = 29.53
p-value < 0.00001.

Virtually the same analysis can be performed with the help of linear regression analysis using the equation.

$$y = a + b \, x \pm se,$$

where

y = the outcome variable = the LDL cholesterol decreases
x = treatment modality   (x = 0 means placebo treatment,
                             x = 1 means pravastatin treatment)
a = intercept
b = regression coefficient
se = standard error of regression.

In SPSS statistical software the data file can be analyzed with the LDL cholesterol values as dependent variable and the treatment modalities, zero for placebo and one for pravastatin as independent variable.

Command:
Analyze....Regression....Linear....Dependent: enter LDL cholesterol values.... Independent (s): enter the treatment modalities....click OK.

In the output sheets the best a and b and e values are given.

a = −0.0376.
b = 1.270.
se = 0.043.

From these values the underneath graph can be drawn.

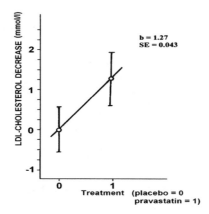

What is the advantage of the regression model? Unlike with t-tests baseline LDL cholesterol can be added to the analytical model and an adjusted analysis can be performed with better precision than the unadjusted analysis.

Command:
Analyze....Regression....Linear....Dependent: enter LDL cholesterol values.... Independent (s): enter (1) baseline LDL cholesterol, (2) the treatment modalities.... click OK.

In the output sheets two instead of one b-value is now produced.

$$b_{\text{LDL cholesterrol decrease}} = 0.41 \ (se \ 0.024)$$
$$b_{\text{treastment}} = 1.27 \ (se \ 0.037)$$

The standard error (se) of the adjusted model is smaller than the unadjusted standard error, namely 0.037 versus 0.043. This would mean that the adjusted model provides a 15% gain in efficiency, and thus, has a better precision. The analysis has some limitations as explained in the Chap. 1, but the potential of regression to take account of more than a single predictive factor remains fascinating, and its benefits has rightly been recognized by the EMA and other drug administrations worldwide.

4   Example of a Multiple Logistic Regression Analysis as Primary Analysis from a Controlled Trial

In a crossover trial of 139 general practitioners the primary scientific question was: is there a significant difference between the numbers of practitioners who give lifestyle advise with or without prior postgraduate education.

		Life style advise with education	
		no	yes
		0	1
Life style advise	no 0	65	28
without education	yes 1	12	34

The above table summarizes the numbers of practitioners giving lifestyle advise with or without postgraduate education. Obviously, before education 65 + 28 = 93 did not give lifestyle, while after education this number fell to 77. It looks as though the education was somewhat successful.

Lifestyle advise-1	lifestyle advise-2
,00	,00
,00	,00
,00	,00
,00	,00
,00	,00
,00	,00
,00	,00
,00	,00
,00	,00
,00	,00

0 = no, 1 = yes

The first 10 patients of the data file is given above. The entire data file is in extras. springer.com, and is entitled "chap3paired binary". It was previously used by the authors in SPSS for starters and 2nd levelers, Chap. 41, Springer Heidelberg Germany, 2016. Start by opening the data file in your computer with SPSS installed. A 3-D Chart of the data was drawn with the help of SPSS statistical software.

Command:
Graphs....3D Bar Chart....X-axis represents: Groups of cases....Z-axis represents: Groups of cases....Define....X Category Axis: lifestyleadvise after....Z Category Axis: lifestyleadvise before....click OK.

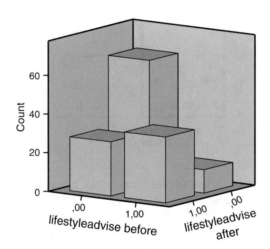

The paired observations show that twice no lifestyleadvise was given by 65 practitioners, twice yes lifestyleadvise by 34 practitioners. Furthermore, 28 practitioners started to give lifestyleadvise after postgraduate education, while, in contrast, 12 stopped giving lifestyleadvise after the education program. McNemar's test is used to statistically test the significance of difference.

Data Analysis with a traditional McNemar's test is performed. For analysis the statistical model Two Related Samples in the module Nonparametric Tests is required.

Command:
Analyze....Nonparametric....Two Related Samples....Test Pairs....Pair 1....Variable 1: enter lifestyleadvise after....Variable 2: enter lifestyleadvise before....mark McNemar....click OK.

lifestyleadvise before & lifestyleadvise after

	lifestyleadvise after	
lifestyleadvise before	,00	1,00
,00	65	28
1,00	12	34

Test Statistics[b]

	lifestyleadvise before & lifestyleadvise after
N	139
Chi-Square[a]	5,625
Asymp. Sig.	,018

a. Continuity Corrected
b. McNemar Test

The above tables show that the McNemar's test is statistically significant at a p-value of 0,018, which is a lot smaller than 0,05. The conclusion can be drawn, that a real difference between the numbers of practitioners giving lifestyle advise after and before postgraduate education is observed. The postgraduate education has, obviously, been helpful. However, the investigators have strong clinical arguments, that the practitioners' age is a concomitant determinant of their willingness to provide postgraduate education, either with or without postgraduate education, and, so, a concomitant assessment of age in the primary analysis seemed vital, and was in the protocol. In order to include age in the analysis binary logistic regression is performed.

Command in SPSS:
Analyze....Regresssion....Binary   Logistic....Dependent: lifestyle....Independent: postgraduate education, age....Click OK.

In the output we view:

Variables in the Equation

		B	S.E.	Wald	df	Sig.	Exp(B)
Step 1[a]	index1	,521	,257	4,105	1	,043	1,683
	age	-,043	,011	16,322	1	,000	,958
	Constant	1,473	,764	3,718	1	,054	4,362

a. Variable(s) entered on step 1: index1, age.

Index1 (postgraduate education) and age are significant, but index1 (postgraduate education) will not be significant anymore after adjustment for multiple testing. A little more sensitive may be the generalized equation model, that allows the predictor postgraduate education to be both ordinal and categorical.

A generalized estimation equation analysis was required. First, we will restructure the data using the Restructure Data Wizard in SPSS.

Command:
click Data....click Restructure....mark Restructure selected variables into cases.... click Next....mark One (for example, w1, w2, and w3)....click Next....Name: id (the patient id variable is already provided)....Target Variable: enter "lifestyleadvise 1, lifestyleadvise 2" ....Fixed Variable(s): enter age....click Next.... How many index variables do you want to create?....mark One....click Next....click Next again....click Next again....click Finish....Sets from the original data will still be in use...click OK.

Return to the main screen and observe, that there are, now, 278 rows instead of 139 in the data file. The first 10 rows are given underneath.

id	age	Index 1	trans 1
1	89,00	1	,00
1	89,00	2	,00
2	78,00	1	,00
2	78,00	2	,00
3	79,00	1	,00
3	79,00	2	,00
4	76,00	1	,00
4	76,00	2	,00
5	87,00	1	,00
5	87,00	2	,00

id: patient identity number
age: age in years
Index 1: 1 = before postgraduate education, 2 = after postgraduate education
trans 1: lifestyleadvise no = 1, lifestyle advise yes = 2

The above data file is adequate to perform a generalized estimation equation analysis. Save the data file. For convenience of the readers it is given in extras. springer.com, and is entitled "chap3pairedbinaryrestructured". Subsequently, a generalized estimation equation analysis will be performed with a binary logistic statistical analysis model. The commands in SPSS are given.

Command:
Analyze....Generalized Linear Models....Generalized Estimation Equations....click Repeated....transfer id to Subject variables....transfer (education) Index1 to Within-subject variables....in Structure enter Unstructured....click Type of Model....mark

Binary logistic....click Response....in Dependent Variable enter (lifestyleadvise) Trans1....click Reference Category....click Predictors....in Factors enter Index1....in Covariates enter age....click Model....in Model enter lifestyleadvise and age.... click OK.

Tests of Model Effects

	Type III		
Source	Wald Chi-Square	df	Sig.
(Intercept)	8,079	1	,004
Index1	6,585	1	,010
age	10,743	1	,001

Dependent Variable: lifestyleadvise before
Model: (Intercept), Index1, age

Parameter Estimates

Parameter	B	Std. Error	95% Wald Confidence Interval		Hypothesis Test		
			Lower	Upper	Wald Chi-Square	df	Sig.
(Intercept)	-2,508	,8017	-4,079	-,936	9,783	1	,002
[Index1=1]	,522	,2036	,123	,921	6,585	1	,010
[Index1=2]	0ª	.	.	.	.	.	.
age	,043	,0131	,017	,069	10,743	1	,001
(Scale)	1						

Dependent Variable: lifestyleadvise before
Model: (Intercept), Index1, age

a. Set to zero because this parameter is redundant.

In the output sheets the above tables are observed. They show that both the index 1 (postgraduate education) and age are significant predictors of lifestyleadvise. The interpretations of the two significant effects are slightly different from one another. The effect of postgraduate education is compared with no postgraduate education at all, while the effect of age is an independent effect of age on lifestyleadvise, the older the doctors the better lifestyle advise given irrespective of the effect of the postgraduate education.

The regression coefficients were 0.522 and 0.043 with standard errors of 0.2036 and 0.013. And the Wald chi-square test statistics with one degree of freedom were 6.585 and 10.743. This would mean that the effects on lifestyleadvise of.

1. postgraduate education and of
2. age

were, independently of one another, larger than zero, with p-values of respectively 0.010 and 0.001. Better sensitivity is provided here than it was with the traditional binary logistic model. Because we tested two null-hypothesis instead of one, the p-values had to be duplicated, 0.020 and 0.002. The p-values after this Bonferroni adjustment are still significant, but 0.02 is not, what you would call highly significant anymore. Interestingly, the adjusted p-value of 0.020 is virtually identical to the above unadjusted McNemar's p-value of 0.018.

5   Example of a Multiple Cox Regression Analysis as Primary Analysis from a Controlled Trial

The underneath graph shows two Kaplan-Meier curves. Estimating the effect on survival of treatment 1 and 2 in two parallel groups of patients with malignancies (33 and 31 patients respectively). The dotted curves present the modeled curves produced by the Cox regression model.

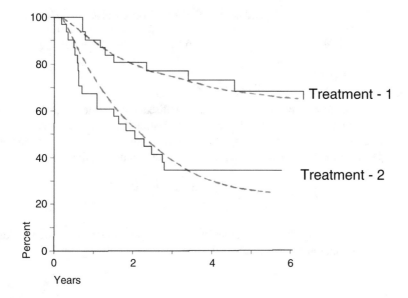

If b is significantly larger than 0, the hazard ratio will be significantly larger than 1, and there will, thus, be a significant difference between treatment-1 and treatment-2. The following results are obtained:

b = 1.1 with a standard error of 0.41
hazard ratio = 3.0
p = 0.01 (t-test)

The Cox regression provides a p-value of 0.01, and, so, it is less sensitive than the traditional summary chi-square test (p-value of 0.002). However, the Cox model has the advantage, that it enables to adjust the data for relevant prognostic factors like disease stage and presence of b-symptoms. The study protocol, indeed, provided, that not only treatment modalities, but also disease stages, and the presence of B-symptoms, being clinical relevant, were included in the primary analysis of this clinical parallel group trial, and so the simple mathematical model needed to be turned into the underneath multiple Cox regression model:

hazard ratio $= e^{b_1x_1+b_2x_2+b_3x_3}$
$x_1 = 0$ (treatment-1) ; $x_1 = 1$ (treatment-2)
$x_2 = 0$ (disease stage I-III); $x_2 = 1$ (disease stage IV)
$x_3 = 0$ (A symptoms); $x_3 = 1$ (B symptoms)

The test for multicollinearity was negative (Pearson correlation coefficient between disease stage and B symptoms <0.85), and, sensitivity analyses with assessing the hazard ratios in the subgroups were statistically insignificant, and, so, the model was deemed appropriate. SPSS produces the following result:

$b_1 = 1.10$ with a standard error of 0.41 (z = 2.68, p-value = 0.007)
$b_2 = 1.38$ "   "   "   0.55 (z = 2.51, p-value = 0.012)
$b_3 = 1.74$ "   "   "   0.69 (z = 2.52, p-value = 0.012)

unadjusted hazard ratio $= 3.0.$
adjusted hazard ratio $= e^{1.10 + 1.38 + 1.74} = 68.0.$
Treatment-2 after adjustment for advanced disease and b-symptoms raises a 68 higher mortality than treatment-1 without adjustments. However, this analysis was not adjusted for multiple testing. When three instead one null-hypothesis are tested, the rejection p-value according to Bonferroni should be adjusted:

rejection p-value $= 0.05 \times 2 / k(k-1)$ where k = 3
$= 0.0166.$

Thus, the three significant p-values unadjusted just under 0.05 were hardly significant anymore after appropriate adjustments.

6   Conclusion

In conclusion, clinical intervention trials of new treatments have been defined as the only form of clinical research that is evidence based, particularly so if they are placebo-controlled and double-blinded. Other lower quality types of research are observational data, the type of data you will observe in epidemiological research. Traditional regression analysis is adequate for epidemiology, but lacks the precision required for clinical investigations. The current chapter shows, however, that regression analyses are adequate for the analysis of prospective controlled clinical trials, and produce the same results as those of traditional variance partition methodologies. In addition, with interventional trials, where we have sound clinical arguments, that baseline characteristics of the patients are important contributors to the outcome, analysis based on multiple regressions may be the only solution. Biases due to multicollinearity and lack of sensitivity must be taken into account (see current chapter, and the Chaps. 1 and 2). A problem with multiple regressions is, that multiple null hypotheses are tested, and the chance of type I errors of finding significances that are chance rather than real findings rapidly rises. Bonferroni adjustments is a traditional remedy, but is overconservative with more than just a few null-hypotheses, leading to no significant effect at all in your data. A more adequate adjustment, then, is the current technique of gate keeping. Gate keeping means that in studies with multiple null hypotheses, the protocol takes care that the family of null hypotheses is given individual alphas, otherwise called boundaries for type I error rates, such that they add-up to 0.05, the traditional overall type I in a single null hypothesis study. Examples are given in Chap. 9, Understanding Clinical Data Analysis, pp. 177–191, Springer Heidelberg Germany, 2017, by the same authors as the current edition.

Reference

To readers requesting more background, theoretical and mathematical information of computations given, several textbooks complementary to the current production and written by the same authors are available: Statistics applied to clinical studies 5th edition, 2012, Machine learning in medicine a complete overview, 2015, SPSS for starters and 2nd levelers 2nd edition, 2015, Clinical data analysis on a pocket calculator 2nd edition, 2016, Understanding clinical data analysis, 2017, all of them edited by Springer Heidelberg Germany.

Chapter 4
Dichotomous Regressions Other than Logistic and Cox

Binary Poisson, Negative Binomial, Probit, Tetrachoric, Quasi-Likelihood Regressions

Abstract Logistic and Cox regressions can model data with binary outcomes. The current chapter reviews 5 additional models for the purpose. Binary Poisson and negative binomial regressions tend to provide better statistics. Probit regression may be more appropriate with very large and very small percentages of responders. Tetrachoric regressions assess dichotomous analyses of continuous outcomes. And quasi-likelihood regressions are adequate for repeated measures models.

Keywords Dichotomous regression · Binary Poisson regression · Negative binomial regression · Probit regression · Tetrachoric regression · Quasi-likelihood regression

1 Introduction, History, and Background

Logistic and Cox regression are immensely popular in clinical research. Instead of a continuous outcome (dependent) variable, they have, in their simplest versions, a binary outcome variable, which could be having had an event or not or any other yes / no variable, or in case of Cox hazard ratios which are much similar to the odds ratios used with logistic regressions. In the Chap. 2 examples have been given of how they work. In the current chapter five more models with binary outcomes will be described.

The first and second are binary Poisson and negative binomial models. They often provide better statistics than a traditional logistic regression does. Unlike with logistic regressions, the output is expressed as chi-square values instead of odds ratios.

The third, the probit model, is, again, from data with binary outcomes, but the results are expressed in response rates or percentages instead of odds ratios. With very large or very small percentages like the responders to different dosages of

Electronic Supplementary Material The online version of this chapter (https://doi.org/10.1007/978-3-030-61394-5_4) contains supplementary material, which is available to authorized users. The videos can be accessed by scanning the related images with the SN More Media App.

pharmacological treatments odds ratios may over- or underestimate the estimated percentages, and probit models may be more appropriate.

Fourth, in practice continuous data can often benefit from dichotomous analyses, where negative means less severe, positive more severe. Tetrachoric regressions are appropriate for the purpose.

With repeated measures binary outcomes a fifth model is adequate: the quasi likelihood model. It is for parallel group data.

We should add, that overdispersion, meaning a frequency distribution wider than compatible with Gaussian models, is a problem with dichotomous outcome data. The above negative binomial modeling is helpful, and software like SPSS has commands in its negative binomial models for adjusting overdispersions.

This chapter will review the five additional models for analyzing data with dichotomous outcomes other than logistic and Cox regressions.

1. Binary Poisson regressions (using the Poisson distribution, after Poisson 1781–1840 Paris)
2. Negative binomial regression (using the Pascal distribution (after Pascal 1623–1662 Paris) and Polya distribution (after Polya 1887–1985 Palo Alto CA)).
3. Probit regressions (using the chi-square distribution, after Bliss, 1899–1979 Ohio USA).
4. Tetrachoric regressions (using a complex model, after Galton, 1822–1911 London UK).
5. Quasi-likelihood regressions (using a complex model, Wedderburn, 1847–1975 Edinburgh UK).

Three of the above five regression models (1, 2, and 5) are in most statistical software packages given the title "generalized linear models". It may be helpful here to provide at this place a condensed overview of history and presence of this category of statistical models.

In statistics, the generalized linear model (Gldm) is a flexible generalization of ordinary linear regression, that allows for response variables that have error distribution models other than a normal distribution. The generalized linear model generalizes linear regression by allowing the linear model to be related to the response variable via a *link function,* and by allowing the magnitude of the variance of each measurement to be a function of its predicted value.

Generalized linear models were formulated by John Nelder (1924–2010 Brushfold UK) and Robert Wedderburn (1947–1975 Edinburgh UK) as a way of unifying various other statistical models, including linear regression, and Poisson regression. They proposed an iteratively reweighted least squares method for maximum likelihood estimation of the model parameters. Maximum-likelihood estimation remains popular, and is the default method on many statistical computing packages. Other approaches, including Bayesian approaches (named after reverend Bayes 1702–1761 from London UK), and least squares fitted to variance stabilized responses, have been developed.

We should emphasize that Gldm, although related with, is much different from, the general linear model (Glm). Glm mainly includes analyses of variance methods like

> Univariate analysis of variance,
> Multivariate analysis of variance,
> Repeated measures analysis,
> and Variance components analysis.

In contrast, Gldm includes many more analysis models like.

> Generalized linear models such as

> Gamma models,
> Tweedie models,
> Poisson models,
> Paired outcome models.

> and

> Generalized estimation models such as

> Paired binary outcome models.

Finally, the generalized linear model methodology is also used within the frame of mixed linear models. The mixed linear model is a statistical model containing both fixed effects and random effects, and is also called multilevel regression, and nested random data analysis. It will be addressed again in this edition several times. Particularly, the generalized linear model with link functions is appropriate for many types of statistical analyses (Machine learning in medicine-a complete overview, Chap. 20, pp 123–130, Springer Heidelberg Germany 2014, from the same authors).

2 Binary Poisson Regression

Poisson regression is the traditional method for rate regressions, but it cannot only be used for counted rates, but also for binary outcome variables. Poisson regression of binary outcome data is different from logistic regression, because it uses a log instead of logit (log odds) transformed dependent variable. It tends to provide better statistics. Can Poisson regression be used to estimate the presence of an illness. Presence means a rate of 1, absence means a rate of 0. If each patient is measured within the same period of time, no weighting variable has to be added to the model. Rates of 0 or 1 after all, do exist in practice. We will see how this approach performs as compared to the logistic regression, traditionally, used for binary outcomes. As an example, in 52 patients with parallel-groups of two different treatments the presence or not of torsades de pointes (brief runs of ventricular tachycardia) was measured. The first 10 patients of the data file is given below. The entire data file is entitled "chap4poissonbinary", and is in extras.springer.com. It was previously used by the authors in SPSS for starters and 2nd levelers, Chap. 47, Springer Heidelberg Germany, 2016. We will start by opening the data file in our computer with SPSS installed. The data from first 10 patients are underneath.

treat	presence of torsade de pointes.
,00	1,00
,00	1,00
,00	1,00
,00	1,00
,00	1,00
,00	1,00
,00	1,00
,00	1,00
,00	1,00
,00	1,00

First, we will perform a traditional binary logistic regression with torsade de pointes as outcome and treatment modality as predictor. For analysis the statistical model Binary Logistic Regression in the module Regression is required.

Command:
Analyze....Regression....Binary Logistic....Dependent: torsade....Covariates: treatment....click OK.

Variables in the Equation

		B	S.E.	Wald	df	Sig.	Exp(B)
Step 1ᵃ	VAR00001	1,224	,626	3,819	1	,051	3,400
	Constant	-,125	,354	,125	1	,724	,882

a. Variable(s) entered on step 1: VAR00001.

The above table shows, that the treatment is not statistically significant. A Poisson regression will be performed subsequently. For analysis the module Generalized Linear Models is required. It consists of two submodules: Generalized Linear Models and Generalized Estimation Models. The first submodule covers many statistical models like gamma regression (Chap. 19), Tweedie regression (Chap. 19), Poisson regression (Chaps. 14 and the current chapter), and the analysis of data files with both paired continuous outcomes and predictors (SPSS for starter and 2nd levelers 2nd edition, Chap. 3, Springer Heidelberg Germany, 2015, from the same authors). The second submodule is for analyzing paired binary outcomes and predictors (SPSS for starter and 2nd levelers, Chap. 42, Springer Heidelberg Germany, 2015, from the same authors).

Command:

Analyze....Generalized Linear Models....Generalized Linear Models ....mark Custom....Distribution: Poisson ....Link Function: Log....Response: Dependent Variable: torsade.... Predictors: Factors: treat....click Model....click Main Effect: enter "treat.....click Estimation: mark Robust Tests....click OK.

Model Information

Dependent Variable	torsade
Probability Distribution	Poisson
Link Function	Log

Goodness of Fit[a]

	Value	df	Value/df
Deviance	31,361	50	,627
Scaled Deviance	31,361	50	
Pearson Chi-Square	22,000	50	,440
Scaled Pearson Chi-Square	22,000	50	
Log Likelihood[b]	-45,681		
Akaike's Information Criterion (AIC)	95,361		
Finite Sample Corrected AIC (AICC)	95,606		
Bayesian Information Criterion (BIC)	99,264		
Consistent AIC (CAIC)	101,264		

Dependent Variable: torsade
Model: (Intercept), treat

a. Information criteria are in small-is-better form.

b. The full log likelihood function is displayed and used in computing information criteria.

Parameter Estimates

Parameter	B	Std. Error	95% Wald Confidence Interval		Hypothesis Test		
			Lower	Upper	Wald Chi-Square	df	Sig.
(Intercept)	-,288	,1291	-,541	-,035	4,966	1	,026
[VAR00001=,00]	-,470	,2282	-,917	-,023	4,241	1	,039
[VAR00001=1,00]	0ª	.	.	.	.	.	.
(Scale)	1ᵇ						

Dependent Variable: torsade
Model: (Intercept), VAR00001

a. Set to zero because this parameter is redundant.

b. Fixed at the displayed value.

The above tables shows the results of the Poisson regression. Regarding the "goodness of fit" table, any statistical model is a simplification of reality and information is lost. The goodness of fit tests give some idea, but they are mainly used for comparing one model versus the other. We will soon address this issue again. Regarding the "parameter estimates" table, the predictor "treatment modality", although insignificant in the above binary logistic model, is now statistically significant at $p = 0.039$. According to the Poisson model the treatment modality is, thus, a significant predictor of torsades de pointes. A 3-D graph will be drawn in order to better clarify the effects of treatments on torsades de pointe.

Command:

Graphs....3D Charts....x-axis treat....z-axis torsade....Define....x-axis treat....z-axis torsade....OK.

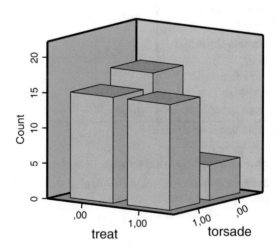

The above graph is in the output sheets. It shows that in the 0-treatment (placebo) group the number of patients with torsades de pointe is virtually equal to that of the patients without. However, in the 1-treatment group the latter number is considerably smaller. The treatment seems to be efficacious. Obviously, Poisson regression is different from linear and logistic regression, because it uses a log transformed dependent variable. For the analysis of yes / no rates Poisson regression is very sensitive and better than standard regression methods.

3 Negative Binomial Regression

The negative binomial model can be used in order to give a more accurate data model than binary Poisson can, because it allows mean and variance of samples to be different, unlike Poisson. Many events have a negative-correlated occurrence. E.g., if you observe many accidents today, you will have more chance of observing less tomorrow, and vice versa. This phenomenon causes a larger variance in the data, than, if the occurrences were entirely independent. The negative binomial model is available in SPSS' Generalized linear models, just like Poisson is. The above data example shall be analyzed once more, but now with negative binomial analysis.

Command:
Analyze....Generalized Linear Models....Generalized Linear Models ....mark Custom....Distribution: Negative Binomial ....Link Function: Log....Response: Dependent Variable: torsade.... Predictors: Factors: treat....click Model....click Main Effect: enter "treat.....click Estimation: mark Robust Tests....click OK.

Model Information

Dependent Variable	torsade
Probability Distribution	Negative binomial (1)
Link Function	Log

Goodness of Fit[a]

	Value	df	Value/df
Deviance	23,491	50	,470
Scaled Deviance	23,491	50	
Pearson Chi-Square	14,432	50	,289
Scaled Pearson Chi-Square	14,432	50	
Log Likelihood[b]	-53,334		
Akaike's Information Criterion (AIC)	110,669		
Finite Sample Corrected AIC (AICC)	110,914		
Bayesian Information Criterion (BIC)	114,571		
Consistent AIC (CAIC)	116,571		

Dependent Variable: torsade
Model: (Intercept), treat

a. Information criteria are in small-is-better form.

b. The full log likelihood function is displayed and used in computing information criteria.

Parameter Estimates

Parameter	B	Std. Error	95% Wald Confidence Interval Lower	Upper	Wald Chi-Square	df	Sig.
(Intercept)	-,288	,1291	-,541	-,035	4,966	1	,026
[treat=,00]	-,470	,2282	-,917	-,023	4,241	1	,039
[treat=1,00]	0[a]	.	.	.	.	.	.
(Scale)	1[b]						
(Negative binomial)	1[b]						

Dependent Variable: torsade
Model: (Intercept), treat

a. Set to zero because this parameter is redundant.

b. Fixed at the displayed value.

The above tables are in the output sheets. The Akaike information criterion is larger than it is in the above Poisson model. This would mean, that Poisson is a better fit model for the data than the negative binomial model is. However, this is not entirely true, because binary count data tend to suffer from overdispersion, and the negative binomial model is adjusted for overdispersion while Poisson is not.

4 Probit Regression

The underneath data are used as example. It was previously used by the authors in SPSS for starters and 2nd levelers, Chap. 49, Springer Heidelberg Germany, 2016. Fourteen populations of mosquitos are exposed to different dosages of repellents. The deaths after exposure are the outcome variable.

Variables

1	2	3	4	5	6	7
1000	18000	1	,02	17000,00	,06	-2,83
1000	18500	1	,03	17500,00	,06	-2,86
3500	19500	1	,03	16000,00	,22	-1,52
4500	18000	1	,04	13500,00	,33	-1,10
9500	16500	1	,07	7000,00	1,36	,31
17000	22500	1	,09	5500,00	3,09	1,13
20500	24000	1	,10	3500,00	5,86	1,77
500	22500	2	,02	22000,00	,02	-3,78
1500	18500	2	,03	17000,00	,09	-2,43
1000	19000	2	,03	18000,00	,06	-2,89
5000	20000	2	,04	15000,00	,33	-1,10
10000	22000	2	,07	12000,00	,83	-,18
8000	16500	2	,09	85000,00	,09	-2,36
13500	18500	2	,10	5000,00	2,70	,99

variable 1 death mosquitos
2 mosquito population
3 presence of repellent nonchemical (burning candles)
4 dilution of repellent chemical
5 nondeath mosquitos
6 odds of death mosquitos (variable 1 / (variable 2 minus variable 1)
7 logodds of death mosquitos

First, in SPSS statistical software a traditional binary logistic regression will be performed. Start by opening the data file entitled "chap4probit" in your computer. The term logodds = logarithm of odds = natural logarithm (ln) of odds.

Command:
Analyze....Regression....Linear Regression....Dependent: logodds of death mosquitos (variable 7)....Independent: enter dilution of repellent chemical (variable 4)....click OK.

The underneath tables are in the output sheets.

Model Summary

Model	R	R Square	Adjusted R Square	Std. Error of the Estimate
1	,831[a]	,690	,664	1,02037

a. Predictors: (Constant), chemical (dilution)

ANOVA[a]

Model		Sum of Squares	df	Mean Square	F	Sig.
1	Regression	27,796	1	27,796	26,697	,000[b]
	Residual	12,494	12	1,041		
	Total	40,290	13			

a. Dependent Variable: lnodds

b. Predictors: (Constant), chemical (dilution)

Coefficients[a]

Model		Unstandardized Coefficients		Standardized Coefficients	t	Sig.
		B	Std. Error	Beta		
1	(Constant)	-3,839	,578		-6,640	,000
	chemical (dilution)	48,689	9,423	,831	5,167	,000

a. Dependent Variable: lnodds

The dilution chemical repellent is a very significant predictor of the logodds of death mosquitos at $p = 0.000$. The regression equation of the binary logistic model is obtained from the coefficients table (burning candles are not assessed):

logodds of death $= -3.839 + 48.689$.(estimate of chemical dilution).

With a dilution of 0.066 the logodds of death $= -0.6255$.

The odds of death $=$ antilog of logodds of death $= 0.53498$.

The percentage of deaths $= 1 / [1 + (1/\text{odds})] = 1/ (1 + 1.87) = 0.35 = 35\%$.

With a dilution of 0.143 the logodds of death $= 3.1235$.

The odds of death $=$ antilog of logodds of death $= 22.726$.

The percentage of deaths $= 1 / [1 + (1/\text{odds})] = 1/ (1 + 1/22.726)) = 1/ 1.044 = 0.96 = 96\%$.

Subsequently, a probit regression for estimating the effect of predictors on yes/no outcomes is performed. If your predictor is multiple pharmacological treatment dosages, then probit regression may be more convenient than logistic regression, because your results will be reported in the form of response rates instead of odds ratios. The dependent variable of the two methods, log odds (otherwise called logit) and log prob (otherwise called probit), are closely related to one another. It can be shown that the log odds of responding $\approx (\pi / \sqrt{3})$ x log probability of responding (see Chap.7, Machine learning in medicine part three, Probit regression, pp 63–68, 2013, Springer Heidelberg Germany, from the same authors). For analysis, in SPSS statistical software the model Probit Regression in the module Regression is required.

Command:
Analyze....Regression....Probit Regression....Response Frequency: enter "mosquitos gone"....Total Observed: enter "n mosquitos"....Covariate(s): enter "chemical".... Transform: select "natural log"....click OK.

Chi-Square Tests

		Chi-Square	df[a]	Sig.
PROBIT	Pearson Goodness-of-Fit Test	7706,816	12	,000[b]

a. Statistics based on individual cases differ from statistics based on aggregated cases.
b. Since the significance level is less than ,150, a heterogeneity factor is used in the calculation of confidence limits.

In the output sheets the above table shows, that the goodness of fit tests of the data is significant, and, thus, the data do not fit the probit model very well. However, SPSS is going to produce a heterogeneity correction factor, and we can proceed. The underneath shows that chemical dilution levels are a very significant predictor of proportions of mosquitos gone.

Parameter Estimates

	Parameter	Estimate	Std. Error	Z	Sig.	95% Confidence Interval	
						Lower Bound	Upper Bound
PROBIT[a]	chemical (dilution)	1,649	,006	286,098	,000	1,638	1,660
	Intercept	4,489	,017	267,094	,000	4,472	4,506

a. PROBIT model: PROBIT(p) = Intercept + BX (Covariates X are transformed using the base 2.718 logarithm.)

Cell Counts and Residuals

	Number	chemical (dilution)	Number of Subjects	Observed Responses	Expected Responses	Residual	Probability
PROBIT	1	-3,912	18000	1000	448,194	551,806	,025
	2	-3,624	18500	1000	1266,672	-266,672	,068
	3	-3,401	19500	3500	2564,259	935,741	,132
	4	-3,124	18000	4500	4574,575	-74,575	,254
	5	-2,708	16500	9500	8405,866	1094,134	,509
	6	-2,430	22500	17000	15410,676	1589,324	,685
	7	-2,303	24000	20500	18134,992	2365,008	,756
	8	-3,912	22500	500	560,243	-60,243	,025
	9	-3,624	18500	1500	1266,672	233,328	,068
	10	-3,401	19000	1000	2498,508	-1498,508	,132
	11	-3,124	20000	5000	5082,861	-82,861	,254
	12	-2,708	22000	10000	11207,821	-1207,821	,509
	13	-2,430	16500	8000	11301,162	-3301,162	,685
	14	-2,303	18500	13500	13979,056	-479,056	,756

The above table shows, that according to chi-square tests the differences between observed and expected proportions of mosquitos gone is several times statistically significant.

It does, therefore, make sense to make some inferences using the underneath confidence limits table, also given in the output sheets.

Confidence Limits

		95% Confidence Limits for chemical (dilution)			95% Confidence Limits for log(chemical (dilution))[b]		
	Probability	Estimate	Lower Bound	Upper Bound	Estimate	Lower Bound	Upper Bound
PROBIT[a]	,010	,016	,012	,020	-4,133	-4,453	-3,911
	,020	,019	,014	,023	-3,968	-4,250	-3,770
	,030	,021	,016	,025	-3,863	-4,122	-3,680
	,040	,023	,018	,027	-3,784	-4,026	-3,612
	,050	,024	,019	,029	-3,720	-3,949	-3,557
	,060	,026	,021	,030	-3,665	-3,882	-3,509
	,070	,027	,022	,031	-3,617	-3,825	-3,468
	,080	,028	,023	,032	-3,574	-3,773	-3,430
	,090	,029	,024	,034	-3,535	-3,726	-3,396
	,100	,030	,025	,035	-3,500	-3,683	-3,365
	,150	,035	,030	,039	-3,351	-3,506	-3,232
	,200	,039	,034	,044	-3,233	-3,368	-3,125
	,250	,044	,039	,048	-3,131	-3,252	-3,031
	,300	,048	,043	,053	-3,040	-3,150	-2,943
	,350	,052	,047	,057	-2,956	-3,059	-2,860
	,400	,056	,051	,062	-2,876	-2,974	-2,778
	,450	,061	,055	,067	-2,799	-2,895	-2,697
	,500	,066	,060	,073	-2,722	-2,819	-2,614
	,550	,071	,064	,080	-2,646	-2,745	-2,529
	,600	,077	,069	,087	-2,569	-2,672	-2,442
	,650	,083	,074	,095	-2,489	-2,598	-2,349
	,700	,090	,080	,105	-2,404	-2,522	-2,251
	,750	,099	,087	,117	-2,313	-2,441	-2,143
	,800	,109	,095	,132	-2,212	-2,351	-2,022
	,850	,123	,106	,153	-2,094	-2,248	-1,879
	,900	,143	,120	,183	-1,945	-2,120	-1,699
	,910	,148	,124	,191	-1,909	-2,089	-1,655
	,920	,154	,128	,200	-1,870	-2,055	-1,608
	,930	,161	,133	,211	-1,827	-2,018	-1,556
	,940	,169	,138	,224	-1,780	-1,977	-1,497
	,950	,178	,145	,239	-1,725	-1,931	-1,430
	,960	,190	,153	,259	-1,661	-1,876	-1,352
	,970	,206	,164	,285	-1,582	-1,809	-1,255
	,980	,228	,179	,324	-1,477	-1,719	-1,126
	,990	,269	,206	,397	-1,312	-1,579	-,923

a. A heterogeneity factor is used.
b. Logarithm base = 2.718.

E.g., one might conclude that a 0,143 dilution of the chemical repellent causes 0,900 (=90%) of the mosquitos to have gone. And 0,066 dilution would mean that 0,500 (=50%) of the mosquitos disappeared.

The probit analysis result is somewhat different from that of the logistic model. What analysis is more realistic. The probit produces percentages, while the logistic model produces odds. An odds of 0.5394 (54%) looks much like a percentage of deaths of about 50%, but with large percentages the odds underestimates the percentage, here by 15%. In contrast, with small percentages it overestimates the estimated percentages, here by 6 % (90% versus 96 % with respectively the logistic and the probit models). Thus the probit model should be more appropriate than the logistic model for the analysis of binary outcomes if your predictor is multiple pharmacological treatment dosages, if the outcome percentages are not very large or very small.

5 Tetrachoric Regression

Tetrachoric regressions make use of tetrachoric correlation coefficients rather than linear correlation coefficients. Linear correlation coefficients can be computed from data with a continuous outcome variable. The linear correlation coefficient was, originally, named product-moment correlation coefficient by its inventor Pearson in the 1900 edition of the Philosophical Transactions of the Royal society of London, Series A, 195, pp 1–147. It is the product of two mean-adjusted variables usually called x and y, and, it is thus, the covariance of the two variables, sometimes written as SPxy (sum of products of x and y variables). Strictly spoken, it is the normalized covariance. The term moment refers to first moment about an origin, which is here the mean. In addition, the correlation coefficient is standardized by dividing it by the square root of the product of the standard deviations (SDs) of x and y:

$$\text{correlation coefficient} = \text{SPxy}/\sqrt{(\text{SD}^2_x.\text{SD}^2_y)}$$

The above construct is pretty messy, but it leaves us with a wonderful functionality. It is, namely, a measure for the strength of linear association between the values from the x variable and the y variable. It varies from -1 to $+1$. the strongest association is either -1 or $+1$, the weakest association is zero.

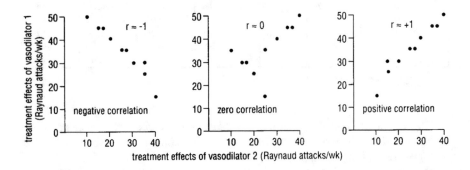

The above graph gives an example of the results of three crossover studies comparing the effect of a test and control treatment in ten patients with Raynaud's phenomenon. Left the association is strong negative, $r = -1$, in the middle the association is zero, $r = 0$, right the association is strong positive, $r = +1$. Correlation coefficients can also be applied for the purpose of analyzing therapeutic treatment comparisons. An example is given underneath.

A parallel-group study of 872 patients studied for cholesterol decreasing capacity of pravastatin and placebo.

	placebo	pravastatin	difference
sample size	434	438	
mean (mmol/l)	-0.04	1.23	1.27
standard deviation (mmol/l)	0.59	0.68	standard error (se) = 0.043

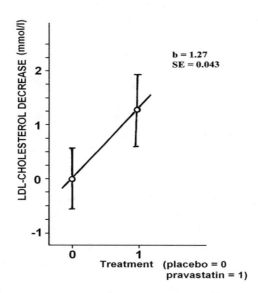

The above graph shows, that the same result of the unpaired t-test for parallel-group analyses is obtained by drawing the best fit regression line for the study data.

The above table and the graph are from the data of the REGRESS study (Jukema, Circulation 1995; 91: 2528), a parallel-group study comparing the cholesterol reducing property of placebo versus pravastatin, the mean difference between the two treatments was 1.27 mmol/l with a standard error (se) of 0.043 mol/l. This result can also be obtained by a linear regression with treatment modality, 0 or 1 for placebo or pravastatin, as x-variable and cholesterol reduction after treatment as outcome. However, the result of the regression analysis is expressed in the regression coefficient and its standard error, b and $se_{b-value}$, 1.27 and 0.043 mmol/l. Here the b-value seems to be useful as the effect size of the study, just like the mean difference between the two treatments is. Even more useful is the standardized b-value:

$$\text{standardized b-value} = \text{b-value}/\text{se}_{\text{b-value}}$$
$$= 1.27/0.043 \text{ standard error units.}$$

The unit of the standardized b-value is often expressed in so-called standard error units. The standardized b-value is equal to the r-value of the linear regression from which it comes.

$$\text{standardized b-value} = \text{correlation coefficient r.}$$

Correlation coefficients of individual studies can be used as the main outcome of comparative therapeutic studies. The meta-analysis of such studies requires information of the standard error of the correlation coefficients (n = sample size of study):

$$\text{standard error}_{\text{correlation coefficient r}} = \left(1 - r^2\right)/\sqrt{n}.$$

Correlation coefficients are calculated from linear regression models with a continuous outcome variable. However, in clinical research, instead of a continuous outcome, a binary outcome is often used, like in a multitude of risk factor studies. They can be analyzed with logistic regression. However, logistic regression does not provide a correlation coefficient. Generally speaking, Pearson correlation cannot deal with categorical variables, mostly, because categorical variables don't have a notion of mean, which Pearson correlation is based on. However, having only 2 binary variables you can consider them as continuous (with values of 1 and 0) and calculate a *kind of* correlation. You may think, nothing is impossible in statistics, but statisticians also thought of ways for obtaining correlation coefficients from binary outcome data. Some approximation methods for computing correlation coefficients from binary data will be given, particularly the ones with help from odds ratios.

An example is given. In a hospital many patients tend to fall out of bed. We wish to find out, whether one department performs better than the other.

fall out of bed	yes	no
department 1	15(a)	20(b)
department 2	15(c)	5(d).

The odds ratio of the odds of falling out of bed in department 1 versus that in department 2 equals:

Odds ratio $= 15/20 \div 15/5 = 1/4 = 0.25 = 25\,\%$.

The correlation coefficient, here otherwise called phi because outcomes are binary (Clinical data analysis on a pocket calculator second edition, Chap. 37, Phi tests for nominal data, 2016, Springer Heidelberg Germany, from the same authors), is computed as shown underneath:

$$
\begin{aligned}
&= (ad - bc)/\sqrt{(a + b)(c + d)(a + c)(b + d)} = \\
&= (75 - 300)/\sqrt{(35 \times 20 \times 30 \times 25)} = \\
&= -0.31.
\end{aligned}
$$

A phi value of -0.31 on a scale from -1 to 0 gives the level of association between the binary variables department and fall out of bed. We might say the correlation coefficient between the two departments. Let us assume, that in a study only the odds ratio (OR) of the departments is given:

$$
OR = a/b \div c/d = 0.25.
$$

The Yule Approximation (Yule 1871–1951 Tuebingen Germany)

Yule (J Roy Stat Soc 1912; 75: 579) provided an equation for approximation of phi from OR:

$$
\text{correlation coefficient} =
$$
$$
\left(OR^{1/2} - 1\right)\left(OR^{1/2} + 1\right) = (-0.5)\,(0.5 + 1) = -0.75
$$

Obviously, this Yule approximation of -0.75 is a poor estimate of the above computed -0.31, but this is due to the large difference in sample size of departments 1 and 2, and will otherwise perform better.

The Ulrich Approximation (Ulrich 1953- Tuebingen Germany)

Suppose the samples sizes are known. Then, the Ulrich approximation (2004, Br J Math Stat Psychol) can be used, and it performs pretty well:

$$
\text{correlation coefficient} =
$$
$$
\log OR/\sqrt{(\log OR^2 + 2.89\,n^2/n_1 n_2)} = -1.386/\sqrt{(1.92 + 12.49)} = -0.36.
$$

The two computations -0.36 for Ulrich and -0.31 for the traditional phi test are very similar. A different approach to estimating correlation coefficients with binary outcome data was given by Sir Galton, a contemporary of Pearson, and will be explained in the next paragraph.

Tetrachoric correlation coefficients are much similar to usual Pearson correlation coefficients, but they have another theoretical background, and provide better sensitivity of testing, if that background is sound and clinically supported.

If we have arguments to assume, that the binary data are from a continuous distribution, as is often the case in practice (e.g., negative means less severe, positive means above a threshold of severity), then a more appropriate approach will be to use a tetrachoric 2 × 2 model, which is best described as an ellipse cut into four parts. The patients in the parts A + C are less severe according to the rater along the x-axis, those in the parts B + D are more severe. The patients in the parts A + B are more severe according to the y-axis rater, those in the C + D parts are less so. The vertical and horizontal lines are thresholds for respectively x- and y-axis raters. It is mathematically pretty hard to calculate the exact chance for patients of being in any of the parts A to D. It requires the concepts of eigenvectors and eigenvalues and bivariate normal distributions. and a lot of arithmetic, which is beyond the scope of the current work.

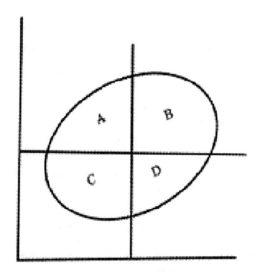

However, like with traditional correlation coefficients, approximation methods are possible. Like with a linear correlation between -1 and +1 as estimate for the strength of association of yes-no data according to two raters, a correlation coefficient can be calculated for the data in an ellipse-form pattern. It is called the tetrachoric correlation coefficient, ranging, just like the linear correlation coefficient, between −1 and + 1. But we should add a very pleasant aspect of the tetrachoric correlation assessments: they produce larger r-values, making the significance testing job more often successful.

Tetrachoric correlation assessments are used for the purpose of finding a correlation coefficient between two raters in a 2x2 interaction matrix of the ellipse type. It was invented by Galton, a contemporary of Pearson, the inventor of the traditional correlation coefficient. The non-tetrachoric Pearson correlation coefficient, also usually called r or R, otherwise sometimes called Cohen's kappa, of the underneath 2 x 2 table is commonly used for estimating the linear correlation between two strictly binary variables, and produces the result given underneath. If , however, an ellipse type interaction matrix is in the data, then the tetrachoric correlation coefficient should produce a larger correlation coefficients, because it better fits the data than does the traditional correlation coefficient.

	rater 1			
	diagnosis yes		no	
rater 2	40 A		10 B	50
	20 C		30 D	50
	60		40	100

$$\text{Pearson correlation coefficient } r = \frac{[A + D - 1/2(A + B + C + D)]}{1/2(A + B + C + D)}$$
$$r = (70 - 50)/(100 - 50) = 0.4.$$

The tetrachoric correlation coefficient (tcc) is calculated much differently.

Approximation methods are usually applied, because exact computations are not available. The underneath approximation by Bonett (Transforming odds ratios into correlations for meta-analytic research, Am Psychologist 2007; 62: 254–55) makes use of the delta method, a mathematical approximation and logarithmic approximation where the variance of log x is equal to the variance of x divided by x squared. This approach using the quadratic approximation and eigenvectors is sufficiently accurate, if samples are not too small.

$$\text{tcc} = (\alpha - 1)/(\alpha + 1)$$
$$\alpha = (AD/BC)^{\pi/4} = (1200/200)^{3.14/4} = 6^{0.785} = 4.08$$
$$\text{tcc} = (4.08 - 1)/(4.08 + 1)$$
$$= 0.606$$

The tetrachoric correlation coefficient on a scale from 0 to 1 (or −1) is, thus, much larger (0.606), than the traditional Pearson correlation coefficient r (0.400). The calculation by Pearson of the tetrachoric correlation coefficient given here is based on a mathematical approximation. A more exact method is computationally also more complex, and Monte Carlo methods are preferred. Also a tetrachoric calculator is available on the internet (free tetrachoric calculator). Tetrachoric correlation is a special case of the polychoric correlation, applicable, when the observed variables are polytomous. The tetrachoric correlation is also called the inferred Pearson Correlation from a two x two table with the assumption of a bivariate normality. The polychoric correlation generalizes this to any larger table, the n x m table, instead of a 2 x 2 table.

In this chapter three approximations are given for the calculation of correlation coefficients from studies with binary outcome variables. Particularly, in such studies where the measure of spread of their outcome, mostly odds ratios (ORs), is often missing in published studies, the Yule, Ulrich and tetrachoric approximations are helpful, and enable to perform relevant meta-analyses of multiple studies (Modern meta-analysis, Chap. 20, Springer Heidelberg Germany, 2017, from the same authors). An OR of 1 means no difference between test treatment and control treatment. An OR > 1 indicates that the test treatment is better than control, and, in addition, it gives an estimate about hów much better. An OR of 2 would mean about twice better, etc. However, if ORs are obtained from a very small data sample, twice better will not mean too much, and a test for statistical significance will be required, and, for that purpose, a measure of spread, e.g., the standard error (se) or the 95% confidence interval is needed. A common procedure is to make an estimate from the equation:

$$\mathrm{se}_{\mathrm{logOR}} = \surd(1/\mathrm{A} + 1/\mathrm{B} + 1/\mathrm{C} + /\mathrm{D}),$$

where A to D are the number of patients per cell in a 2×2 interaction matrix.

	responder yes	no
test treatment	A	B
control treatment	C	D.

In many published studies ORs are reported without information about the numbers of patients per cell, A to D. ORs without spread information are called unweighted ORs, and, those obtained from small studies, will disproportionally determine the final result of average ORs in a meta-analysis. Fortunately, an appropriate estimation of the se of ORs is obtained by the underneath equation:

$$\text{se of correlation coefficient r} = \left(1 - r^2\right)/\surd\mathrm{n},$$

where r = correlation coefficient and n = sample size of study.

6 Quasi-Likelihood Regressions

Overdispersion of a data file means, the data spread is wider than compatible with a normal frequency distribution. This phenomenon is for example a major problem with the random effect analysis of meta-analyses. Wedderburn (Quasi-likelihood functions, Biometrika 1974; 61: 439–47) proposed a quasi-likelihood procedure for the purpose, and this works pretty well (Modern Meta-Analysis, Chap. 5, Springer Heidelberg Germany, 2017, from the same authors). In this chapter we will review the quasi-likelihood method for the analysis of paired proportional data. Log likelihood ratio tests are wonderful, because with large and small numbers they are more sensitive than traditional z- and chi-square tests, and they do not suffer from overdispersion.

Frequency distribution

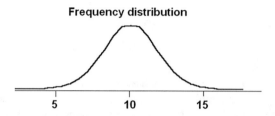

Assume that on average in a population 10 out of a random sample of 15 is sleepy at day time. Then, with multiple sampling 10 / 15 will be encountered most frequently. The chance of observing <10 or > 10 will get gradually smaller. The above normal pattern is likely to be observed. The graph has at the x-(=z-) axis the results of many samples of 15, at the y-axis it has "how often". The proportion 10 / 15 has the maximal likelihood, all other likelihoods are smaller.

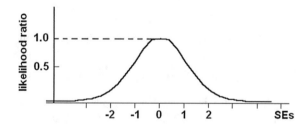

A better approach is the above standardized likelihoods taking ratios:
Likelihood **ratios** = likelihood various proportions / likelihood of 10/15.
Likelihood ratio for 10 / 15 = 1, all other proportions are smaller.
If the proportion 10/15 gets place 0 at z-axis with SE units instead of proportions as unit, then the above graph is a *likelihood ratio* curve with equation $e^{-\frac{1}{2} Z^2}$.
The z-value = the variable of x-axis of normal distribution.
Now transform y-axis to *log* y (means **ln = natural log** here).

y = likelihood ratio $= e^{-1/2\ z\text{-squared}}$
log likelihood ratio $= -\tfrac{1}{2}\ z^2$
-2 log likelihood ratio $= z^2$

With a "normal-like" data frequency distribution

> z-values > 2 means
> there is a difference from zero in normal data at $p < 0.05$,
>
> $z^2 > 4$ equally means
> "   "   "   "   "   " $p < 0.05$.

Thus, if -2 log likelihood ratio > 4, then again there is a difference from zero in normal data at $p < 0.05$, meaning a significant difference between the observed likelihood and the maximal likelihood at the point zero of the z-axis.

This may sound awkward, but it is a lovely method for testing 2×2 contingency tables, and not only these, but also many more continuous or discontinuous data, that have a touch of normality.

Testing two x two tables works as follows:

	events	no events	proportion
group 0	a	b	a / (a+b) = p0
group 1	c	d	c / (c+d) = p1

You find the "log likelihood ratio" equation with binomial formula:

$$\text{log likelihood ratio} = c \ \log\ (p_0/p_1) + d \ \log\ [(1 - p_0)/(1 - p_1)]$$

If -2 log likelihood ratio is larger than 4, then there will be a significant difference between group 0 and group 1 in the above 2×2 table at $p < 0.05$. Two by two tables involve unpaired groups, and the above log likelihood ratio method is unadjusted for paired groups. Often in clinical research paired groups have to be assessed, and, instead of the above test, adjustment for the covariance of the paired observations has to be included. It is then called a quasi-log likelihood test. It is in SPSS statistical software in the module generalized linear models, and will be used in this chapter.

McNemar's chi-square test is appropriate for analysis (SPSS for Starters and second Levelers, Chap. 41, Springer Heidelberg Germany, 2016, from the same authors). However, McNemar's test can not include predictor variables. The analysis of paired outcome proportions including predictor variables in SPSS statistical software requires the module generalized estimating equations. The difference between the two outcomes and the independent effects of the predictors variables on the outcomes are simultaneously tested using quasi-log likelihood tests. First question, is the numbers of yes-responders of outcome-1 significantly different from that of outcome-2. Second question, are the predictor variables significant predictors of the outcomes. For example, in a study of 139 physicians, general practitioners, the primary scientific question was: is there a significant difference between the numbers of practitioners who give lifestyle advise in the periods before and after (postgraduate) education. The second question was, is age an independent predictor of the outcomes.

Lifestyle advise-1	lifestyle advise-2	age (years)
,00	,00	89,00
,00	,00	78,00
,00	,00	79,00
,00	,00	76,00
,00	,00	87,00
,00	,00	84,00
,00	,00	84,00
,00	,00	69,00
,00	,00	77,00
,00	,00	79,00

0 = no, 1 = yes

The first 10 patients of the data file is given above. The entire data file is in extras. springer.com, and is entitled "chap3paired binary". It was already used in the Chap. 3 of this edition for illustrating confirmative regression analyses.

		Lifestyleadvise after education	
		no	yes
		0	1
Lifestyleadvise	no	0	65	28
before education	yes	1	12	34

The above paired 2 × 2 table summarizes the numbers of practitioners giving lifestyle advise in the periods prior to and after postgraduate education. Obviously, before education 65 + 28 = 93 did not give lifestyle, while after education this number fell to 77. It looks as though the education was somewhat successful. According to the McNemar's test this effect was statistically significant $p = 0.018$.

In this chapter we will assess, if the effect still exists after adjustment for doctors' ages.

Start by opening the data file in SPSS statistical software. Prior to a generalized estimation equation analysis, which includes additional predictors to a model with paired binary outcomes, the data will have to be restructured. For that purpose the Restructure Data Wizard will be used.

Command:
click Data....click Restructure....mark Restructure selected variables into cases....
click Next....mark One (for example, w1, w2, and w3)....click Next....Name: id
(the patient id variable is already provided)....Target Variable: enter "lifestyleadvise
1, lifestyleadvise 2"....Fixed Variable(s): enter age....click Next.... How many index
variables do you want to create?....mark One....click Next....click Next again....click
Next again....click Finish....Sets from the original data will still be in use...click OK.

Return to the main screen and observe that there are now 278 rows instead of 139 in the data file. The first 10 rows are given underneath.

id	age	Index 1	trans 1
1	89,00	1	,00
1	89,00	2	,00
2	78,00	1	,00
2	78,00	2	,00
3	79,00	1	,00
3	79,00	2	,00
4	76,00	1	,00
4	76,00	2	,00
5	87,00	1	,00
5	87,00	2	,00

id: patient identity number
age: age in years
Index 1: 1 = before postgraduate education, 2 = after postgraduate education
trans 1: lifestyleadvise no = 1, lifestyle advise yes = 2

The above data file is adequate to perform a generalized estimation equation analysis. Save the data file. For convenience of the readers it is given in extras. springer.com, and is entitled "chap3pairedbinaryrestructured".

For analysis the module Generalized Linear Models is required. It consists of two submodules: Generalized Linear Models and Generalized Estimation Models. The second is for analyzing paired binary outcomes (current chapter).

Command:
Analyze....Generalized Linear Models....Generalized Estimation Equations....click
Repeated....transfer id to Subject variables....transfer Index 1 to Within-subject
variables....in Structure enter Unstructured....click Type of Model....mark Binary

logistic....click Response....in Dependent Variable enter lifestyleadvise....click Reference Category....click Predictors....in Factors enter Index 1....in Covariates enter age....click Model....in Model enter lifestyleadvise and age....click OK.

Tests of Model Effects

	Type III		
Source	Wald Chi-Square	df	Sig.
(Intercept)	8,079	1	,004
Index1	6,585	1	,010
age	10,743	1	,001

Dependent Variable: lifestyleadvise before
Model: (Intercept), Index1, age

Parameter Estimates

Parameter	B	Std. Error	95% Wald Confidence Interval		Hypothesis Test		
			Lower	Upper	Wald Chi-Square	df	Sig.
(Intercept)	-2,508	,8017	-4,079	-,936	9,783	1	,002
[Index1=1]	,522	,2036	,123	,921	6,585	1	,010
[Index1=2]	0ª	.	.	.	.	.	.
age	,043	,0131	,017	,069	10,743	1	,001
(Scale)	1						

Dependent Variable: lifestyleadvise before
Model: (Intercept), Index1, age

a. Set to zero because this parameter is redundant.

In the output sheets the above tables are observed. They show that both the index 1 (postgraduate education) and age are significant predictors of lifestyleadvise. The interpretations of the two significant effects are slightly different from one another. Postgraduate education is compared with no postgraduate education at all, while the effect of age is an independent effect of age on lifestyleadvise, the older the doctors the better lifestyle advise given irrespective of the effect of the postgraduate education.

In the current assessment an analysis is used based on approximations of likelihood, rather than true likelihoods. This limits the meaningfulness of the p-values obtained, and appropriate goodness of fit tests are welcome. Therefore, a goodness of fit with and without age will be computed.

Goodness of Fit[a]

	Value
Quasi Likelihood under Independence Model Criterion (QIC)[b]	356,621
Corrected Quasi Likelihood under Independence Model Criterion (QICC)[b]	355,619

Dependent Variable: lifestyleadvise before

Model: (Intercept), Index1, age

  a. Information criteria are in small-is-better form.

  b. Computed using the full log quasi-likelihood function.

"Quasi-likelihood under independence model criterion (QIC)"
"Corrected quasi-likelihood under independence model criterion (QICC)"

The above terms can be used for respectively estimating the structure of the predictive model and the terms in the predictive model. The smaller the criteria values the better the goodness of fit. This is particularly relevant, if compared with different models and model terms. For assessment we will compute the goodness of fit criteria for the model without the predictor age. The above commands are given once more and the predictor age is removed from the computations. In the output sheets statistics are pretty much the same, but the goodness of fit values are considerably different.

Goodness of Fit[a]

	Value
Quasi Likelihood under Independence Model Criterion (QIC)[b]	371,558
Corrected Quasi Likelihood under Independence Model Criterion (QICC)[b]	371,558

Dependent Variable: lifestyleadvise before

Model: (Intercept), Index1

  a. Information criteria are in small-is-better form.

  b. Computed using the full log quasi-likelihood function.

The above table shows that without age the goodness of fit is considerable smaller than that of the full model. The variable age has managed, that the amount of information loss from the predictive model is less.

In the above section a single predictor variable was assessed, namely patient age. Now, we will use the same methodology for assessing multiple predictors. The same example including a second predictor, namely type of doctor, country/city doctor, will be applied.

patient id	age	educ ation	life style advise	type of doctor
1	89,00	1	,00	1,00
1	89,00	2	,00	1,00
2	78,00	1	,00	1,00
2	78,00	2	,00	1,00
3	79,00	1	,00	1,00
3	79,00	2	,00	1,00
4	76,00	1	,00	1,00
4	76,00	2	,00	1,00
5	87,00	1	,00	1,00
5	87,00	2	,00	1,00

education: 1 = before education, 2 = after education
lifestyle advise: 0 = no lifestyle advise given, 1 = yes lifestyle advise given
type of doctor: 1 = country practice, 2 = city practice.

Command:
Analyze....Generalized Linear Models....Generalized Estimation Equations....click Repeated....transfer id to Subject variables....transfer Education 1 to Within-subject variables....in Structure enter Unstructured....click Type of Model....mark Binary logistic....click Response....in Dependent Variable enter lifestyleadvise....click Reference Category....click Predictors....in Factors enter Education 1....in Covariates enter age and type of doctor....click Model....in Model enter lifestyleadvise and age and type of doctor....click OK.

Tests of Model Effects

Source	Type III		
	Wald Chi-Square	df	Sig.
(Intercept)	,012	1	,912
education	5,892	1	,015
age	2,870	1	,090
countrydoctor	,363	1	,547

Dependent Variable: lifestyleadvise given
Model: (Intercept), education, age, countrydoctor

Parameter Estimates

Parameter	B	Std. Error	95% Wald Confidence Interval		Hypothesis Test		
			Lower	Upper	Wald Chi-Square	df	Sig.
(Intercept)	,195	4,1683	-7,975	8,365	,002	1	,963
[education=1]	,542	,2232	,104	,979	5,892	1	,015
[education=2]	0[a]	.	.	.	.	.	.
age	,031	,0181	-,005	,066	2,870	1	,090
countrydoctor	-1,144	1,8989	-4,866	2,578	,363	1	,547
(Scale)	1						

Dependent Variable: lifestyleadvise given
Model: (Intercept), education, age, countrydoctor

a. Set to zero because this parameter is redundant.

The output sheets are shown above. The statistics are less powerful than in the single predictor model. Type of doctor is not a significant predictor of giving or not giving lifestyle advise. For comparing the goodness of fit of the model with that of the single predictor model the underneath goodness of fit table is given.

Goodness of Fit[a]

	Value
Quasi Likelihood under Independence Model Criterion (QIC)[b]	415,523
Corrected Quasi Likelihood under Independence Model Criterion (QICC)[b]	340,120

Dependent Variable: lifestyleadvise given
Model: (Intercept), education, age, countrydoctor

a. Information criteria are in small-is-better form.

b. Computed using the full log quasi-likelihood function.

The QIC and QICC criteria of the two predictor model are different from those of the one predictor model. The QIC is larger, the QICC slightly smaller. Obviously, the two predictor model is worse, but its terms are better, whatever that means.

Paired proportions have to be assessed when e.g. different diagnostic procedures are performed in one subject. McNemar's chi-square test is appropriate for analysis.

McNemar's test can not include predictor variables, and is not feasible for more than 2 outcomes. For that purpose Cochran's tests are required (SPSS for Starters and 2nd Levelers, Chap. 43, Springer Heidelberg Germany, 2016, from the same authors). The analysis of paired outcome proportions including predictor variables requires quasi-log likelihood models available in the module generalized estimating equations as reviewed in the current chapter. Goodness of fit of the quasi-log likelihood is recommended.

7 Conclusion

In the current chapter five more regression models and their main purposes are explained.

The first, the binary Poisson model, often provides better statistics than traditional logistic regression does. Unlike with logistic regressions, the output is expressed as chi-square values instead of odds ratios. Instead of Poisson models for binary outcomes the second model, negative binomial model, may be more appropriate, because it can adjust for overdispersion of event rates.

With the third model, the probit model, results are expressed in response rates or percentages instead of odds ratios. With very large or very small percentages like the responders to different dosages of pharmacological treatments, odds ratios may over- or underestimate the estimated percentages, and probit models may be more appropriate.

The fourth model, the tetrachoric regression, is much similar to usual Pearson correlation coefficients, but it has another theoretical background, and provides better sensitivity of testing, if that background is sound and clinically supported. That is, if we have arguments to assume, that the binary data are from a continuous distribution, as is often the case in practice (e.g., negative means less severe, positive means above a threshold of severity), then a more appropriate approach will be to use a tetrachoric 2 x 2 model, which is best described as an ellipse cut into four parts. The maths requires the concepts of eigenvectors and eigenvalues and bivariate normal distributions, and is beyond the scope of the current work.

Fifth, the quasi-likelihood regression, is a method for assessing the goodness of fit of repeated measures logistic regressions, and can, thus, be used for estimating, whether repeated measures binary data are largely different or not. The quasi-likelihood under independence model (QIC) criterion, an binary equivalent of the Akaike information criterion, is appropriate for the purpose. The Akaike information criterion is the subtraction of observed regression coefficients and likelihood measures. More information of the Akaike, otherwise Bayesian, information criteria is given in the Chap. 5 of Efficacy Analysis in Clinical Trials an Update, edited 2019 by Springer Heidelberg Germany, from the same authors.

Reference

To readers requesting more background, theoretical and mathematical information of computations given, several textbooks complementary to the current production and written by the same authors are available: Statistics applied to clinical studies 5th edition, 2012, Machine learning in medicine a complete overview, 2015, SPSS for starters and 2nd levelers 2nd edition, 2015, Clinical data analysis on a pocket calculator 2nd edition, 2016, Understanding clinical data analysis, 2017, all of them edited by Springer Heidelberg Germany.

Chapter 5
Polytomous Outcome Regressions

Multinomial, Ordinal, Negative Binomial and Poisson, Random Intercepts, General and Logit Loglinear, Hierarchical Loglinear Regressions

Abstract In clinical research, outcome categories like disease severity levels are very common, and statistics has great difficulty to analyze categories instead of continuously measured outcomes. Polytomous outcome regressions are regressions with categorical rather than continuous outcomes. Five methods of analysis are reviewed in this chapter. With multinomial regression the outcome categories are equally present in the data. With ordinal regression one or two outcome categories are underpresented. With negative binomial and Poisson regressions data are assessed as multivariate models with multiple dummy outcome variables. Random intercept regression is like multinomial, but provides better power. With logit loglinear regression first and second order interactions of the predictors on the outcome categories are assessed. With hierarchical loglinear regression third and fourth order interactions of the predictors on the outcome categories are so.

Keywords Polytomous regressions · Multinomial · Ordinal · Negative binomial · Poisson · Random intercept · General and logit loglinear · Hierarchical loglinear

1 Introduction, History, and Background

If clinical studies have categories as outcome like for example various levels of health or disease, then linear regression will not be adequate for analysis. Linear regression assumes continuous outcomes, where, with each level, severity increases by the same quantity. Outcome categories are very common in medical research. Examples include age classes, income classes, education levels, drug dosages, diagnosis groups, disease severities, etc. Statistics has generally difficulty to assess categories, and traditional models require either binary or continuous variables. Outcome categories are sometimes called polytomous outcomes. They are discrete

Electronic Supplementary Material The online version of this chapter (https://doi.org/10.1007/978-3-030-61394-5_5) contains supplementary material, which is available to authorized users. The videos can be accessed by scanning the related images with the SN More Media App.

outcome data with more than two categories. Five methods of analysis will be reviewed in this chapter.

1. Multinomial regression (McFadden 1937- . . ., from Raleigh North Carolina) will be adequate, if none of the outcome categories are underpresented.
2. Ordinal regression (Winship 1950- . . ., from Chicago University Harvard) will be adequate, if one or two categories are underpresented.
3. Negative binomial and Poisson regressions (Pascal distribution (after Pascal 1623–1662 Paris) and Polya distribution (after Polya 1887–1985 Palo Alto CA)) are rather for binary outcomes than polytomous outcomes, but they can be used for data with more than two outcome categories. The data are then assessed as multivariate models with multiple dummy outcome variables.
4. Random intercept regression often provides better power than multinomial regression. It is also called multilevel regression or nested random data analysis (Yates 1902–1994 from Manchester UK, and Youden 1900–1971 from the University of Rochester NY). See also the Chap. 4.
5. General loglinear regression analyzes all kinds of first and second order interactions of the predictors on the outcome categories. Multinomial regression is adequate for identifying the main predictors of certain outcome categories, like different levels of injury or quality of life (QOL). An alternative approach is logit (= logodds) loglinear modeling available in the SPSS module Loglinear subsection Logit. The latter method does not use continuous predictors on a case by case basis, but rather the weighted means of these predictors. This approach may allow for relevant additional conclusions from your data. This and the next method are pretty new (Simkiss, 2012 in J Trop Pediatics, Analyzing categorical data, loglinear analysis).
6. Hierarchical loglinear regression can analyze higher order interactions of the predictors on the outcome categories. This is not available in the SPSS menu program, but can be easily activated through a few syntax commands, see "Machine learning in medicine – a complete overview", Chap. 39, Loglinear modeling for outcome categories, 233–240, Springer Heidelberg Germany, 2015).

We should add, that data files with polytomous outcomes are rapidly pretty complex. Also many p-values are produced, including increasing numbers of false positive results due to type I errors of finding an effect, where there is none. However, with current big data, study files are generally large, and, therefore, often considered clinically relevant even so, particularly so, when the results match prior hypotheses.

Loglinear models can also be used for assessing incident rates with varying incident risks. This method is available, for example, in the SPSS statistical software module Loglinear section General.

2 Multinomial Regression

Multinomial regression is suitable for analysis. However, if one or two outcome categories in a study are severely underpresented, multinomial regression is flawed, and ordinal regression including specific link functions will provide a better fit for the data. Strictly, ordinal data are, like nominal data, discrete data, however, with a stepping pattern, like severity scores, intelligence levels, physical strength scores. They are usually assessed with frequency tables and bar charts. Unlike scale data, that also have a stepping pattern, they do not necessarily have to have steps with equal intervals. This causes some categories to be underpresented compared to others. The effect of the levels of satisfaction with the doctor on the levels of quality of life (qol) is assessed. In 450 patients with coronary artery disease the satisfaction level of patients with their doctor was assumed to be an important predictor of patient qol (quality of life).

Variable

1	2	3	4
qol	treatment	counseling	sat doctor
4	3	1	4
2	4	0	1
5	2	1	4
4	3	0	4
2	2	1	1
1	2	0	4
4	4	0	1
4	3	0	1
4	4	1	4
3	2	1	4

1. qol	= quality of life score (1 = very low, 5 = vey high)
2. treatment	= treatment modality (1 = cardiac fitness, 2 = physiotherapy, 3 = wellness, 4 = hydrotherapy, 5 = nothing)
3. counseling	= counseling given (0 = no, 1 = yes)
4. sat doctor	= satisfaction with doctor (1 = very low, 5 = very high).

The above table gives the first 10 patients of a 450 patients study of the effects of doctors' satisfaction level and qol. The entire data file is in extras.springer.com, and is entitled "chap5ordinalregression". It was previously used by the authors in SPSS for starters and 2nd levelers, Chap. 48, Springer Heidelberg Germany, 2016. Start by opening the data file in your computer with SPSS installed.

The table shows, that the frequencies of the qol scores are pretty heterogeneous with 111 patients very high scores and only 71 patients medium scores. This could mean that multinomial regression is somewhat flawed and that ordinal regression including specific link functions may provide a better fit for the data.

We will start with a traditional multinomial regression. For analysis the statistical model Multinomial Logistic Regression in the SPSS module Regression is required.

Command:
Analyze....Regression....Multinomial Regression....Dependent: enter qol.... Factor (s): enter treatment, counseling, sat (satisfaction) with doctor....click OK.

In the output sheets a reduced model, testing that all effects of a single predictor are zero, is shown. It is a pretty rough approach, and more details would be in place.

Likelihood Ratio Tests

Effect	Model Fitting Criteria -2 Log Likelihood of Reduced Model	Likelihood Ratio Tests		
		Chi-Square	df	Sig.
Intercept	483,058[a]	,000	0	.
satdoctor	525,814	42,756	16	,000
treatment	496,384	13,326	12	,346
counseling	523,215	40,157	4	,000

The chi-square statistic is the difference in -2 log-likelihoods between the final model and a reduced model. The reduced model is formed by omitting an effect from the final model. The null hypothesis is that all parameters of that effect are 0.

a. This reduced model is equivalent to the final model because omitting the effect does not increase the degrees of freedom.

The next page table is in the output sheets. It shows that the effects of several factors on different qol scores are very significant, like the effect of counseling on very low qol, and the effects of satisfaction with doctor levels 1 and 2 on very low qol. However, other effects were insignificant, like the effects of treatments on very low qol, and the effects of satisfaction with doctor levels 3 and 4 on very low qol. In order to obtain a more general overview of what is going-on an ordinal regression will be performed.

Parameter Estimates

qol score[a]		B	Std. Error	Wald	df	Sig.	Exp(B)	95% Confidence Interval for Exp (B)	
								Lower Bound	Upper Bound
very low	Intercept	-1,795	,488	13,528	1	,000			
	[treatment=1]	-,337	,420	,644	1	,422	,714	,314	1,626
	[treatment=2]	,573	,442	1,678	1	,195	1,773	,745	4,216
	[treatment=3]	,265	,428	,385	1	,535	1,304	,564	3,015
	[treatment=4]	0[b]	.	.	0	.	.	.	.
	[counseling=0]	1,457	,328	19,682	1	,000	4,292	2,255	8,170
	[counseling=1]	0[b]	.	.	0	.	.	.	.
	[satdoctor=1]	2,035	,695	8,579	1	,003	7,653	1,961	29,871
	[satdoctor=2]	1,344	,494	7,413	1	,006	3,834	1,457	10,089
	[satdoctor=3]	,440	,468	,887	1	,346	1,553	,621	3,885
	[satdoctor=4]	,078	,465	,028	1	,867	1,081	,435	2,687
	[satdoctor=5]	0[b]	.	.	0	.	.	.	.
low	Intercept	-2,067	,555	13,879	1	,000			
	[treatment=1]	-,123	,423	,084	1	,771	,884	,386	2,025
	[treatment=2]	,583	,449	1,684	1	,194	1,791	,743	4,320
	[treatment=3]	-,037	,462	,006	1	,936	,964	,389	2,385
	[treatment=4]	0[b]	.	.	0	.	.	.	.
	[counseling=0]	,846	,323	6,858	1	,009	2,331	1,237	4,392
	[counseling=1]	0[b]	.	.	0	.	.	.	.
	[satdoctor=1]	2,735	,738	13,738	1	,000	15,405	3,628	65,418
	[satdoctor=2]	1,614	,581	7,709	1	,005	5,023	1,607	15,698
	[satdoctor=3]	1,285	,538	5,704	1	,017	3,614	1,259	10,375
	[satdoctor=4]	,711	,546	1,697	1	,193	2,036	,699	5,933
	[satdoctor=5]	0[b]	.	.	0	.	.	.	.
medium	Intercept	-1,724	,595	8,392	1	,004			
	[treatment=1]	-,714	,423	2,858	1	,091	,490	,214	1,121
	[treatment=2]	,094	,438	,046	1	,830	1,099	,465	2,594
	[treatment=3]	-,420	,459	,838	1	,360	,657	,267	1,615
	[treatment=4]	0[b]	.	.	0	.	.	.	.
	[counseling=0]	,029	,323	,008	1	,929	1,029	,546	1,940
	[counseling=1]	0[b]	.	.	0	.	.	.	.
	[satdoctor=1]	3,102	,790	15,425	1	,000	22,244	4,730	104,594
	[satdoctor=2]	2,423	,632	14,714	1	,000	11,275	3,270	38,875
	[satdoctor=3]	1,461	,621	5,534	1	,019	4,309	1,276	14,549
	[satdoctor=4]	1,098	,619	3,149	1	,076	2,997	,892	10,073
	[satdoctor=5]	0[b]	.	.	0	.	.	.	.
high	Intercept	-,333	,391	,724	1	,395			
	[treatment=1]	-,593	,371	2,562	1	,109	,552	,267	1,142
	[treatment=2]	-,150	,408	,135	1	,713	,860	,386	1,916
	[treatment=3]	,126	,376	,113	1	,737	1,135	,543	2,371
	[treatment=4]	0[b]	.	.	0	.	.	.	.
	[counseling=0]	-,279	,284	,965	1	,326	,756	,433	1,320
	[counseling=1]	0[b]	.	.	0	.	.	.	.
	[satdoctor=1]	1,650	,666	6,146	1	,013	5,208	1,413	19,196
	[satdoctor=2]	1,263	,451	7,840	1	,005	3,534	1,460	8,554
	[satdoctor=3]	,393	,429	,842	1	,359	1,482	,640	3,432
	[satdoctor=4]	,461	,399	1,337	1	,248	1,586	,726	3,466
	[satdoctor=5]	0[b]	.	.	0	.	.	.	.

a. The reference category is: very high.
b. This parameter is set to zero because it is redundant.

The multinomial regression computes for each outcome category compared to a defined reference category the effect of each of the predictor variables separately without taking interactions into account. The ordinary regression, in contrast, computes the effect of the separate predictors on a single outcome as a variable consistent of categories with a stepping pattern.

3   Ordinal Regression

For analysis the statistical model Ordinal Regression in the SPSS module Regression is required. The above example is used once more.

Command:
Analyze....Regression....Ordinal Regression....Dependent: enter qol....Factor(s): enter "treatment", "counseling", "sat with doctor"....click Options....Link: click Complementary Log-log....click Continue....click OK.

Model Fitting Information

Model	-2 Log Likelihood	Chi-Square	df	Sig.
Intercept Only	578,352			
Final	537,075	41,277	8	,000

Link function: Complementary Log-log.

Parameter Estimates

		Estimate	Std. Error	Wald	df	Sig.	95% Confidence Interval	
							Lower Bound	Upper Bound
Threshold	[qol = 1]	-2,207	,216	103,925	1	,000	-2,631	-1,783
	[qol = 2]	-1,473	,203	52,727	1	,000	-1,871	-1,075
	[qol = 3]	-,959	,197	23,724	1	,000	-1,345	-,573
	[qol = 4]	-,249	,191	1,712	1	,191	-,623	,124
Location	[treatment=1]	,130	,151	,740	1	,390	-,167	,427
	[treatment=2]	-,173	,153	1,274	1	,259	-,473	,127
	[treatment=3]	-,026	,155	,029	1	,864	-,330	,277
	[treatment=4]	0ª	.	.	0	.	.	.
	[counseling=0]	-,289	,112	6,707	1	,010	-,508	-,070
	[counseling=1]	0ª	.	.	0	.	.	.
	[satdoctor=1]	-,947	,222	18,214	1	,000	-1,382	-,512
	[satdoctor=2]	-,702	,193	13,174	1	,000	-1,081	-,323
	[satdoctor=3]	-,474	,195	5,935	1	,015	-,855	-,093
	[satdoctor=4]	-,264	,195	1,831	1	,176	-,646	,118
	[satdoctor=5]	0ª	.	.	0	.	.	.

Link function: Complementary Log-log.

a. This parameter is set to zero because it is redundant.

The above tables are in the output sheets of the ordinal regression. The model fitting information table tells, that the ordinal model provides an excellent overall fit for the data. The parameter estimates table gives an *overall* function of all predictors on the outcome categories. Treatment is not a significant factor, but counseling, and the satisfaction with doctor levels 1–3 are very significant predictors of the quality of life of these 450 patients. The negative values of the estimates can be interpreted as follows: the less counseling, the less effect on quality of life, and the less satisfaction with doctor, the less quality of life.

4   Negative Binomial and Poisson Regressions

The first 10 patients from the above 450 patient file assessing the effect of various factors on qol scores (scores 1–5, very low to very high) is in the underneath table. Analyses have already been performed with multinomial regression and ordinal regression. Multinomial computed for each score levels of qol the effects of each of the predictor variable levels separately. Ordinal computed these effects on the outcome levels as a single overall outcome, the outcome is not 4 outcomes but rather a five level single function. Next, a negative binomial and a Poisson regression will be used for analysis. Like Poisson negative binomial is not for categorical outcomes but rather binary outcomes, but it can be used for categorical outcomes. The data are then assessed as multivariate models with multiple dummy outcome variables.

qol	satdoctor	treatment	counseling
4	4	3	1
2	1	4	0
5	4	2	1
4	4	3	0
2	1	2	1
1	4	2	0
4	1	4	0
4	1	3	0
4	4	4	1
3	4	2	1

Command:
Analyze....Generalized Linear Models....Generalized Linear Models....mark Custom....Distribution: Negative binomial (1)....Link Function: Log....Response: Dependent Variable: qol score....Predictors: Factors: satdoctor, treatment, counseling....click Model....click Main Effects: enter satdoctor, treatment, counseling.... click Estimation: mark Robust Tests...click OK.

The underneath tables are in the output.

Model Information

Dependent Variable	qol score
Probability Distribution	Negative binomial (1)
Link Function	Log

Goodness of Fit[a]

	Value	df	Value/df
Deviance	82,834	441	,188
Scaled Deviance	82,834	441	
Pearson Chi-Square	69,981	441	,159
Scaled Pearson Chi-Square	69,981	441	
Log Likelihood[b]	-1032,531		
Akaike's Information Criterion (AIC)	2083,062		
Finite Sample Corrected AIC (AICC)	2083,471		
Bayesian Information Criterion (BIC)	2120,046		
Consistent AIC (CAIC)	2129,046		

Dependent Variable: qol score
Model: (Intercept), satdoctor, treatment, counseling

a. Information criteria are in small-is-better form.

b. The full log likelihood function is displayed and used in computing information criteria.

Tests of Model Effects

	Type III		
Source	Wald Chi-Square	df	Sig.
(Intercept)	2357,874	1	,000
satdoctor	16,226	4	,003
treatment	3,007	3	,391
counseling	24,063	1	,000

Dependent Variable: qol score
Model: (Intercept), satdoctor, treatment, counseling

Parameter Estimates

			95% Wald Confidence Interval		Hypothesis Test		
Parameter	B	Std. Error	Lower	Upper	Wald Chi-Square	df	Sig.
(Intercept)	1,393	,0541	1,287	1,499	664,630	1	,000
[satdoctor=1]	-,277	,0868	-,448	-,107	10,212	1	,001
[satdoctor=2]	-,184	,0685	-,318	-,050	7,237	1	,007
[satdoctor=3]	-,122	,0656	-,251	,006	3,465	1	,063
[satdoctor=4]	-,036	,0617	-,157	,084	,350	1	,554
[satdoctor=5]	0^a	.	.	.	.	.	.
[treatment=1]	,015	,0534	-,089	,120	,082	1	,775
[treatment=2]	-,092	,0631	-,216	,032	2,129	1	,145
[treatment=3]	-,024	,0578	-,138	,089	,175	1	,676
[treatment=4]	0^a	.	.	.	.	.	.
[counseling=0]	-,212	,0433	-,297	-,127	24,063	1	,000
[counseling=1]	0^a	.	.	.	.	.	.
(Scale)	1^b						
(Negative binomial)	1^b						

Dependent Variable: qol score
Model: (Intercept), satdoctor, treatment, counseling
a. Set to zero because this parameter is redundant.
b. Fixed at the displayed value.

Next, the Poisson model is applied.

Command: Analyze....Generalized Linear Models....Generalized Linear Models....mark Custom....Distribution: Poisson....Link Function: Log....Response: Dependent Variable: qol score....Predictors: Factors: satdoctor, treatment, counseling....click Model....click Main Effects: enter satdoctor, treatment, counseling.... click Estimation: mark Robust Tests...click OK.

The underneath tables are in the output.

Model Information

Dependent Variable	qol score
Probability Distribution	Poisson
Link Function	Log

Goodness of Fit[a]

	Value	df	Value/df
Deviance	298,394	441	,677
Scaled Deviance	298,394	441	
Pearson Chi-Square	279,166	441	,633
Scaled Pearson Chi-Square	279,166	441	
Log Likelihood[b]	-807,965		
Akaike's Information Criterion (AIC)	1633,929		
Finite Sample Corrected AIC (AICC)	1634,338		
Bayesian Information Criterion (BIC)	1670,913		
Consistent AIC (CAIC)	1679,913		

Dependent Variable: qol score
Model: (Intercept), satdoctor, treatment, counseling

a. Information criteria are in small-is-better form.

b. The full log likelihood function is displayed and used in computing information criteria.

Tests of Model Effects

	Type III		
Source	Wald Chi-Square	df	Sig.
(Intercept)	2302,055	1	,000
satdoctor	21,490	4	,000
treatment	4,549	3	,208
counseling	30,512	1	,000

Dependent Variable: qol score
Model: (Intercept), satdoctor, treatment, counseling

Parameter Estimates

Parameter	B	Std. Error	95% Wald Confidence Interval		Hypothesis Test		
			Lower	Upper	Wald Chi-Square	df	Sig.
(Intercept)	1,415	,0503	1,316	1,513	792,213	1	,000
[satdoctor=1]	-,305	,0829	-,467	-,142	13,526	1	,000
[satdoctor=2]	-,206	,0639	-,331	-,081	10,368	1	,001
[satdoctor=3]	-,126	,0610	-,245	-,006	4,250	1	,039
[satdoctor=4]	-,044	,0567	-,155	,067	,597	1	,440
[satdoctor=5]	0[a]	.	.	.	.	.	.
[treatment=1]	,023	,0504	-,075	,122	,214	1	,643
[treatment=2]	-,106	,0607	-,225	,013	3,063	1	,080
[treatment=3]	-,019	,0544	-,126	,087	,127	1	,722
[treatment=4]	0[a]	.	.	.	.	.	.
[counseling=0]	-,235	,0425	-,318	-,152	30,512	1	,000
[counseling=1]	0[a]	.	.	.	.	.	.
(Scale)	1[b]						

Dependent Variable: qol score
Model: (Intercept), satdoctor, treatment, counseling

a. Set to zero because this parameter is redundant.

b. Fixed at the displayed value.

In the above output sheets it is observed, that the Akaike information index and other goodness of fit indices of the Poisson model are better than those of the negative binomial, but both provide pretty similar test statistics. However, the negative binomial adjusts for overdispersion, while Poisson does not. As overdispersion is a common phenomenon with categorical outcome data, this would mean, that the negative binomial model is here slightly more appropriate than the Poisson model. More information of the two methods using binary outcomes has been given in the Chap. 4.

5 Random Intercepts Regression

Outcome categories can be assessed with multinomial regression. With categories both in the outcome and as predictors, random intercept models may provide better sensitivity of testing. The latter models assume that for each predictor category or combination of categories x_1, x_2,... slightly different a-values are computed with a better fit for the outcome category y than a single a-value.

$$y = a + b_1 x_1 + b_2 x_2 + ...$$

We should add that, instead of the above linear equation, even better results were obtained with log-transformed outcome variables (log = natural logarithm).

$$\log y = a + b_1 x_1 + b_2 x_2 + \dots$$

As an example, in a study, three hospital departments (no surgery, little surgery, lot of surgery), and three patient age classes (young, middle, old) were the predictors of the risk class of falling out of bed (fall out of bed no, yes but no injury, yes and injury). Are the predictor categories significant determinants of the risk of falling out of bed with or without injury. Does a random intercept provide better statistics.

outcome fall out of bed	predictor department	predictor ageclass	patient_id
1	0	1,00	1,00
1	0	1,00	2,00
1	0	2,00	3,00
1	0	1,00	4,00
1	0	1,00	5,00
1	0	,00	6,00
1	1	2,00	7,00
1	0	2,00	8,00
1	1	2,00	9,00
1	0	,00	10,00

department =   department class (0 = no surgery, 1 = little surgery, 2 = lot of surgery)
falloutofbed =   risk of falling out of bed (0 = fall out of bed no, 1 = yes but no injury, 2 = yes and injury)
ageclass =   patient age classes (young, middle, old)
patient_id =   patient identification

Only the first 10 patients of the 55 patient file is shown above. The entire data file is in extras.springer.com, and is entitled "chap5randomintercept.sav". It is previously used by the authors in SPSS for starters and 2nd levelers, Chap. 45, Springer Heidelberg Germany, 2016. SPSS version 20 and up can be used for analysis. First, we will perform a fixed intercept model.

Command:
click Analyze....Mixed Models....Generalized Linear Mixed Models....click Data Structure....click "patient_id" and drag to Subjects on the Canvas....click Fields and Effects....click Target....Target: select "fall with/out injury"....click Fixed Effects ....click "agecat", and "department", and drag to Effect Builder:....mark Include intercept....click Run.

The underneath results show that both the various regression coefficients as well
as the overall correlation coefficients between the predictors and the outcome are,
generally, statistically significant.

Source	F	df1	df2	Sig.
Corrected Model ▼	9,398	4	10	,002
agecat	6,853	2	10	,013
department	9,839	2	10	,004

Probability distribution:Multinomial
Link function:Cumulative logit

Model Term		Coefficient ▶	Sig.
Threshold for falloutofbed=	0	2,140	,028
	1	7,229	,000
agecat=0		5,236	,005
agecat=1		-0,002	,998
agecat=2		0,000ª	
department=0		3,660	,008
department=1		4,269	,002
department=2		0,000ª	

Probability distribution:Multinomial
Link function:Cumulative logit

ªThis coefficient is set to zero because it is redundant.

Second a random effect analysis will be performed.

Command:
Analyze....Mixed Models....Generalized Linear Mixed Models....click Data Structure....click "patient_id" and drag to Subjects on the Canvas....click Fields and Effects....click Target....Target: select "fall with/out injury"....click Fixed Effects ....click "agecat", and "department", and drag to Effect Builder:....mark Include intercept....click Random Effects....click Add Block...mark Include intercept ....Subject combination: select patient_id....click OK.....click Run.

Source	F	df1	df2	Sig.
Corrected Model ▼	7,935	4	49	,000
agecat	5,513	2	49	,007
department	7,602	2	49	,001

Probability distribution:Multinomial
Link function:Cumulative logit

Model Term		Coefficient ▶	Sig.
Threshold for falloutofbed=	0	2,082	,015
	1	5,464	,000
agecat=0		3,869	,003
agecat=1		0,096	,900
agecat=2		0,000[a]	
department=0		3,228	,004
department=1		3,566	,000
department=2		0,000[a]	

Probability distribution:Multinomial
Link function:Cumulative logit

[a]This coefficient is set to zero because it is redundant.

The underneath results show the test-statistics of the random intercept model.

The underneath results show the test-statistics of the random intercept model.
p = 0.007 and 0.013	overall for age,
p = 0.001 and 0.004	overall for department,
p = 0.003 and 0.005	regression coefficients for age class 0 versus 2,
p = 0.900 and 0.998	for age class 1 versus 2,
p = 0.004 and 0.008	for department 0 versus 2, and
p = 0.0001 and 0.0002	for department 1 versus 2.

Generalized linear mixed models are, obviously, suitable for analyzing data with multiple categorical variables. Random intercept versions of these models provide better sensitivity of testing than fixed intercept models.

6   Logit Loglinear Regression

Multinomial regression is adequate for identifying the main predictors of outcome categories, like levels of injury or quality of life (QOL). An alternative approach is logit loglinear modeling. It assesses all kinds of first and second order interactions of predictors on outcome categories. This approach may allow for relevant additional conclusions from your data. An example is given.

qol	gender	married	lifestyle	age
2	1	0	0	55
2	1	1	1	32
1	1	1	0	27
3	0	1	0	77
1	1	1	0	34
1	1	0	1	35
2	1	1	1	57
2	1	1	1	57
1	0	0	0	35
2	1	1	0	42
3	0	1	0	30
1	0	1	1	34

age (years)
gender (0 = female)
married (0 = no)
lifestyle (0 = poor)
qol (quality of life levels, 1 = low, 3 = high)

The above table shows the data of the first 12 patiens of a 445 patient data file of qol (quality of life) levels and patient characteristics. The characteristics are the predictor variables of the qol levels (the outcome variable). The entire data file is in extras.springer.com, and is entitled "chap5logitlinear". It is previously used by the authors in SPSS for starters and 2nd levelers, Chap. 51, Springer Heidelberg Germany, 2016. We will first perform a traditional multinomial regression in order to test the linear relationship between the predictor levels and the chance (actually the odds, or to be precise logodds) of having one of the three qol levels. Start by opening SPSS, and entering the data file. For analysis the statistical model Multinomial Logistic Regression in the module Regression is required.

Command:
Analyze....Regression....Multinomial Logistic Regression....Dependent: enter "qol".... Factor(s): enter "gender, married, lifestyle"....Covariate(s): enter "age".... click OK.

The underneath table shows the main results.

Parameter Estimates

qol[a]		B	Std. Error	Wald	df	Sig.	Exp(B)	95% Confidence Interval for Exp (B)	
								Lower Bound	Upper Bound
low	Intercept	28,027	2,539	121,826	1	,000			
	age	-,559	,047	143,158	1	,000	,572	,522	,626
	[gender=0]	,080	,508	,025	1	,875	1,083	,400	2,930
	[gender=1]	0[b]	.	.	0	.			
	[married=0]	2,081	,541	14,784	1	,000	8,011	2,774	23,140
	[married=1]	0[b]	.	.	0	.			
	[lifestyle=0]	-,801	,513	2,432	1	,119	,449	,164	1,228
	[lifestyle=1]	0[b]	.	.	0	.			
medium	Intercept	20,133	2,329	74,743	1	,000			
	age	-,355	,040	79,904	1	,000	,701	,649	,758
	[gender=0]	,306	,372	,674	1	,412	1,358	,654	2,817
	[gender=1]	0[b]	.	.	0	.			
	[married=0]	,612	,394	2,406	1	,121	1,843	,851	3,992
	[married=1]	0[b]	.	.	0	.			
	[lifestyle=0]	-,014	,382	,001	1	,972	,987	,466	2,088
	[lifestyle=1]	0[b]	.	.	0	.			

a. The reference category is: high.
b. This parameter is set to zero because it is redundant.

First, the unmarried subjects have a greater chance of QOL level 1 than the married ones (the b-value is positive here). Second, the higher the age, the less chance of having the low QOL levels 1 and 2 (the b-values (regression coefficients) are negative here). If you wish, you may also report the odds ratios (Exp (B) values) here. Subsequently, we will now perform a logit loglinear analysis. For analysis the statistical model Logit in the module Loglinear is required.

Command:

Analyze.... Loglinear....Logit....Dependent: enter "qol"....Factor(s): enter "gender, married, lifestyle"....Cell Covariate(s): enter: "age"....Model: Terms in Model: enter: "gender, married, lifestyle, age"....click Continue....click Options....mark Estimates....mark Adjusted residuals....mark normal probabilities for adjusted residuals....click Continue....click OK.

The underneath table shows the observed frequencies per cell, and the frequencies to be expected, if the predictors had no effect on the outcome.

Cell Counts and Residuals[a],[b]

gender	married	lifestyle	qol	Observed Count	Observed %	Expected Count	Expected %	Residual	Standardized Residual	Adjusted Residual	Deviance
Male	Unmarried	Inactive	low	7	23,3%	9,111	30,4%	-2,111	-,838	-1,125	-1,921
			medium	16	53,3%	14,124	47,1%	1,876	,686	,888	1,998
			high	7	23,3%	6,765	22,6%	,235	,103	,127	,691
		Active	low	29	61,7%	25,840	55,0%	3,160	,927	2,018	2,587
			medium	5	10,6%	10,087	21,5%	-5,087	-1,807	-2,933	-2,649
			high	13	27,7%	11,074	23,6%	1,926	,662	2,019	2,042
	Married	Inactive	low	9	11,0%	10,636	13,0%	-1,636	-,538	-,826	-1,734
			medium	41	50,0%	43,454	53,0%	-2,454	-,543	-1,062	-2,183
			high	32	39,0%	27,910	34,0%	4,090	,953	2,006	2,958
		Active	low	15	23,8%	14,413	22,9%	,587	,176	,754	1,094
			medium	27	42,9%	21,336	33,9%	5,664	1,508	2,761	3,566
			high	21	33,3%	27,251	43,3%	-6,251	-1,590	-2,868	-3,308
Female	Unmarried	Inactive	low	12	26,1%	11,119	24,2%	,881	,303	,627	1,353
			medium	26	56,5%	22,991	50,0%	3,009	,887	1,601	2,529
			high	8	17,4%	11,890	25,8%	-3,890	-1,310	-1,994	-2,518
		Active	low	18	54,5%	19,930	60,4%	-1,930	-,687	-,978	-1,915
			medium	6	18,2%	5,799	17,6%	,201	,092	,138	,639
			high	9	27,3%	7,271	22,0%	1,729	,726	1,064	1,959
	Married	Inactive	low	15	18,5%	12,134	15,0%	2,866	,892	1,670	2,522
			medium	27	33,3%	29,432	36,3%	-2,432	-,562	-1,781	-2,158
			high	39	48,1%	39,434	48,7%	-,434	-,097	-,358	-,929
		Active	low	16	25,4%	17,817	28,3%	-1,817	-,508	-1,123	-1,855
			medium	24	38,1%	24,779	39,3%	-,779	-,201	-,882	-1,238
			high	23	36,5%	20,404	32,4%	2,596	,699	1,407	2,347

a. Model: Multinomial Logit
b. Design: Constant + qol + qol * gender + qol * married + qol * lifestyle + qol * age

The next page table shows the results of the statistical tests of the data.

Parameter Estimates[c],[d]

Parameter		Estimate	Std. Error	Z	Sig.	95% Confidence Interval	
						Lower Bound	Upper Bound
Constant	[gender = 0] * [married = 0] * [lifestyle = 0]	-7,402[a]					
	[gender = 0] * [married = 0] * [lifestyle = 1]	-7,409[a]					
	[gender = 0] * [married = 1] * [lifestyle = 0]	-6,088[a]					
	[gender = 0] * [married = 1] * [lifestyle = 1]	-6,349[a]					
	[gender = 1] * [married = 0] * [lifestyle = 0]	-6,825[a]					
	[gender = 1] * [married = 0] * [lifestyle = 1]	-7,406[a]					
	[gender = 1] * [married = 1] * [lifestyle = 0]	-5,960[a]					
	[gender = 1] * [married = 1] * [lifestyle = 1]	-6,567[a]					
[qol = 1]		5,332	8,845	,603	,547	-12,004	22,667
[qol = 2]		4,280	10,073	,425	,671	-15,463	24,022
[qol = 3]		0[b]	.	.	.	.	.
[qol = 1] * [gender = 0]		,389	,360	1,079	,280	-,317	1,095
[qol = 1] * [gender = 1]		0[b]	.	.	.	.	.
[qol = 2] * [gender = 0]		-,140	,265	-,528	,597	-,660	,380
[qol = 2] * [gender = 1]		0[b]	.	.	.	.	.
[qol = 3] * [gender = 0]		0[b]	.	.	.	.	.
[qol = 3] * [gender = 1]		0[b]	.	.	.	.	.
[qol = 1] * [married = 0]		1,132	,283	4,001	,000	,578	1,687
[qol = 1] * [married = 1]		0[b]	.	.	.	.	.
[qol = 2] * [married = 0]		-,078	,294	-,267	,790	-,655	,498
[qol = 2] * [married = 1]		0[b]	.	.	.	.	.
[qol = 3] * [married = 0]		0[b]	.	.	.	.	.
[qol = 3] * [married = 1]		0[b]	.	.	.	.	.
[qol = 1] * [lifestyle = 0]		-1,004	,311	-3,229	,001	-1,613	-,394
[qol = 1] * [lifestyle = 1]		0[b]	.	.	.	.	.
[qol = 2] * [lifestyle = 0]		,016	,271	,059	,953	-,515	,547
[qol = 2] * [lifestyle = 1]		0[b]	.	.	.	.	.
[qol = 3] * [lifestyle = 0]		0[b]	.	.	.	.	.
[qol = 3] * [lifestyle = 1]		0[b]	.	.	.	.	.
[qol = 1] * age		,116	,074	1,561	,119	-,030	,261
[qol = 2] * age		,114	,054	2,115	,034	,008	,219
[qol = 3] * age		,149	,138	1,075	,282	-,122	,419

a. Constants are not parameters under the multinomial assumption. Therefore, their standard errors are not calculated.
b. This parameter is set to zero because it is redundant.
c. Model: Multinomial Logit
d. Design: Constant + qol + qol * gender + qol * married + qol * lifestyle + qol * age

The following conclusions are appropriate.

1. The unmarried subjects have a greater chance of QOL 1 (low QOL) than their married counterparts.
2. The inactive lifestyle subjects have a greater chance of QOL 1 (low QOL) than their adequate-lifestyle counterparts.
3. The higher the age the more chance of QOL 2 (medium level QOL), which is neither very good nor very bad, nut rather in-between (as you would expect).

We may conclude that the two procedures produce similar results, but the latter method provides some additional information about the lifestyle.

7 Hierarchical Loglinear Regression

All of the results from the above logit loglinear analysis can also be obtained from a hierarchical loglinear analysis, but the latter can produce more. Not only first order interactions but also second – third – fourth etc. order interactions can be readily computed. For analysis no menu commands are available in SPSS. However, the syntax commands to be given for the purpose are easy.

Command:
click File....click New....click Syntax....Syntax Editor....enter: hiloglinear qol(1,3) lifestyle (0,1) / criteria = delta (0) / design = qol*lifestyle/ print=estim....click Run....click All.

K-Way and Higher-Order Effects

	K	df	Likelihood Ratio		Pearson		Number of Iterations
			Chi-Square	Sig.	Chi-Square	Sig.	
K-way and Higher Order Effects[a]	1	5	35,542	,000	35,391	,000	0
	2	2	24,035	,000	23,835	,000	2
K-way Effects[b]	1	3	11,507	,009	11,556	,009	0
	2	2	24,035	,000	23,835	,000	0

a. Tests that k-way and higher order effects are zero.
b. Tests that k-way effects are zero.

Parameter Estimates

Effect	Parameter	Estimate	Std. Error	Z	Sig.	95% Confidence Interval	
						Lower Bound	Upper Bound
qol*lifestyle	1	-,338	,074	-4,580	,000	-,483	-,193
	2	,246	,067	3,651	,000	,114	,378
qol	1	-,206	,074	-2,789	,005	-,351	-,061
	2	,149	,067	2,208	,027	,017	,281
lifestyle	1	,040	,049	,817	,414	-,057	,137

The above tables in the output sheets show the most important results of the loglinear analysis.

1. There is a significant interaction "qol times lifestyle" at $p = 0,0001$, meaning that the qol levels in the inactive lifestyle group is different from those of the active lifestyle group.
2. There is also a significant qol effect at $p = 0,005$, meaning that medium and high qol is observed significantly more often than low qol.
3. There is no significant lifestyle effect, meaning that inactive and active lifestyles are equally distributed in the data.

For third order hierarchical loglinear modeling the underneath commands are required.

Command:
click File....click New....click Syntax....Syntax Editor....enter: hiloglinear qol(1,3) lifestyle (0,1) married(0,1) / criteria = delta (0) / design = qol*lifestyle*married/ print=estim....click Run....click All.

K-Way and Higher-Order Effects

	K	df	Likelihood Ratio		Pearson		Number of Iterations
			Chi-Square	Sig.	Chi-Square	Sig.	
K-way and Higher Order Effects[a]	1	11	120,711	,000	118,676	,000	0
	2	7	68,839	,000	74,520	,000	2
	3	2	15,947	,000	15,429	,000	3
K-way Effects[b]	1	4	51,872	,000	44,156	,000	0
	2	5	52,892	,000	59,091	,000	0
	3	2	15,947	,000	15,429	,000	0

a. Tests that k-way and higher order effects are zero.
b. Tests that k-way effects are zero.

Parameter Estimates

Effect	Parameter	Estimate	Std. Error	Z	Sig.	95% Confidence Interval	
						Lower Bound	Upper Bound
qol*lifestyle*married	1	-,124	,079	-1,580	,114	-,278	,030
	2	,301	,079	3,826	,000	,147	,456
qol*lifestyle	1	-,337	,079	-4,291	,000	-,491	-,183
	2	,360	,079	4,573	,000	,206	,514
qol*married	1	,386	,079	4,908	,000	,232	,540
	2	-,164	,079	-2,081	,037	-,318	-,010
lifestyle*married	1	-,038	,056	-,688	,492	-,147	,071
qol	1	-,110	,079	-1,399	,162	-,264	,044
	2	,110	,079	1,398	,162	-,044	,264
lifestyle	1	,047	,056	,841	,401	-,062	,156
married	1	-,340	,056	-6,112	,000	-,449	-,231

The above tables give the main results, and show that the analysis allows for some wonderful conclusions.

1. In the married subjects the combined effect of qol and lifestyle is different at p = 0,0001.
2. In the active lifestyle subjects qol scores are significantly different from those of the inactive lifestyle subjects at p = 0,0001.
3. In the married subjects the qol scores are significantly different from those of the unmarried ones at p = 0,037.
4. In the married subjects the lifestyle is not different from that of the unmarried subjects (p = 0,492).
5. The qol scores don't have significantly different counts (p = 0,162).

6. Lifestyles don't have significantly different counts (p = 0,401).
7. The married status is significantly more frequent than the unmarried status (p = 0,0001).

The many p-values need not necessarily be corrected for multiple testing, because of the hierarchical structure of the overall analysis. It start with testing first order models. If significant, then second order. If significant, then third order etc. For fourth order hierarchical loglinear modeling the underneath commands are required.

Command:
click File....click New....click Syntax....Syntax Editor....enter: hiloglinear qol(1,3) lifestyle (0,1) married (0,1) gender (0,1) / criteria = delta (0) / design = qol*lifestyle*married*gender/ print=estim....click Run....click All.

K-Way and Higher-Order Effects

	K	df	Likelihood Ratio		Pearson		Number of Iterations
			Chi-Square	Sig.	Chi-Square	Sig.	
K-way and Higher Order Effects[a]	1	23	133,344	,000	133,751	,000	0
	2	18	81,470	,000	90,991	,000	2
	3	9	25,896	,002	25,570	,002	3
	4	2	,042	,979	,042	,979	3
K-way Effects[b]	1	5	51,874	,000	42,760	,000	0
	2	9	55,573	,000	65,421	,000	0
	3	7	25,855	,001	25,528	,001	0
	4	2	,042	,979	,042	,979	0

a. Tests that k-way and higher order effects are zero.
b. Tests that k-way effects are zero.

Parameter Estimates

Effect	Parameter	Estimate	Std. Error	Z	Sig.	95% Confidence Interval	
						Lower Bound	Upper Bound
qol*lifestyle*married*gender	1	-,006	,080	-,074	,941	-,163	,151
	2	-,010	,080	-,127	,899	-,166	,146
qol*lifestyle*married	1	-,121	,080	-1,512	,130	-,278	,036
	2	,297	,080	3,726	,000	,141	,453
qol*lifestyle*gender	1	-,096	,080	-1,202	,229	-,254	,061
	2	,086	,080	1,079	,281	-,070	,242
qol*married*gender	1	,071	,080	,887	,375	-,086	,228
	2	-,143	,080	-1,800	,072	-,300	,013
lifestyle*married*gender	1	-,065	,056	-1,157	,247	-,176	,045
qol*lifestyle	1	-,341	,080	-4,251	,000	-,498	-,184
	2	,355	,080	4,455	,000	,199	,511
qol*married	1	,382	,080	4,769	,000	,225	,540
	2	-,162	,080	-2,031	,042	-,318	-,006
lifestyle*married	1	-,035	,056	-,623	,533	-,146	,075
qol*gender	1	-,045	,080	-,565	,572	-,203	,112
	2	,018	,080	,223	,823	-,138	,174
lifestyle*gender	1	-,086	,056	-1,531	,126	-,197	,024
married*gender	1	-,007	,056	-,123	,902	-,118	,104
qol	1	-,119	,080	-1,488	,137	-,276	,038
	2	,111	,080	1,390	,164	-,045	,267
lifestyle	1	,041	,056	,720	,472	-,070	,151
married	1	-,345	,056	-6,106	,000	-,455	-,234
gender	1	-,034	,056	-,609	,543	-,145	,076

The above tables show, that the results of the 4th order model are very much similar to that of the 3rd order model, and that the interaction gender*lifestyle* married*qol was not statistically significant. And, so, we can conclude here.

1. In the separate genders the combined effects of lifestyle, married status and quality of life were not significantly different.
2. In the married subjects the combined effect of qol and lifestyle is different at p = 0,0001.
3. In the active lifestyle subjects qol scores are significantly different from those of the inactive lifestyle at p = 0,0001.
4. The difference in married status is significant a p = 0,0001.
5. The qol scores don't have significantly different counts (p = 0,164).

Like with third order hierarchical loglinear modeling, the many p-values need not necessarily be corrected for multiple testing, because of its hierarchical structure. It starts with testing first order models. If significant, then second order. If significant, then third order etc.

Pearson chi-square test can answer questions like: is the risk of falling out of bed different between the departments of surgery and internal medicine. The analysis is very limited, because the interaction between two variables is assessed only. However, we may also be interested in the effect of the two variables separately.

Also, higher order contingency tables do exist. E.g., we may want to know, whether variables like ageclass, gender, and other patient characteristics interact with the former two variables. Pearson is unable to assess higher order contingency tables.

Hiloglinear modeling enables to assess both main variable effects, and higher order (multidimensional) contingency tables. For SPSS hiloglinear modeling the syntax commands are given in this chapter.

Hiloglinear modeling is the basis of a very new and broad field of data analysis, concerned with the associations between multidimensional categorical inputs.

8 Conclusion

Clinical studies often have categories (cats) as outcome, like various levels of health or disease. Multinomial regression is suitable for analysis, but, if one or two outcome categories in a study are severely underpresented, ordinal regression including specific link functions may better fit the data. The underneath table shows that in the example of this chapter the high qol cats were much more often present than the lower qol cats.

Command:
Analyze....Descriptive Statistics....Frequencies....Variable(s): enter "qol score".... click OK.

qol score

		Frequency	Percent	Valid Percent	Cumulative Percent
Valid	very low	86	19,1	19,1	19,1
	low	73	16,2	16,2	35,3
	medium	71	15,8	15,8	51,1
	high	109	24,2	24,2	75,3
	very high	111	24,7	24,7	100,0
	Total	450	100,0	100,0	

The current chapter also shows that, unlike multinomial regression, ordinal regression tests the outcome categories as an overall function.

We should add that with loglinear models many null hypotheses are tested with these pretty complex models and that the risk of type one errors is large. If a more confirmative result is required, then multiplicity adjustments are required, and little statistical significance if any will be left in these kinds of analyses, Nonetheless, they are fascinating, and particularly with large samples often clinically relevant even so. An advantage of hierarchical loglinear modeling is that multiplicity adjustment is less necessary than it is with the other polytomous models.

SPSS Version 22 has started to provide an automated model for association analysis of multiple categorical inputs, and for producing multiway contingency tables. However, the syntax commands, already available in earlier versions, are pretty easy, and SPSS minimizes the risk of typos by providing already written commands.

Reference

More background, theoretical and mathematical information of ordinal regression and ordinal data is given in Machine learning in medicine a complete overview, Chaps. 11 and 37, Springer Heidelberg Germany, 2015, from the same authors. More information about the principles of hierarchical loglinear regressions, otherwise called hiloglinear regressions are in Chap. 52, SPSS for starters and 2nd levelers, Springer Heidelberg Germany, 2015, from the same authors.

Chapter 6
Time to Event Regressions Other than Traditional Cox

Cox with Time Dependent Predictors, Segmented Cox, Interval Censored Regressions, Autocorrelations, Polynomial Regressions

Abstract Time to event regressions are used to describe the percentages of patients having an event in longitudinal studies. However, in survival studies the numbers of events are often disproportionate with time, or checks are not timely made, or changes are sinusoidally due to seasonal effect. Many more mechanisms may bias exponential event patterns. Examples are given as well as appropriate methods for analyses.

Keywords Time dependent predictors · Segmented Cox · Interval censored regression · Autocorrelations · Polynomial regression

1 Introduction, History, and Background

In the Chap. 2 the traditional proportional hazard regression method of Cox is explained. It is underlined, that Cox regression is based on an exponential model with, per time unit, the same percentage of patients having an event, a pretty strong assumption for complex creatures like human beings. Yet it has been widely used for the assessment of time – to – event analyses, even if the data did not very well match an exponential model. Fortunately, several alternative models are currently available, and have proven to often better fit longitudinal clinical data, than the traditional Cox model does. In this chapter the underneath models will be explained with the help of data examples.

1. Cox with Time Dependent Predictors. Special form of Cox regression first published in reviews by the scientific editor Prentice-Hall NJ 1978.
2. Segmented Cox. Special form of Cox regression first published in reviews by the scientific editor Prentice-Hall NJ 1978.

Electronic Supplementary Material The online version of this chapter (https://doi.org/10.1007/978-3-030-61394-5_6) contains supplementary material, which is available to authorized users. The videos can be accessed by scanning the related images with the SN More Media App.

3. Interval Censored Regression. It is one of many generalized linear models with a so – called interval censored link function (see also the Chaps. 4 and 5).
4. Autocorrelations (see also the Chap. 21). It uses linear correlation coefficients for seasonal observations, as developed by Yule (1871–1951 Tuebingen Germany), Box (1932–1982 London UK), and Jenkins (1919–2013 Madison Wisconsin).
5. Polynomial Regression. Polynomial regression models are usually fitted with the help of the method of least squares, as published in 1805 by Legendre (Chap. 1) and in 1809 by Gauss (Chap. 2). The first design of an experiment for polynomial regression appeared in an 1815 paper of Gergonne (1771–1859 Montpellier France).

2   Cox with Time Dependent Predictors

Cox regression assumes, that the proportional hazard of a predictor regarding survival works time-independently. However, in practice time-dependent disproportional hazards are not uncommon. E.g., the level of LDL cholesterol is a strong predictor of cardiovascular survival. However, in a survival study virtually no one will die from elevated values in the first decade of observation. LDL cholesterol may be, particularly, a killer in the second decade of observation, and in the third decade those with high levels may all have died, and other reasons for dying may occur. In other words the deleterious effect of 10 years elevated LDL-cholesterol may be different from that of 20 years. The traditional Cox regression model is not appropriate for analyzing the effect of LDL cholesterol on survival, because it assumes that the relative hazard of dying is the same in the first, second and third decade. Thus, there seems to be a time-dependent disproportional hazard, and if you want to analyze such data, an extended Cox regression model allowing for non-proportional hazards must be applied, and is available in SPSS. We will use the underneath data example. The first 10 patients are given in the table underneath. The entire data file is entitled "chap6coxtimedependent", and is in extras.springer.com (treat = treatment modality).

Variables 1-6

time to event	event 1 = yes	treat 0 or 1	age years	gender 0 = fem	LDL-cholesterol 1 = > 3,9 mmol/l
1	2	3	4	5	6
1,00	1	0	65,00	,00	1,00
1,00	1	0	66,00	,00	1,00
2,00	1	0	73,00	,00	1,00
2,00	1	0	54,00	,00	1,00
2,00	1	0	46,00	,00	1,00
2,00	1	0	37,00	,00	1,00
2,00	1	0	54,00	,00	1,00
2,00	1	0	66,00	,00	1,00
2,00	1	0	44,00	,00	1,00
3,00	0	0	62,00	,00	1,00

Start by opening the data file in your computer with SPSS installed. First, a time-independent Cox analysis will be performed.

Command:
Analyze....Survival....Cox Regression....time: follow years....status: event....Define Event: enter 1....Covariates....click Categorical....Categorical Covariates: enter elevated LDL-cholesterol....click Continue....click Plots....mark Survival....mark Hazard....click Continue....click OK.

Variables in the Equation

	B	SE	Wald	df	Sig.	Exp(B)
cholesterol	-,544	,332	2,682	1	,102	,581

Var 00006 is a binary variable for LDL-cholesterol. It is not a significant predictor of survival with a p-value and a hazard ratio of only 0,102 and 0.581 respectively, as demonstrated above by a simple Cox regression with event as outcome variable and LDL cholesterol as predictor. The investigators believe that the presence of LDL-cholesterol must be a determinant of survival. And if we look at the data, we will observe that something very special is going on: in the first decade virtually no one with elevated LDL-cholesterol dies. In the second decade virtually everyone with an elevated LDL-cholesterol does: LDL-cholesterol seems to be particularly a killer in the second decade. Then, in the third decade other reasons for dying seem to have taken over. In order to assess whether elevated LDL-cholesterol adjusted for time has a significant effect on survival, a time-dependent Cox regression will be performed. For that purpose the time–dependent covariate is defined as a function of both the variable time (called "T_" in SPSS) and the LDL-cholesterol variable, while using the product of the two. This product is applied as "time-dependent predictor of survival", and a usual Cox model is, subsequently, performed (Cov = covariate). For analysis the statistical model Cox Time Dependent in the module Survival is required.

Command:
Analyze....Survival....Cox w/Time-Dep Cov....Compute Time-Dep Cov....Time (T_) transfer to box Expression for T_Cov....add the sign *....add the LDL-cholesterol variable....Model....Time: follow months....Status: event - ?: Define Event: enter 1....click Continue....T_Cov transfer to box Covariates....click OK.

Variables in the Equation

	B	SE	Wald	df	Sig.	Exp(B)
T_COV_	-,131	,033	15,904	1	,000	,877

The above results table of the "Cox regression with time-dependent variables" shows that the presence of an elevated LDL-cholesterol adjusted for differences in time is a highly significant predictor of survival. Time dependent Cox regression is convenient, if some of your predictors are time dependent, like in the above data example.

3   Segmented Cox

Segmented time-dependent Cox regression goes one step further than the above time dependent Cox, and it assesses, whether the interaction with time is different at different periods of the study. An example is given. The primary scientific question: is frailty a time-dependently changing variable in patients admitted to hospital for exacerbation of chronic obstructive pulmonary disease (COPD). A simulated data file of 60 patients admitted to hospital for exacerbation of COPD is given underneath. All of the patients are assessed for frailty scores once a week. The frailty scores run from 0 to 100 (no frail to very frail).

Variables

1 days to discharge	2 cured	3 gender	4 frailty 1st	5 frailty 2nd	6 frailty 3rd week
1,00	1,00	1,00	15,00		
1,00	1,00	1,00	18,00		
1,00	1,00	1,00	16,00		
1,00	1,00	1,00	17,00		
2,00	1,00	1,00	15,00		
2,00	1,00	1,00	20,00		
2,00	1,00	1,00	16,00		
2,00	1,00	1,00	15,00		
3,00	1,00	,00	18,00		
3,00	1,00	,00	15,00		
3,00	1,00	1,00	16,00		
4,00	1,00	1,00	15,00		
4,00	1,00	1,00	18,00		
5,00	1,00	1,00	19,00		
5,00	1,00	1,00	19,00		
5,00	1,00	1,00	19,00		
6,00	1,00	1,00	18,00		
6,00	1,00	1,00	17,00		
6,00	1,00	,00	19,00		
7,00	1,00	,00	16,00		
8,00	1,00	,00	60,00	15,00	
8,00	1,00	,00	69,00	16,00	
8,00	1,00	,00	67,00	17,00	
9,00	1,00	1,00	60,00	19,00	
9,00	1,00	1,00	86,00	24,00	
10,00	1,00	1,00	87,00	16,00	
10,00	1,00	,00	75,00	10,00	
10,00	1,00	,00	76,00	20,00	
10,00	1,00	,00	67,00	32,00	
11,00	1,00	1,00	56,00	24,00	
11,00	1,00	1,00	78,00	25,00	
12,00	1,00	1,00	58,00	26,00	
12,00	1,00	,00	59,00	25,00	
13,00	1,00	,00	77,00	20,00	
13,00	1,00	1,00	66,00	16,00	
13,00	1,00	1,00	65,00	18,00	
13,00	1,00	1,00	68,00	10,00	
14,00	1,00	1,00	85,00	16,00	
14,00	1,00	,00	65,00	23,00	
14,00	1,00	,00	65,00	20,00	
15,00	1,00	,00	54,00	60,00	14,00
16,00	1,00	,00	43,00	68,00	15,00

variable 1 = days to discharge from hospital
variable 2 = cured or lost from observation (1 = cured)
variable 3 = gender
variable 4 = frailty index first week (0-100)
variable 5 = frailty index second week (0-100)
variable 6 = frailty index third week (0-100).
The missing values in the variables 5 and 6 are those from patients already discharged from hospital.

The above table gives the first 42 patients of 60 patients assessed for their frailty scores after 1, 2 and 3 weeks of clinical treatment. It can be observed that in the first week frailty scores at discharge were 15–20, in the second week 15–32, and in the third week 14–24. Patients with scores over 32 were never discharged. Frailty scores were probably a major covariate of time to discharge. The entire data file is in extras. springer.com, and is entitled "chap6segmentedcox". We will first perform a simple time dependent Cox regression. Start by opening the data file in your computer. For analysis the statistical model Cox Time Dependent in the module Survival is required.

Command:
Analyze....Survival....Cox w/Time-Dep Cov....Compute Time-Dep Cov.... Time (T_); transfer to box Expression for T_Cov....add the sign *....add the frailty variable third week....Model....Time: day of discharge....Status: cured or lost.... Define: cured = 1....Continue....T_Cov: transfer to Covariates....click OK.

Variables in the Equation

	B	SE	Wald	df	Sig.	Exp(B)
T_COV_	,000	,001	,243	1	,622	1,000

The above table shows the result: frailty is not a significant predictor of day of discharge. However, patients are generally not discharged from hospital, until they are non-frail at a reasonable level, and this level may be obtained at different periods of time. Therefore, a segmented time dependent Cox regression may be more adequate for analyzing these data.

For analysis the statistical model Cox Time Dependent in the module Survival is again required.

Command:
Survival.....Cox w/Time-Dep Cov....Compute Time-Dependent Covariate....
Expression for T_COV_: enter (T_ >= 1 & T_ < 11) * VAR00004 + (T_ >= 11 & T_ < 21) * VAR00005 + (T_ >= 21 & T_ < 31)....Model....Time: enter Var 1....Status: enter Var 2 (Define events enter 1)....Covariates: enter T_COV_ .... click OK).

Variables in the Equation

	B	SE	Wald	df	Sig.	Exp(B)
T_COV_	-,056	,009	38,317	1	,000	,945

The above table shows that the independent variable, segmented frailty variable T_COV_, is, indeed, a very significant predictor of the day of discharge. We will, subsequently, perform a multiple segmented time dependent Cox regression with treatment modality as second predictor variable. For analysis a multiple segmented time dependent Cox regression must be performed.

Command:
same commands as above, except for Covariates: enter T_COV and treatment.…
click OK.

Variables in the Equation

	B	SE	Wald	df	Sig.	Exp(B)
T_COV_	-,060	,009	41,216	1	,000	,942
VAR00003	,354	,096	13,668	1	,000	1,424

The above table shows that both the frailty (variable T_COV_) and treatment (variable 3) are very significant predictors of the day of discharge with hazard ratios of 0,942 and 1,424. The new treatment is about 1,4 times better and the patients are doing about 0,9 times worse per frailty score point.

4   Interval Censored Regression

In survival studies often time to first outpatient clinic check instead of time to event is measured. Somewhere in the interval between the last and current visit an event may have taken place. For simplicity such data are often analyzed using the proportional hazard model of Cox. However, this analysis is not entirely appropriate here. It assumes that time to first outpatient check is equal to time to relapse. Instead of a time to relapse, an interval is given, in which the relapse has occurred, and so this variable is somewhat more loose than the usual variable time to event. An appropriate statistic for the current variable would be the mean time to relapse inferenced from a generalized linear model with an interval censored link function, rather than the proportional hazard method of Cox.

A data example is given. In 51 patients in remission their status at the time-to-first-outpatient-clinic-control was checked (mths = months, treat = treatment).

time to 1st check (month)	result relapse 0 = no (0 or 1)	treat modality 1 or 2
11	0	1
12	1	0
9	1	0
12	0	1
12	0	0
12	0	1
5	1	1
12	0	1
12	0	1
12	0	0

The first 10 patients are above. The entire data file is entitled "chap6intervalcensored", and is in extras.springer.com. It is previously used by the authors in SPSS for starters and 2nd levelers, Chap. 59, Springer Heidelberg Germany, 2016. Cox regression was first applied. Start by opening the data file in SPSS statistical software. For analysis the module Generalized Linear Models is required. It consists of two submodules: Generalized Linear Models and Generalized Estimation Models. For the censored data analysis the Generalized Linear Models submodule of the Generalized Linear Models module is required.

Command:

Analyze....click Generalized Linear Models....click once again Generalized Linear Models....Type of Model....mark Interval censored survival....click Response.... Dependent Variable: enter Result....Scale Weight Variable: enter "time to first check"....click Predictors....Factors: enter "treatment"....click Model....click once again Model: enter once again "treatment"....click Save.... mark Predicted value of mean of response....click OK.

Parameter Estimates

			95% Wald Confidence Interval		Hypothesis Test		
Parameter	B	Std. Error	Lower	Upper	Wald Chi-Square	df	Sig.
(Intercept)	.467	.0735	.323	.611	40.431	1	.000
[treatment=0]	-.728	.1230	-.969	-.487	35.006	1	.000
[treatment=1]	0ª	.	.	.	.	.	.
(Scale)	1ᵇ						

Dependent Variable: Result
Model: (Intercept), treatment

a. Set to zero because this parameter is redundant.
b. Fixed at the displayed value.

The generalized linear model shows, that, after censoring the intervals, the treatment 0 is, as compared to treat 1, a very significant maintainer of remission.

5   Autocorrelations

Seasonal data are data that are repetitive by season. The assessment of seasonality requires a measure of repetitiveness. Levels of linear autocorrelation is such a measure. It may not be regression analysis, but linear correlation coefficients are closely related to linear regression. Linear correlation coefficients of the values between slightly different seasonal curves are used for making predictions from seasonal data about the presence of seasonality.

Average C-reactive protein in group of healthy subjects (mg/l)	Month
1,98	1
1,97	2
1,83	3
1,75	4
1,59	5
1,54	6
1,48	7
1,54	8
1,59	9
1,87	10

The entire data file is in extras.springer.com, and is entitled "chap6seasonality". It was previously used by the authors in SPSS for starters and 2nd levelers, Chap. 58, Springer Heidelberg Germany, 2016. Start by opening the data file in your computer with SPSS installed.

Command:
Graphs....Chart Builder....click Scatter/Dot....click mean C-reactive protein level and drag to the Y-Axis....click time and drag to the X-Axis....click OK..... double-click in Chart Editor....click Interpolation Line....Properties: click Straight Line.

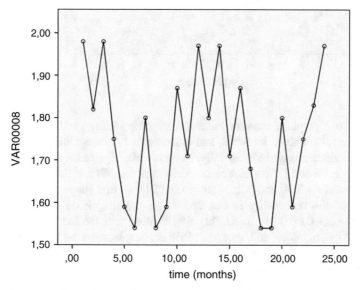

The above graph shows that the average monthly C-reactive protein levels look seasonal.

We will now assess seasonality with autocorrelations.

For analysis the statistical model Autocorrelations in the module Forecasting is required.

Command:

Analyze....Forecasting....Autocorrelations....move monthly percentages into Variable Box....mark Autocorrelations....mark Partial Autocorrelations....OK.

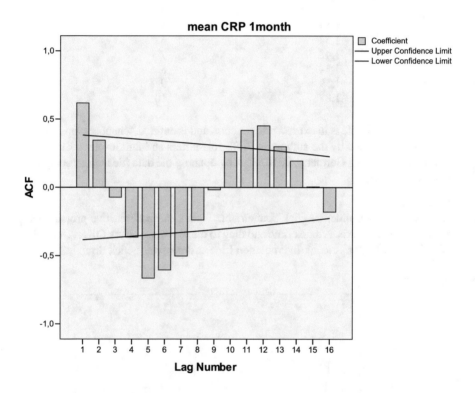

The above graph of monthly autocorrelation coefficients with their 95% confidence intervals is given by SPSS, and it shows that the magnitude of the monthly autocorrelations changes sinusoidally. For example, significant autocorrelations at the month no. 5 and 12 (correlation coefficients of -0.664 and 0.452 (SE 0,174 and 0.139, t-values -3.80 and 3.25, both $p < 0.001$)) further support seasonality. The strength of the seasonality is assessed using the magnitude of r^2. For example $= r^2 = (-0.664)^2 = 0.44$. This would mean that the lagcurve predicts the datacurve by only 44%, and, thus, that 56% is unexplained. And so, the seasonality may be statistically significant, but it is still pretty weak, and a lot of unexplained variability, otherwise called noise, is present.

6 Polynomial Regression

Underneath polynomial regression equations of the first to fifth order are given with y as dependent and x as independent variables.

$$y = a + bx \text{ first order (linear) relationship}$$

$$y = a + bx + cx^2 \text{ second order (parabolic) relationship}$$

$$y = a + bx + cx^2 + dx^3 \text{ third order (hyperbolic) relationship}$$

$$y = a + bx + cx^2 + dx^3 + ex^4 \text{ fourth order (sinusoidal) relationship}$$

$$y = a + bx + cx^2 + dx^3 + ex^4 + fx^5 \text{ fifth order polynomial relationship}$$

Higher order polynomes can visualize longitudinal observations in clinical research.

As an example, in a patient with mild hypertension ambulatory blood pressure measurement was performed using a portable equipment every 30 min for 24 h. The first 10 measurements are underneath, the entire data file is entitled "chap6polynomes", and is in extras.springer.com. It is previously used by the authors in SPSS for starters and 2nd levelers, Chap. 60, Springer Heidelberg Germany, 2016. Start by opening tha data file in your computer with SPSS installed.

Blood   Time
pressure (30 min intervals)
mm Hg

205,00	1,00
185,00	2,00
191,00	3,00
158,00	4,00
198,00	5,00
135,00	6,00
221,00	7,00
170,00	8,00
197,00	9,00
172,00	10,00
188,00	11,00
173,00	12,00

SPSS statistical software will be used for polynomial modeling of these data. Open the data file in SPSS. For analysis the module General Linear Model is required. We will use the submodule Univariate here.

Command:
Analyze....General Linear Model....Univariate....Dependent: enter y (mm Hg)....
Covariate(s): enter x (min)....click: Options....mark: Parameter Estimates....click
Continue....click Paste....in "/Design=x" replace x with a 5th order polynomial
equation tail (* is sign of multiplication)

$$x.x * x.x * x * x.x * x * x * x.x * x * x * x * x$$

$$\left(\text{meaning } x.x^2.x^3.x^4.x^5\right)$$

....then click the green triangle in the upper graph row of your screen.

The underneath table is in the output sheets, and gives you the partial regression
coefficients (B values) of the 5th order polynomial with blood pressure as outcome
and with time as independent variable (-7,135E-6 indicates 0.000007135, which is a
pretty small B value. However, in the equation it will have to be multiplied with x^5,
and a large, very large term will result even so.

Parameter Estimates

Dependent Variable:y

Parameter	B	Std. Error	t	Sig.	95% Confidence Interval	
					Lower Bound	Upper Bound
Intercept	206,653	17,511	11,801	,000	171,426	241,881
x	-9,112	6,336	-1,438	,157	-21,858	3,634
x*x	,966	,710	1,359	,181	-,463	2,395
x*x*x	-,047	,033	-1,437	,157	-,114	,019
x*x*x*x	,001	,001	1,471	,148	,000	,002
x*x*x*x*x	-7,135E-6	4,948E-6	-1,442	,156	-1,709E-5	2,819E-6

Parameter Estimates

Dependent Variable:yy

Parameter	B	Std. Error	t	Sig.	95% Confidence Interval	
					Lower Bound	Upper Bound
Intercept	170,284	11,120	15,314	,000	147,915	192,654
x	-7,034	4,023	-1,748	,087	-15,127	1,060
x*x	,624	,451	1,384	,173	-,283	1,532
x*x*x	-,027	,021	-1,293	,202	-,069	,015
x*x*x*x	,001	,000	1,274	,209	,000	,001
x*x*x*x*x	-3,951E-6	3,142E-6	-1,257	,215	-1,027E-5	2,370E-6

The entire equations can be written from the above B values:

$$y = 206.653 - 9,112x + 0.966x^2 - 0.47x^3 + 0.001x^4 + 0.000007135x^5$$

This equation is entered in the polynomial grapher of David Wees available on the internet at "davidwees.com/polygrapher/", and the underneath graph is drawn. This graph is speculative as none of the x terms is statistically significant. Yet, the actual data have a definite patterns with higher values at daytime and lower ones at night. Sometimes even better fit curve are obtained by taking higher order polynomes like 5th order polynomes. We should add, that, in spite of the insignificant p-values in the above tables, the two polynomes are not meaningless. The first one suggests some white coat effect, the second one suggests normotension and a normal dipping pattern. With machine learning meaningful visualizations can sometimes be produced of your data, even if statistics are pretty meaningless.

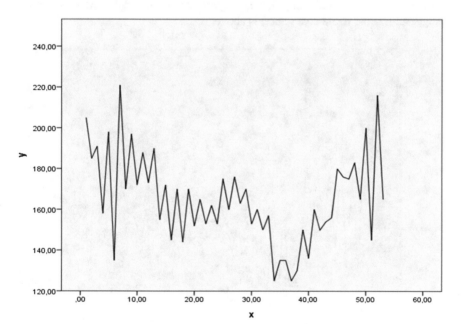

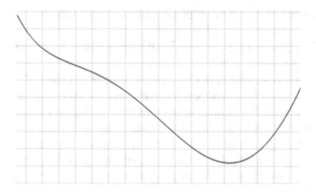

24 hour ABPM (ambulatory blood pressure measurements) recording (30 min measures) of untreated subject with hypertension and 5th order polynome (suggesting some white coat effect) is above.

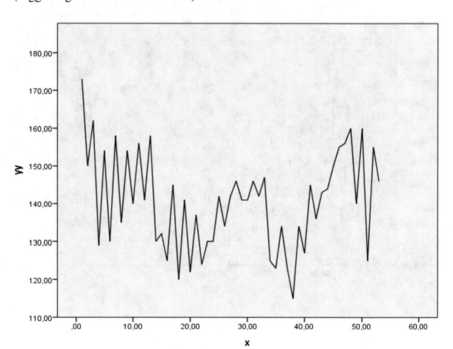

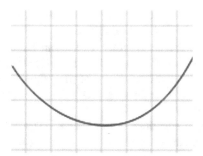

A 24 h ABPM recording (30 min measures) of the above subject treated and 5th order polynome (suggesting normotension and a normal dipping pattern) is above. Polynomes of ambulatory blood pressure measurements can, thus, be applied for visualizing hypertension types and treatment effects, as well as other circadian response patterns in clinical research.

7 Conclusion

Cox regression is based on an exponential model with per time unit the percentage of patients have an event, a pretty strong assumption for complex creatures like human beings. Yet it has been widely used for the assessment of time to event analyses, even if the data do not very well match an exponential model. Fortunately, several alternative models are currently available, and have proven to often better fit longitudinal clinical data than the traditional Cox model does. In this chapter the underneath models have been explained with the help of data examples.

1. Cox with Time Dependent Predictors is particularly suitable with a time-dependent disproportional hazard. Cox regression assumes, that the proportional hazard of a predictor regarding survival works time-independently. However, in practice time-dependent disproportional hazards are not uncommon. E.g., the level of LDL cholesterol is a strong predictor of cardiovascular survival. However, in a survival study, virtually no one will die from elevated values in the first decade of observation. LDL cholesterol may be, particularly, a killer in the second and third decades of observation.
2. Segmented time-dependent Cox regression goes one step further than the above time dependent Cox, and assesses, whether the interaction with time is different at different periods of the study.
3. Traditional Cox assumes, that time to first outpatient check is equal to time to relapse. With interval censored regressions, instead of a time to relapse, an interval is given, in which the relapse has occurred, and, so, this variable is, somewhat, more loose than the usual variable time to event. An appropriate statistic for the current variable would be the mean time to relapse, as inferenced

from a generalized linear model with an interval censored link function, rather than the proportional hazard method of Cox.

4. Seasonal data are data that are repetitive by season. The assessment of seasonality requires a measure of repetitiveness. Levels of linear autocorrelation is such a measure. Linear correlation coefficients of the values between slightly different seasonal curves are used for making predictions from seasonal data about the presence of seasonality.

5. Polynomial regression equations of increasing orders, with y as dependent time variable and x as predictor variable, give linear (= first order) relationship, parabolic (= second order) relationship, hyperbolic (= third order) relationship, sinusoidal (= fourth order) relationship, etc. Higher order polynomes can be used to visualize the effects of predictors like treatments, and patient characteristics, on circadian cardiovascular, hormonal, and other biological rhythms.

Reference

To readers requesting more background, theoretical and mathematical information of computations given, several textbooks complementary to the current production and written by the same authors are available: Statistics applied to clinical studies 5th edition, 2012, Machine learning in medicine a complete overview, 2015, SPSS for starters and 2nd levelers 2nd edition, 2015, Clinical data analysis on a pocket calculator 2nd edition, 2016, Understanding clinical data analysis, 2017, all of them edited by Springer Heidelberg Germany.

Chapter 7
Analysis of Variance (Anova)

Just Another Regression Model with Intercepts and Regression Coefficients

Abstract Data with a continuous outcome and a categorical predictor can be analyzed either with multiple linear regression or with one way Anova (analysis of variance). The results and all of the computations are identical. Analysis of variance is just another regression model with intercepts and regression coefficients.

Keywords Analysis of variance (Anova) · Anova versus regression analysis · Paired and unpaired Anovas

1 Introduction, History, and Background

Not only unpaired t-tests, but also linear regression can be used for analyzing the effect of a binary predictor on a continuous outcome. An example is given in the Chap. 1, Sect. 17. Just like unpaired t-tests, one way analyses of variance (Anovas) are for comparing parallel-groups. However, they allow for more than two parallel-groups. They can not include more than a single predictor in the analysis, often consistent of three or more parallel treatment modalities. The outcome data of the parallel-groups are assumed to be normally distributed. As an example, ethnic race data with strength score as continuous outcome and the presence of different races as multiple mutually elusive categories as predictors can be analyzed, either with multiple linear regression, or with one way Anova. The results and all of the computations are identical. The multiple linear regression of the race data is given as an example in the next section.

Analysis of variance (Anova) is a collection of statistical models and their associated estimation procedures (such as the "variation" among and between groups), as used to analyze the differences among group means in a sample. Anova was developed by the statistician and evolutionary biologist Ronald Fisher 1890–1962. Sir Ronald Aylmer Fisher FRS was a British statistician / geneticist. For

Electronic Supplementary Material The online version of this chapter (https://doi.org/10.1007/978-3-030-61394-5_7) contains supplementary material, which is available to authorized users. The videos can be accessed by scanning the related images with the SN More Media App.

his work in statistics, he has been described as "a genius who almost single-handedly created the foundations for modern statistical science" and as "the single most important figure in 20th century statistics".

In genetics, his work used mathematics to combine Mendelian genetics and natural selection. And this contributed to the revival of Darwinism in the early twentieth-century revision of the theory of evolution known as the "modern synthesis". For his contributions to biology, Fisher has been called "the greatest of Darwin's successors".

From 1919 onward, he worked at the Rothamsted Experimental Station for 14 years; there, he analysed the immense Rothamsted data from crop experiments, since the 1840s, and developed the analysis of variance methodology (Anova). He established his reputation there in the subsequent years as a biostatistician.

He is known as one of the three principal founders of population genetics. He outlined "Fisher's principle", the "Fisherian runaway" and the sexy son hypothesis theories of sexual selection. His contributions to statistics include the maximum likelihood, fiducial inference, the derivation of various sampling distributions, founding principles of the design of experiments, and much more.

Fisher held strong views on race. Throughout his life, he was a prominent supporter of eugenics, an interest which led to his work on statistics and genetics. Notably, he was a dissenting voice in the 1950 UNESCO statement "The Race Question", which very much insisted on racial differences.

He was born in East Finchley, London, England, in a middle-class household; his father, George, was a successful partner in Robinson & Fisher, auctioneers and fine art dealers. He was one of twins, with the other twin being still-born, and he grew up the youngest, with three sisters and one brother. From 1896 until 1904 they lived at Inverforth House in London (where English Heritage installed a blue plaque in 2002), before moving to Streatham. His mother, Kate, died from acute peritonitis, when he was 14, and his father lost his business 18 months later.

Lifelong poor eyesight caused his rejection by the British Army for World War I, but, probably, also helped developing his ability to visualize problems in geometrical terms, rather than mathematical models or proofs. He entered Harrow School age 14, and won the school's Neeld Medal in Mathematics. In 1909, aged 19, he won a scholarship to study Mathematics at Gonville and Caius College, Cambridge. In 1912, he gained a First in Mathematics. In 1915 he published a paper entitled "The evolution of sexual preference on sexual selection and mate choice".

During 1913–1919, Fisher worked for 6 years as a statistician in the City of London and taught physics and maths at a sequence of public schools, at the Thames Nautical Training College, and at Bradfield College. There he settled with his new bride, Eileen Guinness, with whom he had two sons and six daughters.

In 1918 he published "The Correlation Between Relatives on the Supposition of Mendelian Inheritance", in which he introduced the term variance and proposed its formal analysis. He put forward a genetics conceptual model showing that continuous variation amongst phenotypic traits measured by biostatisticians could be produced by the combined action of many discrete genes and thus be the result of Mendelian inheritance. This was the first step towards establishing population genetics and quantitative genetics, which demonstrated that natural selection could change allele frequencies in a population, resulting in reconciling its discontinuous nature with gradual evolution. Joan Box, Fisher's biographer and daughter, says, that Fisher had resolved this problem already in 1911. She not only wrote a biography of her father, but also married the statistician George E. P. Box.

Fisher was noted for being loyal to the English monarchy, and was seen as a patriot. He was a member of the Church of England, politically conservative, as well as a scientific rationalist. He developed a reputation for carelessness in his dress and was the archetype of the absent-minded professor.

Allen Orr describes him in the "Boston Review" as a "deeply devout Anglican who, between founding modern statistics and population genetics, penned articles for church magazines". In a 1955 BBC broadcast on Science and Christianity, he said: "the custom of making abstract dogmatic assertions is not, certainly, derived from the teaching of Jesus, but has been a widespread weakness among religious teachers in subsequent centuries. I do not think, that the word for the Christian virtue of faith should be prostituted to mean the credulous acceptance of all such piously intended assertions. Much self-deception in the young believer is needed to convince himself, that he knows that of which in reality he knows himself to be ignorant. That surely is hypocrisy, against which we have been most conspicuously warned."

2 Little if any Difference Between Anova and Regression Analysis

As an example, a study of four races used to predict strength scores in 60 Patients is applied. The SPSS data file is in extras.springer.com, and is entitled "chap7categorical predictors". It is previously used by the authors in SPSS for starters and 2nd levelers, Chap. 8, Springer Heidelberg Germany, 2016.

patient number	physical strength	race	age	gender	race 1 hispanics	race 2 blacks	race 3 asians	race 4 whites
1	70,00	1,00	35,00	1,00	1,00	0,00	0,00	0,00
2	77,00	1,00	55,00	0,00	1,00	0,00	0,00	0,00
3	66,00	1,00	70,00	1,00	1,00	0,00	0,00	0,00
4	59,00	1,00	55,00	0,00	1,00	0,00	0,00	0,00
5	71,00	1,00	45,00	1,00	1,00	0,00	0,00	0,00
6	72,00	1,00	47,00	1,00	1,00	0,00	0,00	0,00
7	45,00	1,00	75,00	0,00	1,00	0,00	0,00	0,00
8	85,00	1,00	83,00	1,00	1,00	0,00	0,00	0,00
9	70,00	1,00	35,00	1,00	1,00	0,00	0,00	0,00
10	77,00	1,00	49,00	1,00	1,00	0,00	0,00	0,00

The first 10 patients of the data file are above. For the analysis we will use multiple linear regression. The analysis is in SPSS statistical software. Start by opening the file in your computer with SPSS installed.

Command:

Analyze....Regression....Linear....Dependent: physical strength score.... Independent: race 2, race 3, race 4, age, gender....click OK.

The output sheets are below (race 1 = absence of race 2, race 3, and race 4, and, therefore, needs not separately analyzed):

Variables Entered/Removed[a]

Model	Variables Entered	Variables Removed	Method
1	race4, race3, race2[b]	.	Enter

a. Dependent Variable: strengthscore

b. All requested variables entered.

Model Summary

Model	R	R Square	Adjusted R Square	Std. Error of the Estimate
1	,760[a]	,578	,555	9,08741

a. Predictors: (Constant), race4, race3, race2

ANOVA[a]

Model		Sum of Squares	df	Mean Square	F	Sig.
1	Regression	6321,650	3	2107,217	25,517	,000[b]
	Residual	4624,533	56	82,581		
	Total	10946,183	59			

a. Dependent Variable: strengthscore

b. Predictors: (Constant), race4, race3, race2

Coefficients[a]

Model		Unstandardized Coefficients		Standardized Coefficients	t	Sig.
		B	Std. Error	Beta		
1	(Constant)	69,067	2,346		29,436	,000
	race2	18,600	3,318	,596	5,605	,000
	race3	-8,400	3,318	-,269	-2,531	,014
	race4	10,667	3,318	,342	3,215	,002

a. Dependent Variable: strengthscore

The analysis of the same data in one way Anova also in SPSS statistical software is given underneath.

Commands:

Analyze....Compare Means....One-way Anova....Dependent lists: Strengthscore ....
Factor: race 2, race 3, race 4, age, gender....click OK.

Output sheets:

ANOVA

strengthscore

	Sum of Squares	df	Mean Square	F	Sig.
Between Groups	6321,650	3	2107,217	25,517	,000
Within Groups	4624,533	56	82,581		
Total	10946,183	59			

The above one way Anova table is identical to the Anova table of the multiple linear regression model. The intercept (Constant) is simply the mean value of the reference group, race 1. The regression coefficients for the other three groups are the differences between the mean values of reference group and the other groups (the races 2, 3, 4).

Some might defend the view, that Anova and linear regression are not the same. Why so? Because in Anova the predictor is compared to grand mean, while in regression the predictor is dummy coded. Dummy coding creates, e.g., four variables, race 1, 2, 3, 4. Race 1 has a 0 value, and this is the reference group.

Anovas may be pretty much similar to regression analyses, but there are differences. Anovas use variance partitionings, while regressions use ordinary least squares. If you perform an Anova by heart, you will rapidly forget, that a regression would have produced the same result. Also, if you perform a regression analysis, you will observe that the overall variance partitioning is tested against a zero effect with the help of Anova. This may sound complicated, and some simple examples will be given to explain, what, computationally, is going on.

3 Paired and Unpaired Anovas

Anova is really complicated. So wake up. Anova is about partitioning of variances. For example, with an *unpaired Anova* of 3 treatment groups the total variation in the data is split into the between and within group variation.

Total variation

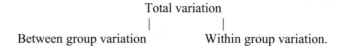

Between group variation　　　　　　Within group variation.

Variations are, subsequently, expressed as sums of squares (SS), and can be added up to obtain the total variation of the data. We will, then, assess, whether the between group variation is large compared to the within group variation (SD = standard deviation, F = Fisher's F Statistic, dfs = degrees of freedom).

1-way Anova

Group	n patients	mean	SD
1	n	-	-
2	n	-	-
3	n	-	-

Grand mean = (mean 1 + 2 +3) / 3

$$SS_{\text{between groups}} = n\,(\text{mean}_1 - \text{grand mean})^2 + n(\text{mean}_2 - \text{grand mean})^2 + \ldots$$

$$SS_{\text{within groups}} = (n - 1)\,SD_1^{\,2} + (n - 1)\,SD_2^{\,2} + \ldots$$

$$F = \frac{SS_{\text{between groups}}/\text{dfs}}{SS_{\text{within groups}}/\text{dfs}}$$

F table gives P-value

With differently sized groups weighted grand mean is required:
weighted mean = $(n_1 \text{ mean}_1 + n_2 \text{ mean}_2) / (n_1 + n_2)$.

Effect of 3 compounds on Haemoglobin levels.

Group	n patients	mean	SD
1	16	8.7125	0.8445
2	16	10.6300	1.2841
3	16	12.3000	0.9419

Grand mean = (mean 1 + 2 + 3)/3 = 10.4926

SS between groups = $16 (8.7125 - 10.4926)^2 + 16(10.6300 - 10.4926)^2 \ldots$

SS within groups = $15 \times 0.84452 + 15 \times 1.28412 + \ldots \ldots$

F = 49.9 and so P < 0.001.

In case of 2 groups: anova = t-test $(F = t^2)$.

Anova is pretty complex. It can make a lot of mess from just a few values. Prior to the computer era statistical analists needed days of hard work for a single Anova. Anova can, indeed, make a lot of mess from only a few numbers. A paired Anova with 3 treatments in a single group is given as an example. Only 4 subjects are in the underneath very small study used as example.

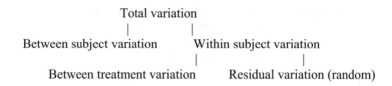

Variations are again expressed as the sums of squares (SS) and can be added up. We will assess whether between treatment variation is large compared to residual variation. For analysis a so-called "two-way Anova, balanced, no replication" is required (SD = standard deviation).

Subject	treatment 1	treatment 2	treatment 3	SD^2
1	-	-	-	-
2	-	-	-	-
3	-	-	-	-
4	-	-	-	-
Treatment mean	-	-	-	

Grand mean = (treatment mean 1 + 2 + 3) / 3= .....

$$SS \text{ within subject} = SD_1{}^2 + SD_2{}^2 + SD_3{}^2$$

$$SS \text{ treatment} = (\text{treatment mean 1} - \text{grand mean})^2 + (\text{treatment mean 2} - \text{grandmean})^2 + \ldots$$

$$SS \text{ residual} = SS \text{ within subject} - SS \text{ treatment}$$

$$F = \frac{SS \text{ treatment}/dfs}{SS \text{ residual}/dfs}$$

F table gives P-value.
Effect of 3 treatments on vascular resistance.

Effect of 3 treatments on vascular resistance

Person	treatment 1	treatment 2	treatment 3	SD^2
1	22.2	5.4	10.6	147.95
2	17.0	6.3	6.2	77.05
3	14.1	8.5	9.3	18.35
4	17.0	10.7	12.3	21.45
Treatment mean	17.58	7.73	9.60	

Grand mean = 11.63

$$\text{SS within subject} = 147.95 + 77.05 + \ldots$$

$$\text{SS treatment} = (17.58 - 11.63)^2 + (7.73 - 11.63)^2 + \ldots$$

$$\text{SS residual} = \text{SS within subject} - \text{SS treatment}$$

$F = 14.31$ and so $P < 0.01$.
In case of 2 treatments: ANOVA $=$ t-test $(F = t^2)$.

4   Conclusion

The effect of a binary predictor on a continuous outcome can be analyzed with unpaired t-tests or with linear regression. The results will be identical. With three or more treatment groups one way analyses of variance or linear regression can likewise be applied for analysis, and will also produce the same results. The results and, indeed, all of the computations are identical. Although Anova is based on variance partitions and linear regressions on ordinary least squares, distinction between the two is hair – splitting, rather than anything else.

Reference

To readers requesting more background, theoretical and mathematical information of computations given, several textbooks complementary to the current production and written by the same authors are available: Statistics applied to clinical studies 5th edition, 2012, Machine learning in medicine a complete overview, 2015, SPSS for starters and 2nd levelers 2nd edition, 2015, Clinical data analysis on a pocket calculator 2nd edition, 2016, Understanding clinical data analysis, 2017, all of them edited by Springer Heidelberg Germany.

Chapter 8
Repeated Outcomes Regression Methods

Repeated Measures Analysis of Variance (Anova), Mixed Linear Models with or Without Random Interaction, Doubly Repeated Measures Multivariate Anova

Abstract In clinical research repeated measures in a single subject are common. In its simplest version it is a statistical model only assessing the differences between within subject differences as outcomes without any predictors. More complex data are increasingly present in clinical research. The current chapter provides analysis methods for repeated outcomes including predictors either continuous or categorical. Also analysis models where within subject comparisons receive fewer degrees of freedom for the benefit of better sensitivity to test between subject differences, are reviewed. Finally, multivariate models with batteries of more than a single set of repeated measures are reviewed. Some of the models reviewed are pretty complex, but increasingly used and sometimes indispensable in modern clinical data analysis.

Keywords Repeated outcome regressions · Within-subject correlation · Repeated measures analysis of variance (Anova) · Repeated measures Anova versus Ancova · Repeated measures with predictors · Mixed linear models · Doubly repeated measures multivariate Anova

1   Introduction, History, and Background

In clinical research repeated measures in a single subject are common. The problem with repeated measures is, that they, usually, are more close to one another than unrepeated measures. If this is not taken into account, then data analysis will lose power. Repeated measures Anova (analysis of variance), in SPSS available in the general linear models module, is the classical model for the analysis of such data. Repeated measures method has multiple outcomes, and are, thus, a multivariate regression methodology.

It has been recently supplemented by a novel method, mixed linear modeling. Mixed modeling was first described by Henderson (Biometrics 1975; 31: 423) in the

---

Electronic Supplementary Material The online version of this chapter (https://doi.org/10.1007/978-3-030-61394-5_8) contains supplementary material, which is available to authorized users. The videos can be accessed by scanning the related images with the SN More Media App.

early 1960s as a linear model for making predictions from longitudinal observations in a single subject. It only became popular in the early 1990s by the work of Robinson (Am Statistician 1991; 45: 54) and McLean (Stat Science 1991; 6: 15), who improved the model by presenting consistent analysis procedures. In the past decade user-friendly statistical software programs like SAS and SPSS have enabled the application of mixed models, even by clinical investigators with limited statistical background.

With mixed models repeated measures *within* subjects receive fewer degrees of freedom than they do with the classical general linear model, because they are nested in a separate layer or subspace. In this way better sensitivity is left in the model for demonstrating differences *between* subjects. Therefore, if the main aim of your research is, to demonstrate differences *between* subjects, then the mixed model should be more sensitive. However, the two methods should be equivalent, if the main aim of your research is, to demonstrate differences between repeated measures, for example different treatment modalities in a single subject. A limitation of the mixed model is, that it includes additional variances, and is, therefore, more complex. More complex statistical models are, ipso facto, more at risk of power loss, particularly, with small data.

The current chapter uses examples to demonstrate how the mixed model will perform in practice. The examples show, that the mixed model, unlike the general linear, produced a very significant effect in a parallel-group study with repeated measures, and that the two models were approximately equivalent for analyzing a crossover study with two treatment modalities.

The underneath statistical models with their special properties and advantages will be explained.

1. Repeated measures Anova (analysis of variance) and repeated measures of covariance (Ancova) are multivariate analysis of variance methodologies. Many statisticians have improved these important methodologies, which were originally primarily based on one way analysis of variance (Fisher, see Chap. 7).

   Mauchly (1907–2004 from Ambler Pennsylvania) recognized the problem of lacking sphericity of variances in 1940.

   Greenhouse (from NY City, currently 101 years of age) and Geisser (1929–2004 from Chapel Hill North Carolina) invented adjustment for the above lack in 1959.

   Huynh (pronounced as Gwin) and Feldt invented the above adjustment with the help of the Box correction in 1940 (see also Chap. 18).

   Pillay (in Ann Math Stat 1955; 19: 57) provided intercept models and exact statistics for multivariate tests.

   Wilks (in Biometrika 1932; 24: 471) also provided intercept models and exact statistics for multivariate tests.

   Hotelling (in Proceedings 2nd Berkeley Symposium on Statistics 1951) also provided intercept models and exact statistics for multivariate tests.

   Roy (Wiley and Son 1947) provided the largest roots method for multivariate tests.

2. Repeated measures Anova with predictors is like ordinary repeated measures analysis of variance, but with a full factorial design. A factorial Anova compares means across two or more independent variables. Again, a one-way Anova has one independent variable, that splits the sample into two or more groups, whereas the factorial Anova has two or more independent variables, that splits the sample into four or more groups.

3. Mixed linear model is the third model. With the mixed model repeated measures, *within* subjects receive fewer degrees of freedom than they do with the classical general linear model, because they are nested in a separate layer or subspace. In this way better sensitivity is left in the model for demonstrating differences *between* subjects. Therefore, if the main aim of your research is, to demonstrate differences *between* subjects, then the mixed model should be more sensitive.

4. Mixed linear model with random interaction is the fourth model. Ronald Fisher (see also Chap. 7) introduced random effects models to study the correlations of trait values between relatives. In the 1950s, Charles Roy Henderson (1911–1989 from Coin Iowa) provided best linear unbiased estimates of fixed effects and best linear unbiased predictions of random effects. Subsequently, mixed modeling has become a major area of statistical research, including work on computation of maximum likelihood estimates, non-linear mixed effects models, missing data in mixed effects models, and Bayesian estimation of mixed effects models. Mixed models are applied in many disciplines where multiple correlated measurements are being made on each unit of interest. They are prominently used in research involving human and animal subjects in fields, ranging from genetics to marketing, and have also been used in sports like baseball and industrial statistics.

5. Doubly repeated measures multivariate Anova is the fifth model. Repeated-measures Anova uses repeated measures of a single outcome variable in a single subject. If a second outcome variable is included, and measured in the same way, the doubly-repeated-measures Anova procedure, available in the general linear models module, will be an adequate method for analysis.

2   Repeated Measures Anova (Analysis of Variance)

In the underneath repeated measures parallel group study each patient is measured for fall of systolic blood pressure for 4 months once a month, while on treatment with treatment 0 or treatment 1.

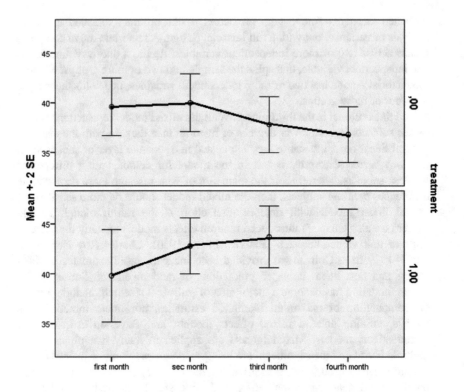

| patient | outcome (fall in systolic blood pressure) | | | |
no	1st	2nd	3rd	4th month
1	33,00	32,00	31,00	30,00
2	33,00	34,00	32,00	31,00
3	38,00	38,00	36,00	35,00
4	39,00	40,00	37,00	36,00
5	39,00	40,00	37,00	36,00
6	40,00	41,00	38,00	37,00
7	40,00	41,00	39,00	38,00
8	41,00	41,00	39,00	38,00
9	46,00	46,00	44,00	43,00
10	47,00	47,00	45,00	44,00
11	31,00	36,00	36,00	36,00
12	32,00	37,00	37,00	37,00
13	36,00	41,00	42,00	42,00
14	37,00	42,00	43,00	43,00
15	56,00	42,00	43,00	43,00
16	38,00	43,00	44,00	44,00
17	39,00	44,00	45,00	44,00
18	39,00	44,00	45,00	45,00
19	44,00	49,00	51,00	50,00
20	46,00	50,00	51,00	51,00

The above table gives the data file. SPSS statistical software will be used for testing whether the differences between the subsequent months of observation will produce different results. Start by entering the data file in your computer with SPSS installed. It is in extras.springer.com, and is entitled "chap8rmsanova". It is previously used by the authors in SPSS for starters and 2nd levelers, Chap. 11, Springer Heidelberg Germany, 2016.

Command:

Analyze....General linear model....Repeated Measurements....Define factors.... Within-subjects factor names: month....number levels: 4....click Add....click Define....enter month 1, 2, 3, 4 in box: "Within-subjects Variables"....click OK.

The underneath tables are in the output sheets.

Multivariate Tests[b]

Effect		Value	F	Hypothesis df	Error df	Sig.
month	Pillai's Trace	,680	12,044[a]	3,000	17,000	,000
	Wilks' Lambda	,320	12,044[a]	3,000	17,000	,000
	Hotelling's Trace	2,125	12,044[a]	3,000	17,000	,000
	Roy's Largest Root	2,125	12,044[a]	3,000	17,000	,000

a. Exact statistic
b. Design: Intercept
  Within Subjects Design: month

Mauchly's Test of Sphericity[a]

Measure: MEASURE_1

Within Subjects Effect	Mauchly's W	Approx. Chi-Square	df	Sig.	Epsilon[b]		
					Greenhouse-Geisser	Huynh-Feldt	Lower-bound
month	,006	91,390	5	,000	,401	,413	,333

Tests the null hypothesis that the error covariance matrix of the orthonormalized transformed dependent variables is proportional to an identity matrix.

a. Design: Intercept
  Within Subjects Design: month

b. May be used to adjust the degrees of freedom for the averaged tests of significance. Corrected tests are displayed in the Tests of Within-Subjects Effects table.

Tests of Within-Subjects Effects

Measure: MEASURE_1

Source		Type III Sum of Squares	df	Mean Square	F	Sig.
month	Sphericity Assumed	32,700	3	10,900	1,645	,189
	Greenhouse-Geisser	32,700	1,204	27,168	1,645	,215
	Huynh-Feldt	32,700	1,240	26,365	1,645	,215
	Lower-bound	32,700	1,000	32,700	1,645	,215
Error(month)	Sphericity Assumed	377,800	57	6,628		
	Greenhouse-Geisser	377,800	22,869	16,520		
	Huynh-Feldt	377,800	23,565	16,032		
	Lower-bound	377,800	19,000	19,884		

Overall statistics is given in the Multivariate Tests table. If significant, then there must be a significant effect somewhere in the data. We do not yet know exactly where. Mauchly's test is also significant. It means, the data do not have homogeneity of variances, and, therefore, the first test for within-subject effects is not appropriate. Instead three other tests are possible. However, none of them is statistically significant, each of them with p-values at 0.215. Instead of repeated measures Anova, a nonparametric test like the Friedman test is performed.

Command:
Analyze....Nonparametric....K-related Samples....Test variables: enter month 1, 2, 3, 4....Click OK.

The underneath result shows, that this test is not statistically significant either. The different months do not produce different falls in systolic blood pressure.

Test Statistics[a]

N	20
Chi-Square	5,851
df	3
Asymp. Sig.	,119

a. Friedman Test

3   Repeated Measures Anova Versus Ancova

The difference between Anova and Ancova can be easily demonstrated by repeated measures Anova using a between-subject factor instead of a covariance as concomitant predictor variable. A 20 patient data file has the effects of three repeated treatment modalities as main outcome. Three sleeping pills are tested for hours of sleep. The data file is in extras.springer.com and is entitled "chap8rmsancova". Start by opening the data file in your computer with SPSS installed. We will start with a repeated measures anova without any between-subjects factor(s) or covariate(s).

Command:

Analyze....General linear model....Repeated Measurements....Define factors.... Within-subjects factor names: treatment....number levels: 3....click Add....click Define....enter treatmenta, treatmentb, treatmentc in box: "Within-subjects Variables"....click OK.

Within-Subjects Factors

Measure: treat

factor1	Dependent Variable
1	treatmenta
2	treatmentb
3	treatmentc

Multivariate Tests[a]

Effect		Value	F	Hypothesis df	Error df	Sig.
factor1	Pillai's Trace	,292	3,706[b]	2,000	18,000	,045
	Wilks' Lambda	,708	3,706[b]	2,000	18,000	,045
	Hotelling's Trace	,412	3,706[b]	2,000	18,000	,045
	Roy's Largest Root	,412	3,706[b]	2,000	18,000	,045

a. Design: Intercept
   Within Subjects Design: factor1

b. Exact statistic

Mauchly's Test of Sphericity[a]

Measure: treat

Within Subjects Effect	Mauchly's W	Approx. Chi-Square	df	Sig.	Epsilon[b]		
					Greenhouse-Geisser	Huynh-Feldt	Lower-bound
factor1	,466	13,744	2	,001	,652	,680	,500

Tests the null hypothesis that the error covariance matrix of the orthonormalized transformed dependent variables is proportional to an identity matrix.

a. Design: Intercept
   Within Subjects Design: factor1

b. May be used to adjust the degrees of freedom for the averaged tests of significance. Corrected tests are displayed in the Tests of Within-Subjects Effects table.

Tests of Within-Subjects Effects

Measure: treat

Source		Type III Sum of Squares	df	Mean Square	F	Sig.
factor1	Sphericity Assumed	4,164	2	2,082	6,685	,003
	Greenhouse-Geisser	4,164	1,304	3,194	6,685	,011
	Huynh-Feldt	4,164	1,361	3,061	6,685	,010
	Lower-bound	4,164	1,000	4,164	6,685	,018
Error(factor1)	Sphericity Assumed	11,836	38	,311		
	Greenhouse-Geisser	11,836	24,772	,478		
	Huynh-Feldt	11,836	25,850	,458		
	Lower-bound	11,836	19,000	,623		

The output sheets are from the above 20 patient data file treated with three different treatment modalities. The Multivariate tests table shows, that the treatment modalities are different in their efficacies at $p = 0.045$. Mauchly' s test is significant indicating that the paired treatment samples are not entirely Gaussian. So, Gaussian patterns, otherwise called sphericity, cannot not be assumed.

The Greenhouse, Huynh, and Lower-bound tests show, that the variable treatment modality is a very significant predictor of the outcome, treatment efficacy.

Now, in the analysis a continuous variable is included, age, which can be used in the analysis for adjusting the outcome effects for the patients' ages.

Command:

Analyze....General linear model....Repeated Measurements....Define factors.... Within-subjects factor names: treatment....number levels: 3....click Add....click Define....enter treatmenta, treatmentb, treatmentc in box: "Within-subjects Variables"....Covariate(s): enter age....click OK.

Multivariate Tests[a]

Effect		Value	F	Hypothesis df	Error df	Sig.
factor1	Pillai's Trace	,284	3,375[b]	2,000	17,000	,058
	Wilks' Lambda	,716	3,375[b]	2,000	17,000	,058
	Hotelling's Trace	,397	3,375[b]	2,000	17,000	,058
	Roy's Largest Root	,397	3,375[b]	2,000	17,000	,058
factor1 * age	Pillai's Trace	,353	4,646[b]	2,000	17,000	,025
	Wilks' Lambda	,647	4,646[b]	2,000	17,000	,025
	Hotelling's Trace	,547	4,646[b]	2,000	17,000	,025
	Roy's Largest Root	,547	4,646[b]	2,000	17,000	,025

a. Design: Intercept + age
   Within Subjects Design: factor1

b. Exact statistic

Mauchly's Test of Sphericity[a]

Measure: treat

Within Subjects Effect	Mauchly's W	Approx. Chi-Square	df	Sig.	Greenhouse-Geisser	Huynh-Feldt	Lower-bound
					Epsilon[b]		
factor1	,467	12,961	2	,002	,652	,721	,500

Tests the null hypothesis that the error covariance matrix of the orthonormalized transformed dependent variables is proportional to an identity matrix.

a. Design: Intercept + age
   Within Subjects Design: factor1

b. May be used to adjust the degrees of freedom for the averaged tests of significance. Corrected tests are displayed in the Tests of Within-Subjects Effects table.

Tests of Within-Subjects Effects

Measure: treat

Source		Type III Sum of Squares	df	Mean Square	F	Sig.
factor1	Sphericity Assumed	1,309	2	,655	2,478	,098
	Greenhouse-Geisser	1,309	1,304	1,004	2,478	,122
	Huynh-Feldt	1,309	1,443	,908	2,478	,117
	Lower-bound	1,309	1,000	1,309	2,478	,133
factor1 * age	Sphericity Assumed	2,324	2	1,162	4,398	,020
	Greenhouse-Geisser	2,324	1,304	1,782	4,398	,038
	Huynh-Feldt	2,324	1,443	1,611	4,398	,033
	Lower-bound	2,324	1,000	2,324	4,398	,050
Error(factor1)	Sphericity Assumed	9,512	36	,264		
	Greenhouse-Geisser	9,512	23,476	,405		
	Huynh-Feldt	9,512	25,966	,366		
	Lower-bound	9,512	18,000	,528		

For the age adjustment it has to be entered as a continuous covariate. The Anova model, now being an Ancova model gives a disappointing result. In the multivariate test the treatment efficacy is no longer 0.050 while an interaction with age at $p = 0.025$ is observed. Also the within-subject tests are no longer <0.050, again with an interaction of treatment with age at 0.038–0.050. Instead of entering the covariate age as a continuous variable, it can also be entered as a between-subject factor. This means that a continuous variable will automatically be transferred into a between-subject factor, which is a categorical variable. The underneath table is in the output, and shows, how it works in the example given.

Between-Subjects Factors

		N
age	44,00	2
	53,00	2
	55,00	2
	56,00	2
	62,00	1
	63,00	1
	65,00	5
	66,00	1
	83,00	1
	84,00	1
	85,00	1
	86,00	1

Thus, in the output the transformed covariate is a categorical variable with 12 categories.

Multivariate Tests[a]

Effect		Value	F	Hypothesis df	Error df	Sig.
factor1	Pillai's Trace	,758	10,949[b]	2,000	7,000	,007
	Wilks' Lambda	,242	10,949[b]	2,000	7,000	,007
	Hotelling's Trace	3,128	10,949[b]	2,000	7,000	,007
	Roy's Largest Root	3,128	10,949[b]	2,000	7,000	,007
factor1 * age	Pillai's Trace	1,513	2,258	22,000	16,000	,050
	Wilks' Lambda	,042	2,469[b]	22,000	14,000	,042
	Hotelling's Trace	9,603	2,619	22,000	12,000	,044
	Roy's Largest Root	7,938	5,773[c]	11,000	8,000	,010

a. Design: Intercept + age
   Within Subjects Design: factor1

b. Exact statistic

c. The statistic is an upper bound on F that yields a lower bound on the significance level.

Tests of Within-Subjects Effects

Measure: treat

Source		Type III Sum of Squares	df	Mean Square	F	Sig.
factor1	Sphericity Assumed	6,070	2	3,035	15,981	,000
	Greenhouse-Geisser	6,070	1,174	5,169	15,981	,002
	Huynh-Feldt	6,070	2,000	3,035	15,981	,000
	Lower-bound	6,070	1,000	6,070	15,981	,004
factor1 * age	Sphericity Assumed	8,797	22	,400	2,105	,065
	Greenhouse-Geisser	8,797	12,917	,681	2,105	,129
	Huynh-Feldt	8,797	22,000	,400	2,105	,065
	Lower-bound	8,797	11,000	,800	2,105	,150
Error(factor1)	Sphericity Assumed	3,039	16	,190		
	Greenhouse-Geisser	3,039	9,394	,323		
	Huynh-Feldt	3,039	16,000	,190		
	Lower-bound	3,039	8,000	,380		

The model with age as categorical predictor provides better statistics than does the above two models. The multivariate tests improved with p-values of 0.58 and 0.45 to a p-value of 0.007. The within subject tests improved from p-values of 0.010 and 0.117 to one of 0.000. In contrast, interaction between age and treatment modalities is no longer statistically significant.

In conclusion, Ancova may cause loss of precision. In contrast, Ancova with the covariate transformed into categorical variable may, rather than loss, provide increase of precision, albeit with some loss of interaction effect.

4   Repeated Measures Anova with Predictors

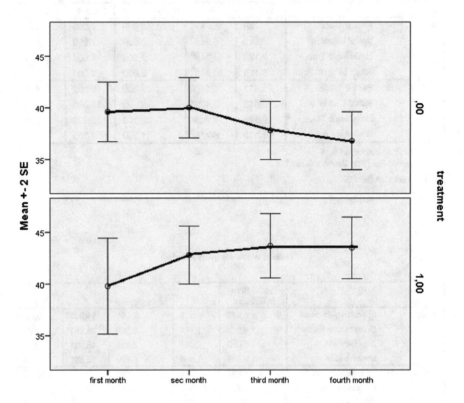

The data file from Sect. 2 of this chapter is used once more. A parallel group study of the treatments 0 and 1 for the treatment of hypertension is assessed with four monthly-repeated measurements. According to the above graph a significant difference between the two treatments is expected. The above graph of the summaries of the data suggests, indeed, that already after 2 months one treatment group seems to perform better than the other. Multiple unpaired t-tests of treatment-0 versus treatment-1 demonstrate a trend to a significant difference with p-values as small as 0.08 at the months 2, 3 and 4 (analysis not shown). However, this analysis is not entirely appropriate, because it does not take the repeated nature of the data into account. A repeated measurements analysis of variance using the classical general linear model will be performed (full factorial design). SPSS statistical software will be used.

Command:
Analyze....General linear model....Repeated Measurements....Define factors....
Within-subjects factor names: month....number levels: 4....click Add....click Define
....enter month 1, 2, 3, 4 in box: "Within-subjects Variables"....enter treatment in box between-subjects covariates....click OK.

The underneath tables are in the output sheets.

Tests of Within-Subjects Effects

Measure: months

Source		Type III Sum of Squares	df	Mean Square	F	Sig.
month	Sphericity Assumed	32,700	3	10,900	2,411	,077
	Greenhouse-Geisser	32,700	1,038	31,501	2,411	,137
	Huynh-Feldt	32,700	1,106	29,564	2,411	,134
	Lower-bound	32,700	1,000	32,700	2,411	,138
month * treatment	Sphericity Assumed	133,700	3	44,567	9,859	,000
	Greenhouse-Geisser	133,700	1,038	128,799	9,859	,005
	Huynh-Feldt	133,700	1,106	120,879	9,859	,004
	Lower-bound	133,700	1,000	133,700	9,859	,006
Error(month)	Sphericity Assumed	244,100	54	4,520		
	Greenhouse-Geisser	244,100	18,685	13,064		
	Huynh-Feldt	244,100	19,909	12,261		
	Lower-bound	244,100	18,000	13,561		

Tests of Between-Subjects Effects

Measure: months ...

	Type III Sum				
Intercept	131220,000	1	131220,000	1501,278	,000
treatment	304,200	1	304,200	3,480	,078
Error	1573,300	18	87,406		

As the test for sphericity (equal standard errors) had to be rejected again, sphericity could not be accepted, and the Huynh-Feldt test was the next best for demonstrating a difference between the repeated measures. With p = 0.134 no significant difference between the repeated measures could be demonstrated (upper part of the above table). The subsequent between-subjects comparison of the two treatments neither showed a significant effect with a p-value of 0.078 (lower part of the above table). The two treatments did not produce significantly different falls in blood pressure, although a trend to significance at 0.078 (= < 0.10) was observed.

5 Mixed Linear Model Analysis

The data example from the above Sects. 2 and 4 is used once more. With mixed models repeated measures *within* subjects receive fewer degrees of freedom than they do with the classical general linear model, because they are nested in a separate layer or subspace. In this way better sensitivity is left in the model to demonstrate differences *between* subjects. Therefore, if the main aim of your research is to demonstrate differences *between* subjects, then the mixed model should be more sensitive. As an alternative, therefore, a mixed linear model is applied using again SPSS. However, for that purpose the data file has to be adapted. Every month must be given a separate row. The restructure data wizard will be used.

Command:
click Data....click Restructure....mark Restructure selected variables into cases.... click Next....mark One (for example, w1, w2, and w3)....click Next....Name: id (the patient id variable is already provided)....Target Variable: enter "firstmonth, secondmonth...... fourthmonth"....Fixed Variable(s): enter treatment....click Next.... How many index variables do you want to create?....mark One....click Next....click Next again....click Next again....click Finish....Sets from the original data will still be in use...click OK.

Return to the main screen, and observe that there are now 80 rows instead of 20 in the data file. The above table is adequate to perform a mixed linear model analysis. For readers' convenience it is saved in extras.springer.com, and is entitled "chap8rmsanova-restructured". SPSS calls the levels "indexes", and the outcome values after restructuring "Trans" values, terms pretty confusing to us.

Click the data screen. It now looks like underneath. Index1 is the month of treatment, trans1 is the outcome value, the fall in systolic blood pressure after treatment in mmHg.

id1	treatment	Index1	trans1
1	,00	1	33,00
1	,00	2	32,00
1	,00	3	31,00
1	,00	4	30,00
2	,00	1	33,00
2	,00	2	34,00
2	,00	3	32,00
2	,00	4	31,00
3	,00	1	38,00
3	,00	2	38,00
3	,00	3	36,00
3	,00	4	35,00
4	,00	1	39,00
4	,00	2	40,00
4	,00	3	37,00
4	,00	4	36,00
5	,00	1	39,00
5	,00	2	40,00
5	,00	3	37,00
5	,00	4	36,00
6	,00	1	40,00
6	,00	2	41,00
6	,00	3	38,00
6	,00	4	37,00
7	,00	1	40,00
7	,00	2	41,00
7	,00	3	39,00
7	,00	4	38,00
8	,00	1	41,00
8	,00	2	41,00
8	,00	3	39,00
8	,00	4	38,00
9	,00	1	46,00
9	,00	2	46,00
9	,00	3	44,00
9	,00	4	43,00
10	,00	1	47,00
10	,00	2	47,00
10	,00	3	45,00
10	,00	4	44,00
11	1,00	1	31,00
11	1,00	2	36,00
11	1,00	3	36,00
11	1,00	4	36,00
12	1,00	1	32,00
12	1,00	2	37,00
12	1,00	3	37,00
12	1,00	4	37,00
13	1,00	1	36,00
13	1,00	2	41,00
13	1,00	3	42,00
13	1,00	4	42,00
14	1,00	1	37,00
14	1,00	2	42,00
14	1,00	3	43,00
14	1,00	4	43,00
15	1,00	1	56,00
15	1,00	2	42,00
15	1,00	3	43,00
15	1,00	4	43,00
16	1,00	1	38,00
16	1,00	2	43,00
16	1,00	3	44,00
16	1,00	4	44,00
17	1,00	1	39,00
17	1,00	2	44,00
17	1,00	3	45,00
17	1,00	4	44,00
18	1,00	1	39,00
18	1,00	2	44,00
18	1,00	3	45,00
18	1,00	4	45,00
19	1,00	1	44,00
19	1,00	2	49,00
19	1,00	3	51,00
19	1,00	4	50,00
20	1,00	1	46,00
20	1,00	2	50,00
20	1,00	3	51,00
20	1,00	4	51,00

The above table is adequate to perform a multilevel modeling analysis with mixed linear models, and adjusts for the positive correlation between the presumably positive correlation between the weekly measurements in one patient. The module Mixed Models consists of two statistical models:

Linear,
Generalized Linear.

For analysis the statistical model Linear is required.

Command:
Analyze....Mixed Models....Linear....Specify Subjects and Repeated....Subject: enter id ....Continue....Linear Mixed Model....Dependent Variables: Trans1 ....Factors: Index1, treatment....Fixed....Build Nested Term....Treatment ....Add ....Index1....Add.... Index1 build term by* treatment....Index1 *treatment.... Add ....Continue....click OK (* = sign of multiplication).

The underneath table shows the result. SPSS has applied the effects of the cluster levels and the interaction between cluster levels and treatment modality for adjusting the effects of the correlation levels between the monthly repeated measurements. The adjusted analysis shows that one treatment now performs much better than the other, while the repeated measures Anova produced an insignificant p-value of only 0,078. One treatment modality performs better than the other at p = 0,001, and this is adjusted for within subjects differences between repeated measures. Akaike criterion is given and will be discussed in the next section.

Information Criteria[a]

-2 Restricted Log Likelihood	452,829
Akaike's Information Criterion (AIC)	460,829
Hurvich and Tsai's Criterion (AICC)	461,426
Bozdogan's Criterion (CAIC)	473,936
Schwarz's Bayesian Criterion (BIC)	469,936

The information criteria are displayed in smaller-is-better forms.

a. Dependent Variable: trans1.

Type III Tests of Fixed Effects[a]

Source	Numerator df	Denominator df	F	Sig.
Intercept	1	66,858	5198,547	,000
treatment	1	66,858	12,052	,001
Index1	3	36,283	,423	,738
Index1 * treatment	3	36,283	1,582	,211

a. Dependent Variable: trans1.

6   Mixed Linear Model with Random Interaction

The within subject effect of the months of treatment of the data example from the previous section was huge and not entirely expected. Therefore, it can be interpreted as a random rather than fixed interaction, that will not happen again next time. For mixed linear models with random interaction the module Generalized Mixed Linear Models is slightly more appropriate than the fixed effect mixed linear module.

It will be applied next.

Command:
Analyze....Mixed Linear....Generalized Mixed Linear Models....click Data Structure....click left mouse and drag patient_id to Subjects part of the canvas ....click left mouse and drag month to Repeated Measures part of the canvas.... click Fields and Effects....click Target....check that the variable outcome is already in the Target window....check that Linear model is marked....click Fixed Effects....drag treatment and month to Effect builder....click Random Effects....click Add Block ....click Add a custom term....move month*treatment (* is symbol multiplication and interaction) to the Custom term window....click Add term....click OK....click Run.

Model Summary
Target: trans1

Target		trans1
Probability Distribution		Normal
Link Function		Identity
Information Criterion	Akaike Corrected	479,867
	Bayesian	490,585

Information criteria are based on the -2 log likelihood
(468,997) and are used to compare models. Models with
smaller information criterion values fit better.

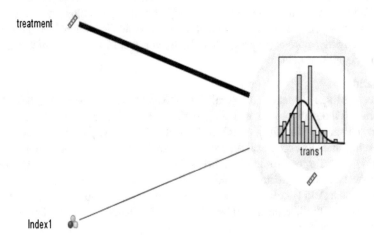

Source	F	df1	df2	Sig.
Corrected Model ▼	2,837	4	75	,030
treatment	10,000	1	75	,002
Index1	0,482	3	75	,696

Probability distribution:Normal
Link function:Identity

In the output sheet tables and a graph is observed with the mean and standard errors of the outcome value displayed with the best fit Gaussian curve. The F-value of 10,000 indicates that one treatment is very significantly better than the other with p <0,002. The thickness of the lines are a measure for level of significance, and so the significance of the index (repeated measures) is very thin and thus very weak. The F-value of this random effect model is slightly smaller than that of the fixed effect mixed model (F = 12,052). The random effect analysis takes more into account than the fixed effect analysis does, namely interaction effect. A statistical model that accounts more, usually, loses some power, but the tiny difference is academic.

Also the Akaike criterion of the random effect model is a bit larger than that of the fixed effect model, 479,... versus 460,... The smaller the Akaike criterion of a statistical model, the better the model fits the data. Again the difference is small and negligible.

7 Doubly Repeated Measures Multivariate Anova

Repeated-measures Anova uses repeated measures of a single outcome variable in a single subject. If a second outcome variable is included and measured in the same way, the doubly-repeated-measures Anova procedure, available in the general linear models module, will be adequate for analysis. A data example is given.

Morning body temperatures in patients with sleep deprivation is lower than in those without sleep deprivation. In 16 patients a three period crossover study of three sleeping pills (treatment levels) were studied. The underneath table gives the data of the data file and is entitled "chap8doublyrmsanova", and is in extras.springer.com. It is previously used by the authors in SPSS for starters and 2nd levelers, Chap. 11, Springer Heidelberg Germany, 2016. Two outcome variables are measured at three levels each. This study would qualify for a doubly multivariate analysis.

Hours			temp			age	gender
a	b	c	a	b	c		
6,10	6,80	5,20	35,90	35,30	36,80	55,00	,00
7,00	7,00	7,90	37,10	37,80	37,00	65,00	,00
8,20	9,00	3,90	38,30	34,00	39,10	74,00	,00
7,60	7,80	4,70	37,50	34,60	37,70	56,00	1,00
6,50	6,60	5,30	36,40	35,30	36,70	44,00	1,00
8,40	8,00	5,40	38,30	35,50	38,00	49,00	1,00
6,90	7,30	4,20	37,00	34,10	37,40	53,00	,00
6,70	7,00	6,10	36,80	36,10	36,90	76,00	,00
7,40	7,50	3,80	37,30	33,90	37,40	67,00	1,00
5,80	5,80	6,30	35,70	36,30	35,90	66,00	1,00
6,10	6,80	5,20	35,90	35,30	36,80	55,00	,00
7,00	7,00	7,90	37,10	37,80	37,00	65,00	,00
8,20	9,00	3,90	38,30	34,00	39,10	74,00	,00
6,90	7,30	4,20	37,00	34,10	37,40	53,00	,00
6,70	7,00	6,10	36,80	36,10	36,90	76,00	,00
8,40	8,00	5,40	38,30	35,50	38,00	49,00	1,00

hours = hours of sleep on sleeping pill
a, b, c = different sleeping pills (levels of treatment)
age = patient age
gen = gender
temp = different morning body temperatures on sleeping pill

We will start by opening the data file in SPSS. For analysis the statistical model Repeated Measures in the module General Linear Model is required.

Command:
Analyze....General Linear Model....Repeated Measures....Within-Subject Factor Name: type treatment....Number of Levels: type 3....click Add....Measure Name: type hours....click Add....Measure Name: type temp....click Add....click Define .... Within-Subjects Variables(treatment): enter hours a, b, c, and temp a, b, c.... Between-Subjects Factor(s): enter gender....click Contrast....Change Contrast .... Contrast....select Repeated....click Change....click Continue....click Plots.... Horizontal Axis: enter treatment....Separate Lines: enter gender....click Add....click Continue....click Options....Display Means for: enter gender*treatment....mark Estimates of effect size....mark SSCP matrices....click Continue....click OK.

The underneath table is in the output sheets.

Multivariate Tests[b]

Effect			Value	F	Hypothesis df	Error df	Sig.	Partial Eta Squared
Between Subjects	Intercept	Pillai's Trace	1,000	3,271E6	2,000	13,000	,000	1,000
		Wilks' Lambda	,000	3,271E6	2,000	13,000	,000	1,000
		Hotelling's Trace	503211,785	3,271E6	2,000	13,000	,000	1,000
		Roy's Largest Root	503211,785	3,271E6	2,000	13,000	,000	1,000
	gender	Pillai's Trace	,197	1,595[a]	2,000	13,000	,240	,197
		Wilks' Lambda	,803	1,595[a]	2,000	13,000	,240	,197
		Hotelling's Trace	,245	1,595[a]	2,000	13,000	,240	,197
		Roy's Largest Root	,245	1,595[a]	2,000	13,000	,240	,197
Within Subjects	treatment	Pillai's Trace	,562	3,525[a]	4,000	11,000	,044	,562
		Wilks' Lambda	,438	3,525[a]	4,000	11,000	,044	,562
		Hotelling's Trace	1,282	3,525[a]	4,000	11,000	,044	,562
		Roy's Largest Root	1,282	3,525[a]	4,000	11,000	,044	,562
	treatment * gender	Pillai's Trace	,762	8,822[a]	4,000	11,000	,002	,762
		Wilks' Lambda	,238	8,822[a]	4,000	11,000	,002	,762
		Hotelling's Trace	3,208	8,822[a]	4,000	11,000	,002	,762
		Roy's Largest Root	3,208	8,822[a]	4,000	11,000	,002	,762

a. Exact statistic

b. Design: Intercept + gender
Within Subjects Design: treatment

Doubly multivariate analysis has two sets of repeated measures plus separate predictor variables. For analysis of such data both between and within subjects tests are performed. We are mostly interested in the within subject effects of the treatment levels, but the above table starts by showing the not so interesting gender effect on hours of sleep and morning temperatures. They are not significantly different between the genders. More important is the treatment effects. The hours of sleep and the morning temperature are significantly different between the different treatment levels at p = 0,044. Also these significant effects are different between males and females at p = 0,002.

Tests of Within-Subjects Contrasts

Source	Measure	treatment	Type III Sum of Squares	df	Mean Square	F	Sig.	Partial Eta Squared
treatment	hours	Level 1 vs. Level 2	,523	1	,523	6,215	,026	,307
		Level 2 vs. Level 3	62,833	1	62,833	16,712	,001	,544
	temp	Level 1 vs. Level 2	49,323	1	49,323	15,788	,001	,530
		Level 2 vs. Level 3	62,424	1	62,424	16,912	,001	,547
treatment * gender	hours	Level 1 vs. Level 2	,963	1	,963	11,447	,004	,450
		Level 2 vs. Level 3	,113	1	,113	,030	,865	,002
	temp	Level 1 vs. Level 2	,963	1	,963	,308	,588	,022
		Level 2 vs. Level 3	,054	1	,054	,015	,905	,001
Error(treatment)	hours	Level 1 vs. Level 2	1,177	14	,084			
		Level 2 vs. Level 3	52,637	14	3,760			
	temp	Level 1 vs. Level 2	43,737	14	3,124			
		Level 2 vs. Level 3	51,676	14	3,691			

The above table shows, whether differences between levels of treatment were significantly different from one another by comparison with the subsequent levels (contrast tests). The effects of treatment levels 1 versus (vs) 2 on hours of sleep were different at p = 0,026, levels 2 vs 3 at p = 0,001. The effects of treatments levels 1 vs 2 on morning temperatures were different at p = 0,001, levels 2 vs 3 on morning temperatures were also different at p = 0,001. The effects on hours of sleep of treatment levels 1 vs 2 accounted for the differences in gender remained very significant at p = 0,004.

gender * treatment

Measure	gender	treatment	Mean	Std. Error	95% Confidence Interval	
					Lower Bound	Upper Bound
hours	,00	1	6,980	,268	6,404	7,556
		2	7,420	,274	6,833	8,007
		3	5,460	,417	4,565	6,355
	1,00	1	7,350	,347	6,607	8,093
		2	7,283	,354	6,525	8,042
		3	5,150	,539	3,994	6,306
temp	,00	1	37,020	,284	36,411	37,629
		2	35,460	,407	34,586	36,334
		3	37,440	,277	36,845	38,035
	1,00	1	37,250	,367	36,464	38,036
		2	35,183	,526	34,055	36,311
		3	37,283	,358	36,515	38,051

The above table shows the mean hours of sleep and mean morning temperatures for the different subsets of observations. Particularly, we observe the few hours of sleep on treatment level 3, and the highest morning temperatures at the same level. The treatment level 2, in contrast, pretty many hours of sleep and, at the same time, the lowest morning temperatures (consistent with longer periods of sleep). The underneath figures show the same.

Estimated Marginal Means of hours

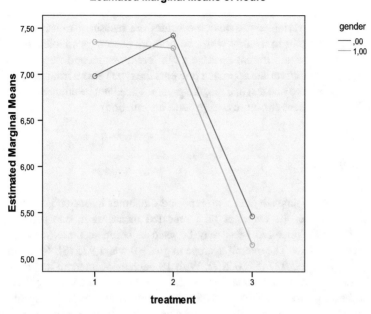

Estimated Marginal Means of temp

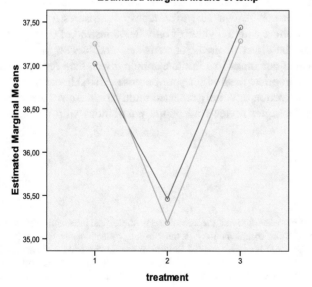

Doubly multivariate Anova is for studies with multiple paired observations with more than a single outcome variable. For example, in a study with two or more different outcome variables the outcome values are measured repeatedly during a period of follow up or in a study with two or more outcome variables the outcome values are measured at different levels, e.g., different treatment dosages or different compounds. The multivariate approach prevents the type I errors from being inflated, because we only have one test and, so, the p-values need not be adjusted for multiple testing (see References for more background information)

8 Conclusion

You might want to analyze data with repeated outcomes in one subject in different ways. For example, the averages of 5 repeated measures in one patient can be calculated, and an unpaired t-test may be used to compare these averages in the two treatment groups. The overall average in group 0 was 1,925 (SE (standard error) 0,0025), in group 1 2,227 (SE 0,227). With 18 degrees of freedom and a t-value of 1,99 the difference did not obtain statistical significance, $0,05 < p < 0,10$. There seems to be, expectedly, a strong positive correlation between the 5 repeated measurements in one patient. In order to take account of this strong positive correlation, a mixed linear model should be used. This model showed, that treatment 1 now performed significantly better than did treatment 0, at $p = 0,0001$.

You might want to analyze data with repeated outcomes in one subject otherwise using a repeated measures Anova. However, repeated-measures Anova produced a treatment modality effect with a p-value of only 0,048 instead of 0,0001. If you are more interested in the effect of predictor variables, and less so in the difference between the repeated outcomes like in the example given, then repeated-measures Anova is not an appropriate method for your purpose. Mixed linear models with or without random interaction or doubly repeated multivariate Anova are more appropriate. The current chapter reviews all of the possibilities with the help of data examples.

Reference

To readers requesting more background, theoretical and mathematical information of computations given, several textbooks complementary to the current production and written by the same authors are available: (1) Statistics applied to clinical studies 5th edition, 2012, (2) Machine learning in medicine a complete overview, 2015, (3) SPSS for starters and 2nd levelers 2nd edition, 2015, (4) Clinical data analysis on a pocket calculator 2nd edition, 2016, (5) Understanding clinical data analysis, 2017, (6) Modern Meta-analysis, 2017, all of them edited by Springer Heidelberg Germany.

Chapter 9
Methodologies for Better Fit of Categorical Predictors

Restructuring Categories into Multiple Binary Variables, Variance Components Regressions, Contrast Coefficients Regressions

Abstract Restructuring categorical predictors in a linear regression with categories into multiple binary variables will be an adequate solution for the lack of fit, if the categories are non-incremental. Other methods to deal with the problem are variance components and contrast coefficients regressions. The latter methods will be particularly suitable, if the categorical predictors are unexpected, otherwise called random. Variance components compares such random effects with those of the residual error of a study, otherwise called the unexplained error of a study. With contrast coefficients regression, arbitrary weights adding up to one are given to subgroups (the categories) in the data. Different models are possible, and the best fit model is chosen for final analysis.

Keywords Categorical predictors · Restructuring categories · Variance components regressions · Contrast coefficients regressions

1 Introduction, History, and Background

The outcome of a linear regression is always a continuous variable. The predictor, however, can be continuous, binary or categorical. The equation of a linear regression is an incremental function in the form of a straight line, $y = a + bx$. If the predictor x increases, the outcome y will increase proportionally. With a binary predictor the x-values are often given the amount of 0 or 1, e.g. in therapeutic trials, a zero for the worse treatment, a one for the better treatment, and, so, this is an incremental function too. Similarly, a categorical predictor may be incremental, e.g. in therapeutic trials with incremental dosages of a drug. But, this is not always true, e.g. in a therapeutic trial with completely different compounds an incremental function may very well be lacking, and, thus, linear regression will be insignificant, and is an inappropriate model for assessment. Restructuring non-incremental

Electronic Supplementary Material The online version of this chapter (https://doi.org/10.1007/978-3-030-61394-5_9) contains supplementary material, which is available to authorized users. The videos can be accessed by scanning the related images with the SN More Media App.

categories into multiple binary variables is a solution for the problem. Other methods to deal with the problem are variance components and contrast coefficients regressions. They will be particularly suitable, if the categorical predictors are unexpected, otherwise called random. Variance components compares such random effect with that of the residual error of a study, otherwise called the unexplained error of a study. With contrast coefficients regression, arbitrary weights, adding up to one, are given to subgroups (the categories) in the data. Different models are possible, and the best fit model is chosen for final analysis.

1. In 1996 Nichols (1944-..., pharmacologist from Covington Kentucky) polled statistical software users, and found, that the proper use of categorical variables was of major concern to them (in: SPSS Keywords, 1996; 56: pp. 1–4). Adequate methods were published soon after. Restructuring non-incremental categories into multiple binary variables is an adequate solution for the problem. Other methods to deal with the problem, are the underneath variance components and contrast coefficients regressions methods. They will be particularly suitable, if the categorical predictors are unexpected, otherwise called random.
2. Variance components compares such random effect with that of the residual error of a study, otherwise called the unexplained error of a study. F Satterthwaite, received his PhD from University of Iowa 1941. He described Variance components in: Biometric Bulletin 1946; 2: 110).
3. With contrast coefficients regression arbitrary weights adding up to one are given to subgroups (the categories) in the data. Different models are possible, and the best fit model is chosen for final analysis. The methodology is pretty new (see Casella (1951–2012 from The Bronx, NY City). He described Contrast coefficients in: Cengage, The Digital Learning Platform, 2001.

2   Restructuring Categories into Multiple Binary Variables

In a study with a categorical predictor like races, the race-values 1–4 have no incremental function, and, therefore, linear regression is not appropriate for assessing their effect on any outcome. Instead, restructuring the data for categorical predictors does the job. As an example, in a study the scientific question was: does race have an effect on physical strength. The variable race has a categorical rather than linear pattern. The effects on physical strength (scores 0–100) were assessed in 60 subjects of different races (hispanics (1), blacks (2), asians (3),and whites (4)), ages (years), and genders (0 = female, 1 = male). The first 10 patients are in the table underneath.

patient number	physical strength	race	age	gender
1	70,00	1,00	35,00	1,00
2	77,00	1,00	55,00	0,00
3	66,00	1,00	70,00	1,00
4	59,00	1,00	55,00	0,00
5	71,00	1,00	45,00	1,00
6	72,00	1,00	47,00	1,00
7	45,00	1,00	75,00	0,00
8	85,00	1,00	83,00	1,00
9	70,00	1,00	35,00	1,00
10	77,00	1,00	49,00	1,00

The entire data file is in extras.springer.com, and is entitled "chap9restructuring categories". It is previously used by the authors in SPSS for starters and 2nd levelers, Chap. 8, Springer Heidelberg Germany, 2016. Start by opening the data file in your computer with SPSS installed.

Command:
click race....click Edit....click Copy....click a new "var"....click Paste....highlight the values 2-4....delete and replace with 0,00 values....perform the same procedure subsequently for the other races.

patient number	physical strength	race	age	gender	race 1 hispanics	race 2 blacks	race 3 asians	race 4 whites
1	70,00	1,00	35,00	1,00	1,00	0,00	0,00	0,00
2	77,00	1,00	55,00	0,00	1,00	0,00	0,00	0,00
3	66,00	1,00	70,00	1,00	1,00	0,00	0,00	0,00
4	59,00	1,00	55,00	0,00	1,00	0,00	0,00	0,00
5	71,00	1,00	45,00	1,00	1,00	0,00	0,00	0,00
6	72,00	1,00	47,00	1,00	1,00	0,00	0,00	0,00
7	45,00	1,00	75,00	0,00	1,00	0,00	0,00	0,00
8	85,00	1,00	83,00	1,00	1,00	0,00	0,00	0,00
9	70,00	1,00	35,00	1,00	1,00	0,00	0,00	0,00
10	77,00	1,00	49,00	1,00	1,00	0,00	0,00	0,00

The result is shown above. For the analysis we will use multiple linear regression. First the inadequate analysis will be performed.

Command:

Analyze....Regression....Linear....Dependent: physical strength score.... Independent: race, age, gender....OK.

The table shows that age and gender are significant predictors but race is not.

Coefficients^a

Model		Unstandardized Coefficients		Standardized Coefficients		
		B	Std. Error	Beta	t	Sig.
1	(Constant)	79,528	8,657		9,186	,000
	race	,511	1,454	,042	,351	,727
	age	-,242	,117	-,260	-2,071	,043
	gender	9,575	3,417	,349	2,802	,007

a. Dependent Variable: strengthscore

The variable race is analyzed as a stepwise rising function from 1 to 4, and the linear regression model assumes, that the outcome variable will rise (or fall) simultaneously and linearly, but this needs not be necessarily so. Next, a categorical analysis will be performed.

The above commands are given once more, but now the independent variables are entered slightly differently.

Command:

Analyze....Regression....Linear....Dependent: physical strength score.... Independent: race 2, race 3, race 4, age, gender....click OK.

Coefficients^a

Model		Unstandardized Coefficients		Standardized Coefficients		
		B	Std. Error	Beta	t	Sig.
1	(Constant)	72,650	5,528		13,143	,000
	race2	17,424	3,074	,559	5,668	,000
	race3	-6,286	3,141	-,202	-2,001	,050
	race4	9,661	3,166	,310	3,051	,004
	age	-,140	,081	-,150	-1,716	,092
	gender	5,893	2,403	,215	2,452	,017

a. Dependent Variable: strengthscore

The above table shows that race 2–4 are significant predictors of physical strength.

The results can be interpreted as follows.

The underneath regression equation is used:

$$y = a + b_1x_1 + b_2x_2 + b_3x_3 + b_4x_4 + b_5x_5$$

a = intercept	
b_1 = regression coefficient for	blacks (0 = no,1 = yes),
b_2 =	asians
b_3 =	whites
b_4 =	age
b_5 =	gender

If an individual is hispanic (race 1), then x_1, x_2, and x_3 will turn into 0, and the regression equation turn into $y = a + b_4 x_4 + b_5 x_5$.

In summary:
if hispanic, $y = a + b_4 x_4 + b_5 x_5$.
if black, $y = a + b_1 + b_4 x_4 + b_5 x_5$.
if asian, $y = a + b_2 + b_4 x_4 + b_5 x_5$.
if white, $y = a + b_3 + b_4 x_4 + b_5 x_5$.

So, e.g., the best predicted physical strength score of a white male of 25 years of age would equal.
    $y = 72.65 + 9.66 - 0.14 * 25 + 5.89 * 1 = 84.7$ (on a linear scale from 0 to 100),
    ($* =$ sign of multiplication).
    Compared to the presence of the hispanic race, the black and white races are significant positive predictors of physical strength ($p = 0.0001$ and 0.004 respectively), the asian race is a significant negative predictor ($p = 0.050$). All of these results are adjusted for age and gender, at least if we used $p = 0.10$ as criterion for statistical significance.

3 Variance Components Regressions

If we have reasons to believe, that, in a study, certain patients due to co-morbidity, co-medication, and other factors will respond differently from others, then the spread in the data will be caused not only by residual effect, but also by the subgroup properties, otherwise called random effects. Variance components analysis is able to assess the magnitudes of the random effects as compared to that of the residual error of a study. Can a variance components analysis by including the random effect in the analysis reduce the unexplained variance in a study, and, thus, increase the accuracy of the analysis model as used.

The data from three parallel-group studies of each 5 patients assessed for hours of sleep during different treatments were analyzed with variance components models. When the levels of a factor such as narcoleptics have been chosen at random, such as for example different treatment dosages in phase I-II studies, investigators are often interested in components of variance. A components variance model is appropriate for the purpose. Five treatment modalities have to be assessed and three parallel group studies of 5 sleeping sessions each will be performed for the purpose. The hours of sleep is the main outcome. The study samples are from a single population and have the same population variance, independent of sample size. The outcome is hours of sleep, the predictor = factor = treatment modalities.

treatment modality	hours of sleep	study number	patient number
1,00	7,40	1,00	1
2,00	6,80	1,00	2
3,00	7,50	1,00	3
4,00	7,20	1,00	4
5,00	7,90	1,00	5
1,00	7,60	2,00	6
2,00	7,10	2,00	7
3,00	7,70	2,00	8
4,00	7,40	2,00	9
5,00	8,10	2,00	10
1,00	7,50	3,00	11
2,00	7,20	3,00	12
3,00	7,70	3,00	13
4,00	7,30	3,00	14
5,00	7,90	3,00	15

The data file is entitled "chap9variance components", and is in extras.springer. com. It is previously used by the authors in Machine learning in medicine a complete overview, Chap. 36, Springer Heidelberg Germany, 2015. We will start with a traditional one way Anova. Open the datafile in your computer mounted with SPSS statistical software.

Command:
Analyze....Compare Means....One Way Anova....Dependent List: hours of sleep.... Factors; treatment modality....click OK.

ANOVA

outcome1

	Sum of Squares	df	Mean Square	F	Sig.
Between Groups	1,477	4	,369	20,519	,000
Within Groups	,180	10	,018		
Total	1,657	14			

The mean squares give the error between treatments (0,369), and the residual error, (0,018). The total error equals $0,369 + 0,018 = 0,953$. Thus $0,953 = 95.3\%$ of the differences in the data is due to between treatment differences, and 4.7% is residual.

We will now perform a random effects analysis. The mean squares give the error between treatments (0,369), and the residual error, (0,018). With random effects the errors are assessed differently. A variance components analysis will be performed.

Command:

Analyze....General Linear Model....Variance Components....Dependent Variable: hours of sleep....Random Factor(s): treatment modality....click OK.

Factor Level Information

		N
treatment modality	1,00	3
	2,00	3
	3,00	3
	4,00	3
	5,00	3

Dependent Variable: VAR00002

Variance Estimates

Component	Estimate
Var(VAR00001)	,117
Var(Error)	,018

Dependent Variable:
VAR00002
Method: Minimum Norm
Quadratic Unbiased
Estimation (Weight = 1 for
Random Effects and
Residual)

The above tables are in the output sheets. The difference in the data due to the between treatment differences is now 0,117. The residual error is 0,018. This would mean that $0,117 / (0,018 + 0,117) = 0,867 = 86.7\%$ of the differences in the data is due to differences between the treatments. This is considerably less than the 95.3% in the one way anova analysis, but, with phase I-II studies, it is more accurate. How does the above computation work. The mathematical model is.

$$y = mean \pm mean\ variance_{population} \left(otherwise\ called\ s^2_{population}\right) \pm residual\ error.$$

The mean variance is independent of sample size, and is an adequate estimate of the spread in the data. A one way analysis of variance is performed with treatment modalities as factor, a categorical predictor.

The mean squares give the error between treatments (0,369), and the residual error, (0,018). However, with random effects the errors are assessed differently:

$$residual\ error = s^2_e = 0.018.$$

The mean square treatment now consists of the residual error plus three time the population error.

$$variance = s^2_e + 3\ s^2_{population} = 0.369.$$

This would mean that:

$$s^2_{population} = \frac{0.369 - 0.018}{3} = 0.117$$

and

$$s^2{}_{total} = s^2{}_{population} + s^2{}_e = 0.117 + 0.018 = 0.135$$

and

$$0.117/0.135 = 86.7\%.$$

Thus, 86.7% of the total variance is due to differences between treatments, and only 13.3% is due to differences within the treatment groups. The 86,7% difference in the data due to treatment effect may be more realistic and accurate than the 95.3% from the one way Anova.

4   Contrast Coefficients Regressions

Contrast coefficients regression is different from multiple binary variables or variance components analyses. With contrast coefficients regression, arbitrary weights, adding up to one, are given to subgroups (the categories) in the data. Different models are possible, and the best fit model is chosen for final analysis.

The underneath data example will be used for explanation. In 4 datasets involving 10 patients per dataset the fall in systolic blood pressure after treatments was the main outcome measure.

4 datasets of 10 patients	individual outcomes fall in systolic (syst) blood pressure (mm Hg)
,00	6,00
,00	7,10
,00	8,10
,00	7,50
,00	6,40
,00	7,90
,00	6,80
,00	6,60
,00	7,30
,00	5,60
1,00	5,10
1,00	8,00
1,00	3,80
1,00	4,40
1,00	5,20
1,00	5,40
1,00	4,30
1,00	6,00
1,00	3,70
1,00	6,20
2,00	4,10
2,00	7,00
2,00	2,80
2,00	3,40
2,00	4,20
2,00	4,40
2,00	3,30
2,00	5,00
2,00	2,70
2,00	5,20
3,00	4,30
3,00	4,30
3,00	6,20
3,00	5,60
3,00	6,20
3,00	6,00
3,00	5,30
3,00	5,40
3,00	5,40
3,00	5,30

The above SPSS data file is in extras.springer.com and is entitled "chap9contrast coefficients". It is previously used by the authors in Modern meta-analysis, Chap.22, Springer Heidelberg germany, 2017. Start by entering the data file in your computer installed with SPSS statistical software. Open the data file.

Command:
click Graphs....click Legacy Dialogs....Error Bars....mark Summaries for groups of cases....click Define....Variable: fall in blood pressure....Category Axis: dataset....click Bars Represent Confidence Intervals....Level: 95%....click OK.

The underneath graph is shown in the output.

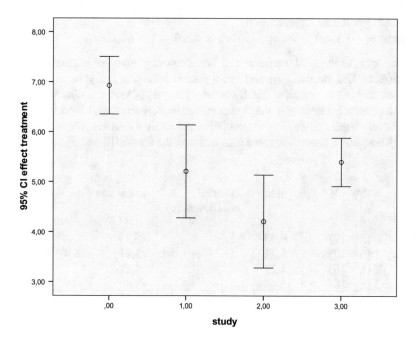

It shows the study means and 95% confidence intervals of four different treatment datasets assessing the fall in systolic blood after treatment. Using the Report commands will give you the means and the standard deviations.

Report

effect treatment

study	Mean	N	Std. Deviation
,00	6,9300	10	,80561
1,00	5,2100	10	1,29910
2,00	4,2100	10	1,29910
3,00	5,4000	10	,67659
Total	5,4375	40	1,41615

We will assess the above datasets for heterogeneity with linear contrast testing. We will first need a null-hypothesis. The table gives the means and standard deviations of the datasets 0, 1, 2, and 3. The datasets 0 and 3 were performed in the country, the other two in cities. The question is, are the results of studies in the country different from those in the cities. The null-hypothesis to be tested will be:

The results of the country studies are not different from the city studies.
The contrast is denoted as follows:
Contrast $= -1$ mean $_0 + 1$ mean $_1 + 1$ mean $_2 - 1$ mean $_3$
The null hypothesis: Contrast $= 0$
Contrast $= -1 \times 6930 + 1 \times 5210 + 1 \times 4210 - 1 \times 5400 = 0$

One way analysis of variance can be adequately used for assessing contrast coefficients. But we will first perform a pocket calculator analysis, and, then, use contrast analysis as option of SPSS one way analysis of variance (Anova).

The underneath table give the pocket calculator computations for one way Anova for a linear contrast analysis of the data from the above 4 datasets. We will assess the hypothesis, that a linear contrast exists between the studies [0 and 3] versus [1 and 2].

Study	mean	contrast coefficient (c)	mean times c	c^2
0 (n = 10)	6,93	-1	-6,93	1
1 (n = 10)	5,21	1	5,21	1
2 (n = 10)	4,21	1	4,21	1
3 (n = 10)	5,40	-1	-5,40	1 +
		0	-2.91	4

Computations are:

mean square $_{\text{between datasets}}$ / sum of squares $_{\text{within datasets}}$ =
mean square $_{\text{enumerator}}$ / sum of squares $_{\text{denominator}}$.

$$\text{sum of squares}_{\text{enumerator}} = \frac{10 \left(\sum (-1 \times 6,93 + 1 \times 5,21 + 1 \times 4,21 - 1 \times 5,40) \right)^2}{\sum \left((-1)^2 + (1)^2 + (1)^2 + (-1)^2 \right)}$$

$$= \frac{10(-2,91^2)}{4} = 84,7/4$$

$$= 21,4$$

For 1 degree of freedom:

$$\text{mean square}_{\text{enumerator}} = 21,4/1 = 21,4$$

$$\text{mean square}_{\text{denominator}} = \text{mean square}_{\text{error}}$$

$$\text{mean square}_{\text{error}} = \frac{(10-1)\,SD_{\text{dataset }0}{}^2 + \ldots + (10-1)\,SD_{\text{dataset }3}{}^2}{\text{degree of freedom } (= 4n - 4)}$$

$$= 9\,(0,65 + 1,69 + 1,69 + 0,46)/36$$

$$= 1,125$$

SD = standard deviation.

$$F\ (\text{Fisher})\ \text{statistic} = \text{mean square}_{\text{enumerator}}/\text{mean square}_{\text{denominator}}$$

$$= 21,4/1,125$$

$$= 19,02$$

This F-statistic for linear contrast has 1 and 36 degrees of freedom, and is, thus, statistically very significant with p = 0.0001 (see, e.g., Free p-Value Calculator for an F-Test at www.danielsoper.com). We conclude, that the datasets 0 and 3, from the country, are very significantly different from the datasets 1 and 2, from the cities.

Now, as an alternative, a linear contrast analysis will be performed using SPSS statistical software.

Command:
click Menu....click Comparing Means....click One Way Analysis of Variance.... Dependent List: enter effect treatment....Factor: enter dataset....click Contrast

Coefficients: enter -1....click Add....click 1....click Add....click 1....click Add.... click -1....click Add.... click continue....click OK.

The underneath tables are in the output file, and show the results of the one way analysis of variance first, and, the contrast coefficients analysis, second. For the contrast coefficients analysis SPSS does not use Anova tests, but rather unpaired t-tests.

ANOVA

VAR00003

	Sum of Squares	df	Mean Square	F	Sig.
Between Groups	37,875	3	12,625	11,267	,000
Within Groups	40,339	36	1,121		
Total	78,214	39			

Contrast Coefficients

	VAR00001			
	,00	1,00	2,00	3,00
Contrast				
1	-1	1	1	-1

Contrast Tests

		Contrast	Value of Contrast	Std. Error	t	df	Sig. (2-tailed)
VAR00003	Assume equal variances	1	-2,9100	,66949	-4,347	36	,000
	Does not assume equal variances	1	-2,9100	,66949	-4,347	28,576	,000

The datasets 1 and 4 have been given the contrast coefficients -1 and -1. The datasets 2 and 3 have been given the contrast coefficients 1 and 1. Together they add up to zero. The test statistics for this contrast model equals $-29,100$, and the t-value is much smaller than -1.96. This means that the outcomes of the datasets 2 and 3 are largely different from those of the datasets 1 and 4. The t-value squared is approximately equal to the F-value from the above pocket calculator computation, namely $(-4.347^2) = 18.896$ versus 19.02. The pocket calculator, using SDs, thus, produced, virtually, the same results, as the SPSS calculation using individual data of the separate datasets did.

If the datasets had been equal in size, a weighted contrast analysis would have been required. The means of the datasets 1 and 4 are, then, estimated by (n_1 mean$_1$ + n_4 mean$_4$) $/(n_1 + n_4)$ and those of the datasets 2 and 3 by (n_2 mean$_2$ + n_3 mean$_3$) $/(n_2 + n_3)$.

SPSS readily supplies the computations.

If you have arguments for contrasting group 0 versus the combined groups 2,3,4, then your model will produce even better statistics. With similar commands given and contrast coefficients of -3, 1, 1, and 1, the underneath results of the contrast coefficients analysis is given.

Contrast Coefficients

	VAR00001			
	,00	1,00	2,00	3,00
Contrast				
1	3	-1	-1	-1

Contrast Tests

		Contrast	Value of Contrast	Std. Error	t	df	Sig. (2-tailed)
VAR00003	Assume equal variances	1	5,9700	1,15958	5,148	36	,000
	Does not assume equal variances	1	5,9700	,98357	6,070	21,045	,000

Obviously, if you take into account less, e.g., equal variances, then your test statistic will further rise. Here the t-value rose from −4347 to 6070.

5 Conclusion

A categorical predictor in a linear regression may be incremental, e.g. in therapeutic trials incremental dosages of a drug. But, this is not always true, e.g. in a therapeutic trial with completely different compounds an incremental function may very well be lacking, and, thus, linear regression will be insignificant, and is an inappropriate model for assessment. Three adequate analysis methods are given.

1. Restructuring non-incremental categories into multiple binary variables is an adequate solution for the problem. Other methods to deal with the problem are variance components and contrast coefficients regressions. They are particularly suitable, if the categorical predictors are unexpected, otherwise called random.
2. Variance components compares such random effects with that of the residual error of a study, otherwise called the unexplained error of a study.
3. With contrast coefficients regression arbitrary weights adding up to one are given to subgroups (the categories) in the data. Different models are possible, and the best fit model is chosen for final analysis.

Reference

To readers requesting more background, theoretical and mathematical information of computations given, several textbooks complementary to the current production and written by the same authors are available: Statistics applied to clinical studies 5th edition, 2012, Machine learning in medicine a complete overview, 2015, SPSS for starters and 2nd levelers 2nd edition, 2015, Clinical data analysis on a pocket calculator 2nd edition, 2016, Understanding clinical data analysis, 2017, Modern Meta-analysis, 2017, all of them edited by Springer Heidelberg Germany.

Chapter 10
Laplace Regressions, Multi- Instead of Mono-Exponential Regressions

Work of Excellence from the San-Francisco Non-Mem (Nonlinear Mixed Effects) Pharmacological Group

Abstract There is an increasing trend towards the use of nonlinear mixed effect models for describing the pharmacokinetics and pharmacodynamics of drugs in humans. The term mixed effect models refers to the random effect statistical regression models applied. More information of mixed effects linear models is given in the Chap. 8. The current chapter addresses mixed effect *nonlinear* models, particularly mixed effects multi-exponential models. Especially, concentration-time relationships in pharmacokinetics do fit multi-exponential modelings pretty well, and these modeling approaches are increasingly an important part of drug approval processes. As logarithmic transformations allow for mono-exponential equations only, Laplace transformations, based on second differentiations, must be used for mathematical analysis. The current chapter gives examples.

Keywords Laplace regressions · Multi-exponential models · Non-mem analyses (nonlinear mixed effects analyses)

1 Introduction, History, and Background

In 1780, 100 years after Newton and Leibnitz invented differentiations, a great mathematician / statistician lived in Normandy France. Pierre-Simon, marquis de Laplace (23 March 1749–5 March 1827) was a French scholar whose work was important to the development of engineering, mathematics, statistics, physics, astronomy, and philosophy. He summarized and extended the work of his predecessors in his five-volume Mécanique Céleste (Celestial Mechanics) (1799–1825). This work translated the geometric study of classical mechanics to one based on calculus, opening up a broader range of problem solutions. In statistics, the Bayesian interpretation of probability was developed mainly by Laplace. Laplace formulated Laplace's equation, and pioneered the Laplace transform, which

© The Author(s), under exclusive license to Springer Nature Switzerland AG 2021
T. J. Cleophas, A. H. Zwinderman, *Regression Analysis in Medical Research*,
https://doi.org/10.1007/978-3-030-61394-5_10

appeared in many branches of mathematical physics, a field that he took a leading role in forming. The Laplaçian differential operator, widely used in mathematics, is also named after him. He restated, and developed the nebular hypothesis of the origin of the Solar System, and was one of the first scientists to postulate the existence of black holes, and the notion of gravitational collapse. Laplace is remembered as one of the greatest scientists of all time. Sometimes referred to as the French Newton or Newton of France, he has been described as possessing a phenomenal natural mathematical faculty, superior to that of any of his contemporaries. He was Napoleon's examiner, when Napoleon attended the École Militaire in Paris in 1784. Laplace became a count of the Empire in 1806, and was named a marquis in 1817, after the Bourbon Restoration. He, thus, also invented, what we currently call Laplace transformations. All of the Laplace transformations are second differentiations, and they enabled computations, that had been impossible before. Laplace transformations namely went one step further than traditional mono-exponential functions, and provided the calculus for multi -, rather than mono – exponential functions. Till then many biological processes had been modeled as mono-exponential models. Mono-exponential models may be adequate for the analysis of the survival of mosquitos, for multicausal and multi-dimensional processes they are not.

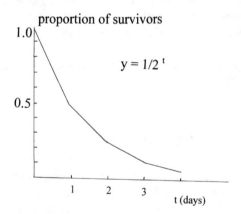

As a hypothesized example, the survival half life of mosquitos was given. As observed in the above graph, mosquitos only die, when colliding against a concrete wall. A survival half life can be pretty easily computed using the above exponential model. In the above example, after one day 50% of the mosquitos are alive, after the 2nd day 25% etc. A mathematical equation for proportion survivors $= (1/2)^t = 2^{-t}$. In true biology the e value, 2.71828, better fits the data, than does the value 2, and k is applied as a constant for the species. In this way the proportion survivors can be expressed as.

$$\text{proportion survivors} = e^{-kt}$$

The above model is used not only with mosquito survival, but also with survival studies of more complex creatures like humans, e.g. in the form of Cox proportional hazard analyses for explaining and making predictions from Kaplan-Meier curves (Chap. 2). This edition will review many regression methodologies based on exponential modeling like logistic regressions (the Chaps. 2 and 4), Poisson regressions (the Chaps. 5 and 14), loglinear regression (Chap. 5), quasi-likelihood regressions (Chap. 4), Markov models (Chap. 11), Cox regressions (Chap. 2 and 5), but all of them are mono-exponential. The current chapter is the only one, that will review models, that apply more than a single exponential term.

Why should we use multi-exponential regression models. Biological processes are often multicausal and multi-dimensional. The survival of mosquitos may be determined by them colliding against a concrete wall, with more complex creatures survival has multiple determinants, and these determinants may better suit exponential, or, even, multi-exponential, than linear or any other type of modeling. Already, Laplace assessed determinants of flower growth accordingly in the year 1780. Nowadays, important and helpful examples of multi-exponential modeling is found in pharmacology, particularly in the development of medicines. This chapter will review simple multi-exponential models in pharmacokinetics, like the ones applied in the Non-Mem Program of the San Francisco Non-Mem Pharmacological Group (Regression Analysis with Laplace transformations, Statistics applied to clinical studies 5th edition, pp 213–8, Springer Heidelberg Germany, 2012, from the same authors).

2 Regression Analysis with Laplace Transformations with Due Respect to Those Clinical Pharmacologists Who Routinely Use it

There is an increasing trend towards the use of nonlinear mixed effect models (commonly called population pharmacokinetics and pharmacodynamics) for describing the pharmacokinetics and pharmacodynamics of drugs in humans. The term mixed effect models refers to the random effect statistical regression models applied. More information of mixed effects linear models is given in the Chap. 8. However, this chapter will address mixed effect nonlinear models, particularly mixed effects multi-exponential models. These models allow for sparse sampling, and, at the same time, can account for multiple effects from associated variables, and even account for errors in samplings (Boeckman, NONMEM user's guide, NONMEM Project Group, University of California, San Francisco, 1992; Davidian, Nonlinear models for repeated measurements data Chapman and Hall, New York, 1995; Lindstrom, Nonlinear mixed effects models for repeated measures data, Biometrics, 1990; 46: 673–8). Particularly concentration-time relationships in pharmacokinetics do fit multi-exponential modelings pretty well, and these modeling approaches are increasingly becoming an important part of drug approval processes. They routinely make use of multi-exponential models, according to equations like the one underneath:

$$f(t) = D/V \left(e^{-at} + e^{-bt} + e^{-ct} .... \right)$$

D = dose drug
V = volume of distribution
a = absorption constant of compartment 1
b = elimination constant of "   1
c = elimination   "   " "   2
t = time

The above equation can also be described as follows.

$$f(t) = C(t) = C(0) \left(e^{-at} + e^{-bt} + e^{-ct} + ... \right)$$

C(t) = drug concentration at time t, C(0) = drug concentration at time 0.

As logarithmic transformations only allow for mono-exponential equations, Laplace transformations, based on second differentiations, must be used for mathematical analysis.

We should add, that NON-MEM is an abbreviation of non-linear mixed effect model. The model is mixed, because some covariates are assumed to have random (= unpredictable) effect, and are analyzed as nested terms (see also Chap. 8).

3   Laplace Transformations: How Does it Work

We will start with a Laplace transformation of a mono-exponential model. The log transformation of a mono-exponential function turns it into a linear function (ft = function of variable t (time), ln = natural logarithm, a = constant of linear function).

$$ft = e^{-at}$$

$$Ln\ ft = -at \text{ (linear function)}$$

Instead of transformation into a linear function, a Laplace procedure is possible. For example, ft. may be the function of a drug concentration in the blood.

Laplace is based on 2nd differentiations. With the underneath initial function (mono-exponential).

$$ft = C(t) = C_{(0)} . e^{-at}$$

the Laplace-transformation will look like:

$$fs = C(0)/(s + a)$$

where s = the Laplace variable, meaning here s = the amount drug cleared per unit of time, while the variable t (time) has gone. It is easy to observe that "1 / fs" is a linear function. If the initial function is not mono- but bi-exponential, then the underneath equation will be adequate.

$$ft = C(t) = C_{(0)} \cdot \left(e^{-at} + e^{-bt} \right)$$

a = absorption constant compartment 1.
b = elimination constant compartment 1.

Laplace-transformation of the above initial function produces the underneath function with again Laplace's "s", instead of "t" as outcome variable.

$$fs = C_{(0)}/(s + a)(s + b),$$

"s" = variable Laplace, "t" again has gone. It is again easy to observe, that here "1 / fs" is a simple quadratic function. Laplace has produced many transformation models for all kinds of multi-exponential equations, that are more complex, but, otherwise, similar to the above two.

4 Laplace Transformations and Pharmacokinetics

With Laplace procedures the variable "time", thus, disappears from the equation. The final data are, then, transformed back to their initial equations. The advantage of the exponential modeling in pharmacokinetics is, that it is very easy to calculate the keystone pharmacokinetic parameters according to which compounds are currently registered: plasma-half-life, volume of distribution, plasma-clearance rate etc. Exponential pharmacokinetic models assume first order kinetics, and it may be true, that many drugs at the therapeutically given concentrations would follow first order kinetics. However, zero order patterns are followed for example by ethyl-alcohol, acetyl-salicylic-acid, and by any drug at higher dosages, while second order elimination-patterns are followed, for example, by drugs, that are hydrolyzed or conjugated before excretion (Keusch, Chemical kinetic, rate laws, Arrhenius equation-experiments, www.uniregensburg.de). The simplest equations and curves for zero, first, and second order kinetics are given underrneath. We should add, that the nonlinear mixed model analyses are impossible without software, and, that software like.

S-plus SAS
Non-mem software (USF)

are pretty expensive.

The principles are straightforward with commands like below:

1. enter data, time, confounders (gender, age, comorbidities),
2. requested Laplace-transformations.

The software, subsequently, computes the best fit concentration-time curves, and adjusts confounders, and tests, whether correlations are statistically significant.

The advantages of Laplace transformations is, that they are based on first order kinetics, and produce important pharmacokinetic parameters, including plasma-half-life, distribution volume, equations for clearances).

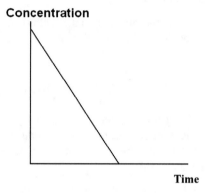

The above graph with time on the x-axis and drug concentration in the blood on the y-axis is an example of a linear zero order drug elimination pattern (k = elimination constant). The equation of a zero order elimination pattern is C (t) = C (0) − kt.

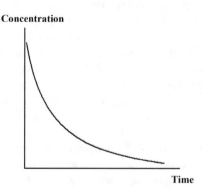

The above graph has again time on the x-axis and drug concentration in the blood on the y-axis is an example of a first order drug elimination pattern. The equation of a first order elimination pattern is C (t) = C (0). e^{-kt}.

Concentration

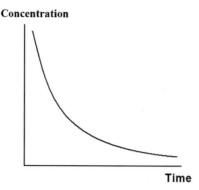

Time

The above graph again time on the x-axis and drug concentration in the blood on the y-axis is an example of a second order drug elimination pattern. The equation of a second order elimination pattern $1 / C\ (t) = 1 / C\ (0) - kt$.

In summary, examples of time-concentration curves following zero-, first-, and second-order pharmacokinetics have different patterns:

1. Zero order $C\ (t) = C\ (0)$ times - kt   linear pattern
2. First order $C(t) = C(0)$ times e^{-kt} exponential pattern
3. Second order $[1 / C\ (t)] = [1 / C(0)]$ times - kt   hyperbolic pattern

   k = elimination constant

The above examples are from hypothetical data.

As shown in the example of the underneath figure, in real practice there may be a wide spread in the data of a pharmacokinetic study. The 95% confidence intervals calculated with the NON-MEM software, which uses the Laplace transformations, assumes a first order pharmacokinetic. In fact both a zero and a second order pattern provided a better fit of the data. However, a problem with either of them is, that it is impossible to derive plasma-half-life and other pharmacokinetic parameters from them. As can be observed only in the above equation 2, plasma-half-life is not dependent on C(0). With equations 1 and 3, we have many plasma-half-lives, with equation 2 we have only one. This is a very elegant advantage of first order kinetics, but it should not mean, that the best fit data models are sacrificed for the purpose of unreliable pharmacokinetic parameters. A second problem with the Laplace models is, that they assume independence of confounders. In pharmacology confounders like gender, age, body mass, renal function, notoriously interact with the treatment modalities.

It is time that we summarized some of the limitations of the nonlinear mixed model analyses:

1. Nature does not always follow an exponential curve (curve = approximation with wide spread).
2. The possibility of interaction is not usually included.

3. Sometimes zero order (ethanol, aspirin, higher dosages of anything) or second order kinetics (methylation / hydrolysis) are more adequate.
4. Non-mem models are actually regression models and regression models are strictly for non-repeated data, and so, regression to the mean biases may be introduced.
5. The plotted data are usually not as precise as desired (see the wide data spread in underneath figure, showing drug elimination measurements of a biphosphonate compound.

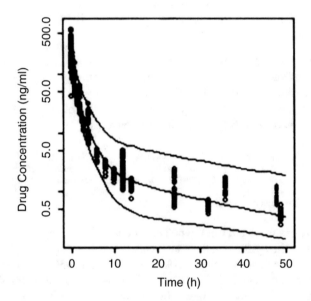

5 Conclusion

There is an increasing trend towards the use of non linear mixed effect models for describing the pharmacokinetics and pharmacodynamics of drugs in humans. The term mixed effect models refers to the random effect statistical regression models applied. More information of mixed effects linear models is given in the Chap. 8. This chapter addressed mixed effect nonlinear models, particularly mixed effects multi-exponential models. These models allow for sparse sampling and at the same time can account for multiple effect associated variables and even account for errors in sampling. Particularly concentration-time relationships in pharmacokinetics do fit multi-exponential modelings pretty well, and these modeling approaches are increasingly an important part of drug approval processes. They routinely make use of multi-exponential models.

As logarithmic transformations allow for mono-exponential equations only, Laplace transformations, based on second differentiations, must be used for mathematical analysis. We should add that NON-MEM is an abbreviation of nonlinear mixed effect model. The model is mixed, because some covariates have random ($=$ unpredictable) effect, and are analyzed as nested terms (see also Chap. 8 for more examples of data analyzed with nested terms).

Reference

To readers requesting more background, theoretical and mathematical information of computations given, several textbooks complementary to the current production and written by the same authors are available: Statistics applied to clinical studies 5th edition, 2012, Machine learning in medicine a complete overview, 2015, SPSS for starters and 2nd levelers 2nd edition, 2015, Clinical data analysis on a pocket calculator 2nd edition, 2016, Understanding clinical data analysis, 2017, all of them edited by Springer Heidelberg Germany.

Chapter 11
Regressions for Making Extrapolations

Gaussian Process Regressions (Kriging Regressions), Markov Regressions

Abstract Kriging, otherwise called cumulative Gaussian regression with an exponential model, is a statistical model where observations occur in a continuous domain, e.g., time or space. It uses matrix algebra to fit correlations between known and unknown places in time or space. A second methodology for making predictions about unmeasured places from measured ones is Markov regressions, which is, just like kriging, an exponential methodology, where, also with the help of matrix algebra, long term predictions can be made about short term observations. The current chapter reviews the two methods for making extrapolations, and uses real and hypothesized data examples for the purpose.

Keywords Regressions for making extrapolations · Kriging regressions · Semi-variography · Markov regressions

1 Introduction, History, and Background

In probability theory and statistics, a Gaussian process is a particular kind of statistical model in which observations occur in a continuous domain, e.g. time or space. It is called Gaussian, because it assumes, that the data are Gaussian-like, otherwise called normally, distributed. Danie Krige predicted the chance of finding gold. The chance was assumed to depend on (1) distances from measured places, (2) amounts of gold, and (3) distances of all of the successful places in the past from one another. Accounting these three requirements already is not a simple task, but it can be done with the help of Gaussian process regression, otherwise called Kriging regression. Currently, Kriging regression has myriads of applications, and this chapter will review how, with the help of some matrix algebra, it can be used to pretty well fit correlations between known and unknown places in time or space. A second methodology for making predictions about unmeasured places from measured ones is Markov regressions, an exponential methodology, where, also with the help of matrix algebra, long term predictions can be made about short term observations. The current chapter will review the two methods for extrapolations using real and hypothesized data examples.

The founders of Gaussian process regressions and Markov regressions are underneath.

Danie Gerhardus Krige (1919–2013 Johannesburg) was a South African statistician and mining engineer, who pioneered the field of geostatistics, and was professor at the University of the Witwatersrand, Republic of South Africa. The technique of kriging is named after him. His teacher, Georges François Paul Marie Matheron (1930–2000 Paris) was a French mathematician and civil engineer of mines, known as the founder of geostatistics and a co-founder of mathematical morphology. In 1968, he created the Centre de Géostatistique et de Morphologie Mathématique at the Paris School of Mines in Fontainebleau. Andrey Andreyevich Markov (1856–1922 Saint Petersburg) was a Russian mathematician best known for his work on stochastic processes. A primary subject of his research later became known as Markov chains and Markov processes. Markov and his younger brother Vladimir Andreevich Markov proved the Markov brothers' inequality to be true (n is constant, P = probability, X = nonnegative random variable, E = expected value)

$$n.P(X \geq n) \leq E(X).$$

The equation may look Greek to non-mathematicians, but the underneath graph may be somewhat helpful. It shows, that the green line, which is a graph of the function E(X), is always larger than the blue line which is a function of n.P(X>n).

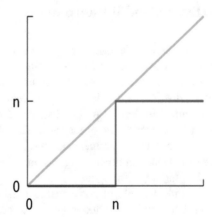

The Markov inequality is of great help to understanding probability assessments, a major goal of statistical analyses. See also the Laboratory for Intelligent Probability Systems text at "princeton.edu/a-geometric-intuition-for-Markovs-inequality", and many more similar communications on the internet.

2 Kriging Regression

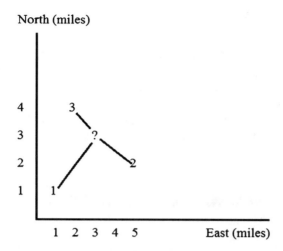

Krige was a student in Matheron's class (Sorbonne Paris 1965) who wrote his thesis on the South African gold search: with just a few bore hole places successful in the past, he tried and predicted the chance of finding gold at places in the neighbourhood. The chance was assumed to depend on distances from measured places, amounts of gold, and distances of all of the successful places in the past. Unlike gold, any other local characteristic can be extrapolated from the data of known places in the neighbourhood. We will assess the predicted % life expectancy at the place with the question mark, and use for the purpose the data of three places in the neighbourhood as shown in the above graph.

2.1 Semi Variography

It can be demonstrated, that, the more distant our place with unknown life expectancy from the known places in the neighbourhood, the more different the life expectancies will be. Matheron (Les variables regionalisees et leur estimation, PhD Thesis, Sorbonne Paris, Edited by Masson, Paris 1965) invented semi variography. He demonstrated, that the correlation between the outcome measures of two places with approximately the same distance will be high, and, of course, one (= 100%), if the two places are identical. The wider, the smaller the level of correlation, often called covariance here. The relationships between distances and levels of correlations is not linear, but either cubic or exponential, and cumulative Gaussian regressions seemed to fit such data pretty well. It is remarkable, but Matheron, and subsequently many others, showed that the underneath equation can be used to adequately predict the level of correlation from distances given in many situations:

correlation $= 0.8\ (1\ -\ 1.5\ (d/4) + 0.5\ (d/4)^3)$

The amount 0.8 in the above equation is called the sill. It is the amplitude of the curve.

With d very large, the curve will approach 0.8.

The term d is the variable for distance intervals, e. g., 1 mile meaning 0.5 to 1.5, 2 miles meaning 1.5 to 2.5 etc.

The amount of 4 in the above equation is called the range. It is the level where the cubic model flattens.

A graph of distance intervals versus the mean correlation levels between the two outcomes of the distance intervals of similar length is given underneath.

2.2 Correlation Levels Between Observed Places and Unobserved Ones

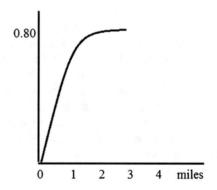

We will now use the above semi variogram equation for computing various correlation levels. The unknown place will be defined as place "?".

Place 1 to place "?"
distance $=\ \sqrt{8} = 2.8$ correlation $= 0.8\ (1\ -\ 1.5(2.8/4) + 0.5(2.8/$
$$4)^3\ ) = 0.096$$

Place 2 to place "?"
distance $=\ \sqrt{5} = 2.2$ correlation $= 0.8\ (1\ -\ 1.5(2.2/4) + 0.5\ (2.2/$
$$4)^3\ ) = 0.146$$

Place 3 to place "?"
distance $=\ \sqrt{2} = 1.4$ correlation $= 0.8\ (1\ -\ 1.5(1.4/4) + 0.5\ (1.4/$
$$4)^3\ ) = 0.505.$$

The matrix of distances between all of the places is given underneath

	1	2	3
1	0	$\sqrt{17}$	$\sqrt{10}$
2	$\sqrt{17}$	0	$\sqrt{13}$
3	$\sqrt{10}$	$\sqrt{13}$	0

$$\sqrt{17} = \sqrt{(4^2 + 1^2)}$$
$$\sqrt{10} = \sqrt{(3^2 + 1^2)}$$
$$\sqrt{13} = \sqrt{(3^2 + 2^2)}$$

Or in decimal terms:

	1	2	3
1	0	4.123	3.162
2	4.123	0	3.606
3	3.162	3.606	0

2.3   The Correlation Between the Known Places and the Place "?"

The correlation between the places and the place "?" is computed below:

place 1 to place 2: correlation $= 0.8\ (1-1.5(4.123/4) + 0.5\ (4.123/4)3) = 0.0016$
place 1 to place 3: correlation $= 0.8\ (1-1.5(1.186/4) + 0.5\ (1.186/4)3) = 0.049$
place 2 to place 3: correlation $= 0.8\ (1-1.5(3.61/4) + 0.5\ (3.61/4)3) = 0.467$

A covariance or correlation matrix can be drawn:

	1	2	3
1	0.80	0.0016	0.0049
2	0.0016	0.80	0.467
3	0.0049	0.467	0.80

Additional matrix algebra is required. The inverse power of the above correlation matrix (correlation-matrix)$^{-1}$ must be computed. It is, however, pretty complex to compute inverse power matrices by heart, and, therefore, the online matrix calculator "matrixcalc.org" will be used. First, enter the values of the correlation matrix at the appropriate place, then press inverse. The inverse power correlation matrix as computed must be multiplied with a column of correlations between the place 1, 2, and 3 versus the place "?".

$$\text{Correlation}^{-1} \times \begin{bmatrix} 0.096 \\ 0.146 \\ 0.80 \end{bmatrix} = \begin{bmatrix} 325/2813 \\ -148/563 \\ 2795/3516 \end{bmatrix} = \begin{bmatrix} 0.116 \\ -0.281 \\ 0.823 \end{bmatrix}$$

$$\begin{bmatrix} 0.116 & -0.281 & 0.823 \end{bmatrix} \begin{bmatrix} 0.07 \\ -0.09 \\ -0.06 \end{bmatrix} = -1597/1\,00\,000 = -0.016$$

With an overall % life expectancy as compared to 100%, of 0.780 = 78%, in practice a plausible percentage for many regions,
and
with a % life expectancy as compared to 100% in the measured places of

place 1   0.85
place 2   0.69
place 3   0.72

what will be the % life expectancy in place "?":

%life expectancy at place "?" = 0.78 − 0.016
$$= 0.764$$
$$= 76.4\%.$$

The %s life expectancies in known places are

place 1   0.85 (85%)
place 2   0.69 (69%)
place 3   0.72 (72%)

The place "?" with a % life expectancy of 76.4% is

> closest to place 3 (difference in % life expectancy only 4.4%),
> farthest from place 1 (difference in % life expectancy 8.6 %,
> In-between is place 2 with a difference of 7.4 %.

The above calculations illustrate, that the life expectancy at an unknown place can be computed with the help of the data from three places in the neighbourhood. With multiple computations at unknown places, the place with the highest life expectancy probability can be identified pretty closely.

3   Markov Regression

Markov modeling, otherwise called stochastic processes, assumes that per time unit the same % of a population will have an event, and it is used for long term predictions from short term observations. This chapter is to assess, whether the method can be applied by non-mathematicians using an online matrix-calculator. If per time unit the same % of patients will have an event like surgery, medical treatment, a complication like a co-morbidity or death, what will be the average time before such events take place.

   As an example, patients with three states of treatment for a disease are checked every 4 months. The underneath matrix is a so-called transition matrix. The states 1–3 indicate the chances of treatment: 1 = no treatment, 2 = surgery, 3 = medicine. If you are in state 1 today, there will be a 0.3 = 30% chance that you will receive no treatment in the next 4 months, a 0.2 = 20% chance of surgery, and a 0.5 = 50% chance of medicine treatment. If you are still in state 1 (no treatment) after 4 months, there will again be a 0.3 chance that this will be the same in the second 4 month period etc. So, after 5 periods the chance of being in state 1 equals $0.3 \times 0.3 \times 0.3 \times 0.3 \times 0.3 = 0.00243$. The chance that you will be in the states 2 or 3 is much larger, and there is something special about these states. Once you are in these states you will never leave them anymore, because the patients who were treated with either surgery or medicine are no longer followed in this study. That this happens can be observed from the matrix: if you are in state 2, you will have a chance of 1 = 100% to stay in state 2 and a chance of 0 = 0% not to do so. The same is true for the state 3.

	State in next period (4 months)		
State in current time	1	2	3
1	0.3	0.2	0.5
2	0	1	0
3	0	0	1

Now we will compute what will happen with the chances of a patient in the state 1 after several 4 month periods.

4 month period	chances of being in state: state 1	state 2	state 3
1st	30%	20%	50%
2nd	30x0.3= 9%	20+0.3x20= 26%	50+0.3x50= 65%
3rd	9x0.3= 3%	26+9x0.2= 27.8%	65+9x0.5= 69.5%
4th	3x0.3= 0.9%	27.8+3x0.2= 28.4%	69.5+3x0.5= 71.0%
5th	0.9x0.3= 0.27%	28.4+0.9x0.2= 28.6%	71.0+0.9x0.5= 71.5..

Obviously, the chances of being in the states 2 or 3 will increase, though increasingly slowly, and the chance of being in state 1 is, ultimately, going to approximate zero. In clinical terms: postponing the treatment does not make much sense, because everyone in the no treatment group will eventually receive a treatment and the ultimate chances of surgery and medicine treatment are approximately 29% and 71%. With larger matrices this method for calculating the ultimate chances is rather laborious. Matrix algebra offers a rapid method.

	State in next period (4 months)			
	1	2	3	
State in current time				
1	[0.3]	[0.2 0.5]	matrix Q	matrix R
2	[0]	[1 0]	matrix O	matrix I
3	[0]	[0 1]		

The states are called transient, if they can change (the state 1), and absorbing if not (the states 2 and 3). The original matrix is partitioned into four submatrices, otherwise called the canonical form:

[0.3]   Upper left corner:
                      This square matrix Q can be sometimes very large with rows and
                      columns respectively presenting the transient states.

[0.2   0.5]   Upper right corner:
                      This R matrix presents in rows the chance of being absorbed from
                      the transient state.

[1   0]   Lower right corner:
[0   1]   This identity matrix I presents rows and columns with chances of
                      being in the absorbing states, the I matrix must b e adjusted to the
                      size of the Q matrix (here it will look like [1] instead of   [1 0]
                                                                                      [0 1]

[0]   Lower left corner.
[0]   This is a matrix of zeros (0 matrix).

From the above matrices a fundamental matrix (F) is constructed.

$$[(\text{matrix I}) - (\text{matrix R})]^{-1} = [0.7]^{-1} = 10/7$$

With larger matrices a matrix calculator, like the Bluebit Online Matrix Calculator can be used to compute the matrix to the -1 power by clicking "Inverse".

The fundamental matrix F equals 10/7. It can be interpreted as the average time, before someone goes into the absorbing state ($10/7 \times 4$ months = 5.714 months). The product of the fundamental matrix F and the R matrix gives more exact chances of a person in state 1 ending up in the states 2 and 3.

$F \times R = (10/7) \times [0.2 \quad 0.5] = [2/7 \quad 5/7] = [0.285714 \quad 0.714286]$.

The two latter values add up to 1.00, which indicates a combined chance of ending up in an absorbing state equal to 100%.

As an second example, patients with three states of treatment for a chronic disease are checked every 4 months.

		State in next period (4 months)		
		1	2	3
State in current time				
1		0.3	0.6	0.1
2		0.45	0.5	0.05
3		0	0	1

The above matrix of three states and second periods of time gives again the chances of different treatment for a particular disease, but it is slightly different from the first example. Here state 1 = no treatment state, state 2 = medicine treatment, state 3 = surgery state. We assume that medicine can be stopped while surgery is irretrievable, and, thus, an absorbing state. We first partition the matrix.

	State in next period (4 months)				
	1	2	3		
State in current time					
1	[0.3	0.6]	[0.1]	matrix Q	matrix R
2	[0.45	0.5]	[0.05]		
3	[0 0]		[1]	matrix O	matrix I

The R matrix $\begin{bmatrix} 0.1 \\ 0.05 \end{bmatrix}$ is in the upper right corner.

The Q matrix $\begin{bmatrix} 0.3 & 0.6 \\ 0.45 & 0.5 \end{bmatrix}$ is in the left upper corner.

The I matrix $[1]$ is in the lower right corner, and must be adjusted, before it can be subtracted from the Q matrix according

to $\begin{bmatrix} 0 & 1 \\ 1 & 0 \end{bmatrix}$

The 0 matrix $[0 \quad 0]$ is in the lower left corner.

$$I - Q = \begin{bmatrix} 1 & 0 \\ 0 & 1 \end{bmatrix} - \begin{bmatrix} 0.3 & 0.6 \\ 0.45 & 0.5 \end{bmatrix} = \begin{bmatrix} 0.7 & -0.6 \\ 0.45 & -0.5 \end{bmatrix}$$

The inverse of [I – Q] is obtained by marking "Inverse" at the online Bluebit Matrix Calculator and equals

$[I - Q]^{-1}$ = fundamental matrix F =

$$\begin{bmatrix} 6.25 & 7.5 \\ 5.625 & 8.75 \end{bmatrix}$$

It is interpreted as the average periods of time before some transient state goes into the absorbing state:

(6.25+7.5=13.75) × 4 months for the patients in state 1 first and state 2 second, (5.625+8.75=14.375) × 4 months for the patients in state 2 first and state 1 second).

Finally, the product of matrix F times matrix R is calculated. It gives the chances of ending up in the absorbing state for those starting in the states 1 and 2.

$$\begin{bmatrix} 6.25 & 7.5 \\ 5.625 & 8.75 \end{bmatrix} \times \begin{bmatrix} 0.1 \\ 0.05 \end{bmatrix} = \begin{bmatrix} 1.00 \\ 1.00 \end{bmatrix}$$

Obviously the chance of both the transient states for ending up in the absorbing state is 1.00 = 100%.

A third example will be given.

State 1 = stable coronary artery disease (CAD),
state 2 = complications,
state 3 = recovery state,
state 4 = death state).

	State in next period (4 months)			
	1	2	3	4
State in current time				
1	0.95	0.04	0	0.01
2	0	0	0.9	0.1
3	0	0.3	0.3	0.4
4	0	0	0	1

If you take higher powers of this transition matrix (P), you will observe long-term trends of this model. For that purpose use the matrix calculator and square the transition matrix (P^2 gives the chances in the 2nd 4 month period etc) and compute also higher powers (P^3 ,P^4 ,P^5, etc).

P^2

0.903 0.038 0.036 0.024
0.270 0.270 0.460
0.090 0.360 0.550
0.000 0.000 1.000

P^6

0.698 0.048 0.063 0.191
0.026 0.064 0.910
0.021 0.047 0.931
0.000 0.000 1.000

The above higher order transition matrices suggest that with rising powers, and, thus, after multiple 4 month periods, there is a general trend towards the absorbing state: in each row the state 4 value continually rises. In the end we all will die, but in order to be more specific about the time, a special matrix like the one described in the previous examples is required. In order to calculate the precise time before the transient states go into the absorbing state, we need to partition the initial transition matrix.

	State in next period (4 months)			
	1	2	3	4
State in current time				
1	$\begin{bmatrix} 0.95 & 0.04 & 0.0 \\ 0.0 & 0.0 & 0.9 \\ 0.0 & 0.3 & 0.3 \end{bmatrix}$		$\begin{bmatrix} 0.01 \\ 0.1 \\ 0.4 \end{bmatrix}$ matrix Q	matrix R
2				
3				
4	$[0 \quad 0 \quad 0]$		$[1]$ matrix O	matrix I

$$F = (I - Q)^{-1}$$

$$I - Q = \begin{bmatrix} 1 & 0 & 0 \\ 0 & 1 & 0 \\ 0 & 0 & 1 \end{bmatrix} - \begin{bmatrix} 0.95 & 0.04 & 0.0 \\ 0.0 & 0.0 & 0.9 \\ 0.0 & 0.3 & 0.3 \end{bmatrix}$$

$$F = \begin{bmatrix} 0.5 & -0.04 & 0 \\ 0.0 & 1.0 & -0.9 \\ 0.0 & -0.3 & 0.7 \end{bmatrix}^{-1}$$

The online Bluebit Matrix calculator (mark inverse) produces the underneath result.

$$F = \begin{bmatrix} 20.0 & 1.302 & 1.674 \\ 0.0 & 1.628 & 2.093 \\ 0.0 & 0.698 & 2.326 \end{bmatrix}$$

The average time before various transient states turn into the absorbing state (dying in this example) is given.

State 1: $(20 + 1.302 + 1.674) \times 4$ months $= 91.904$ months.
State 2: $(0.0 + 1.628 + 2.093) \times 4$ months $= 14.884$ months.
State 3: $(0.0 + 0.698 + 2.326) \times 4$ months $= 12.098$ months.

The chance of dying for each state is computed from matrix F times matrix R (click multiplication, enter the data in the appropriate fields and click calculate.

$$
F.R = \begin{bmatrix} 20.0 & 1.302 & 1.672 \\ 0.0 & 1.628 & 2.093 \\ 0.0 & 0.698 & 2.326 \end{bmatrix} \times \begin{bmatrix} 0.01 \\ 0.1 \\ 0.4 \end{bmatrix} = \begin{bmatrix} 1.0 \\ 1.0 \\ 1.0 \end{bmatrix}.
$$

Like in the previous examples again the products of the matrices F and R show that all of the states end up with death. However, in the state 1 this takes more time than it does in the other states.

4   Conclusion

A Gaussian process is a particular kind of statistical model where observations occur in a continuous domain, e.g. time or space. It is called Gaussian, because it assumes, that the data are Gaussian-like, otherwise called normally distributed. Gaussian process modeling, otherwise called Kriging regression, has myriads of applications for the purpose, and this chapter reviewed how, with the help of some matrix algebra it can be used to pretty well fit correlations between known and unknown places in time or space. A second methodology for making predictions about unmeasured places from measured ones is Markov regressions, an exponential methodology, where, also with the help of matrix algebra, long term predictions can be made about short term observations. The current chapter reviewed the two methods for extrapolations using real and hypothesized data examples. We should add, that there is no mathematical explanation, why the two methods perform well in statistical practice. But an important assumption of both of them is, that the data – to – be – extrapolated should have a Gaussian or invert – Gaussian distance from the reference data.

Reference

To readers requesting more background, theoretical and mathematical information of computations given, several textbooks complementary to the current production and written by the same authors are available: Statistics applied to clinical studies 5th edition, 2012, Machine learning in medicine a complete overview, 2015, SPSS for starters and 2nd levelers 2nd edition, 2015, Clinical data analysis on a pocket calculator 2nd edition, 2016, Understanding clinical data analysis, 2017, all of them edited by Springer Heidelberg Germany.

Chapter 12
Standardized Regression Coefficients (SEMs)

Path Analysis, Structural Equation Modeling, Bayesian Networks, Extending the Hypothesis of Correlation to That of Causality

Abstract In clinical efficacy studies the outcome is often influenced by multiple causal factors like drug-noncompliance, frequency of counseling, and many more factors. Path analysis, structural equation modeling (SEM), and Bayesian networks are able to account not only direct but also indirect relationships across a network of variables. Unfortunately, p-values are not provided, but Akaike information indexes can be used to assess and compare better and worse networks. In this chapter the results of traditional regression analyses were compared with those of the novel networks. It is shown, that worthwhile additional information is given by the network methodologies.

Keywords Standardized regression coefficients · Path statistics · Structural equation modeling · Bayesian networks · Hypothesis of causality

1 Introduction, History, and Background

Instead of multiple linear regressions, currently, multistage regressions using path statistics are increasingly applied. It produces often better estimation of multi-step relationships than does standard linear regression. In clinical efficacy studies the outcome is often influenced by multiple causal factors like drug-noncompliance, frequency of counseling, and many more factors. Path analysis, structural equation modeling, and Bayesian networks are based on standardized rather than the traditional non-standardized regression coefficients, also called path statistics.

Structural equation modeling, as a term, is, currently, used in sociology and psychology, although, so far, little in medicine. It evolved from the earlier methods in genetic path modeling of Sewall Wright, an evolutionary geneticist from Harvard University MA. More modern forms came about with computer intensive

---

Electronic Supplementary Material The online version of this chapter (https://doi.org/10.1007/978-3-030-61394-5_12) contains supplementary material, which is available to authorized users. The videos can be accessed by scanning the related images with the SN More Media App.

T. J. Cleophas, A. H. Zwinderman, *Regression Analysis in Medical Research*,
https://doi.org/10.1007/978-3-030-61394-5_12

implementations in the 1960s and 1970s. SEM (structural equation modeling) evolved in three different streams:

1. Systems of equation regression methods were developed mainly by the Cowles Foundation for Research in Economics is an economic research institute at Yale University, New Haven Connecticut. It was created at the Cowles Commission for Research in Economics at Colorado Springs in 1932 by businessman and economist Alfred Cowles.
2. Iterative maximum likelihood algorithms for path analysis developed mainly by Karl Gustav Jöreskog (1935 – ...) at the Educational Testing Service Stockholm, and, subsequently, at the Uppsala University in Sweden.
3. Iterative canonical correlations, for fitting algorithms of path analysis, also developed at the Uppsala University through Hermann Wold (1908–1992 from Stockholm University). Much of this development occurred at a time, that automated computing was offering substantial upgrades over the existing calculator, and that analogue computing methods were available.

They were the spin-off of the proliferation of office equipments in the late 20th century. The 2015 edition *"Structural Equation Modeling: From Paths to Networks"* by Christopher Westland, from Illinois University Chicago, edited by Springer Heidelberg Germany, provides a thorough history of the methods.

Loose and confusing terminologies have been used to obscure weaknesses in the methods. In particular, PLS-PA (the Lohmoller algorithm) has been conflated with partial least squares regression (PLSR), which is a substitute for ordinary least squares regression, and has nothing to do with path analysis. PLS-PA has been falsely promoted as a method, that works with small datasets, when other estimation approaches fail. J Christopher Westland decisively showed this not to be true, and developed an algorithm for sample sizes in SEM. Since the 1970s, the "small sample size" assertion has been recognized to be false (see, for example, papers of Dhrymes, 1972, 1974; Dhrymes & Erlat, 1972; Dhrymes et al., 1972; Gupta, 1969; Sobel, 1982).

The founders of the methods of path analysis, structural equation modeling, and Bayesian networks are underneath. First, path analysis was invented by Sewall Wright (1889–1988 from Madison Wisconsin), while he developed his shifting balance theory of population genetics. He needed a technique, that could be used to track effects of forces on changes in one variable over time, or the effect of different variables on one another. Second, various structural equation models were invented by Judea Pearl (1936–..., Tel Aviv). Third, Bayesian networks were invented by many, but have also been commonly credited to Pearl.

Pearl was an Israeli-American computer scientist and philosopher, best known for championing the probabilistic approach to artificial intelligence. He is also credited for developing a theory of causal and counterfactual inference based on structural equation models.

2 Path Analysis

As an example, the effects of counseling and non-compliance (pills not used) on the efficacy of a novel laxative drug was studied in 35 patients. The first 10 patients are in the data file underneath. The entire data file is in extras.springer.com, and is entitled "chap12amos1.sav". It is previously used by the authors in Machine learning in medicine a complete overview, Chap. 48, Springer Heidelberg Germany, 2015. Start by opening the data file in your computer with SPSS installed (var = variable).

Var 1	var 2	var 3 (var = variable)
Outcome	Pt Instrumental variable	Problematic predictor
Frequency counseling	Pills not used	Efficacy estimator (ther eff) of new laxative (stools/month)
1. 24	8	25
2. 30	13	30
3. 25	15	25
4. 35	14	31
5. 39	9	36
6. 30	10	33
7. 27	8	22
8. 14	5	18
9. 39	13	14
10. 42	15	30

First, two linear regressions will be performed. Start by opening the data file in your computer.

Command:
Analyze....Regression....Linear....Dependent Variable: ther eff....Independent Variable(s): counseling, non-compliance....click OK.

The underneath table is from the output sheet.

Coefficients^a

Coefficients[a]

Model		Unstandardized Coefficients		Standardized Coefficients	t	Sig.
		B	Std. Error	Beta		
1	(Constant)	2,270	4,823		,471	,641
	counseling	1,876	,290	,721	6,469	,000
	non-compliance	,285	,167	,190	1,705	,098

a. Dependent Variable: ther eff

Next, with similar commands, the linear effect of non-compliance on counseling is assessed.

Coefficients^a

Model		Unstandardized Coefficients		Standardized Coefficients		
		B	Std. Error	Beta	t	Sig.
1	(Constant)	4,228	2,800		1,510	,141
	non-compliance	,220	,093	,382	2,373	,024

a. Dependent Variable: counseling

The two tables show the results of two linear regressions assessing (1) the effects of counseling and non-compliance on therapeutic efficacy, and (2) the effect of non-compliance on counseling. With p = 0.10 as cut-off p-value for statistical significance all of the effects were statistically significant. Non-compliance was a significant predictor of counseling, and, at the same time, a significant predictor of therapeutic efficacy. This would mean, that non-compliance worked two ways: it predicted therapeutic efficacy *directly* and *indirectly* through counseling. However, the indirect way is not taken into account in the usual one step linear regression. An adequate approach for assessing both ways simultaneously is path statistics. Path analysis uses add-up sums of regression coefficients for better estimation of multiple step relationships. Because regression coefficients have the same unit as their variable, they can not be added up unless they are standardized by dividing them by their own variances. SPSS routinely provides the standardized regression coefficients, otherwise called path statistics, in its regression tables as shown above. The underneath figure gives a diagram of path statistics from the data.

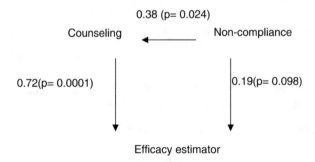

The standardized regression coefficients are added to the arrows. Single step path analysis gives a standardized regression coefficient of 0.19. This underestimates the real effect of non-compliance. Two step path analysis is more realistic, and, shows, that the add-up path statistic is larger, and equals

$$0.19 + 0.38 \times 0.72 = 0.46$$

The two-path statistic of 0.46 is a lot better than the single path statistic of 0.19 with an increase of 60%. The advantage of multistep regressions using path statistics is increasingly recognized, and it is the basis of many expert systems and bioinformatics for identifying networks with particular eye towards causal relationships, although it is not yet widely applied in biomedicine. In biomedicine causality is a difficult subject. Often effects are multicausal, and confounders rather than causal effects cannot be ruled out. Multistep path statistics, however, brings the search for causalities one step further. If you have multiple arrows like the ones in the above graph, and if they point directly or indirectly to the same center, and if your add-up regressions coefficients are increasingly large, then causal relationships will get increasingly probable.

3 Structural Equation Modeling

In clinical efficacy studies the outcome is often influenced by multiple causal factors, like drug – noncompliance, frequency of counseling, and many more factors. Structural equation modeling (SEM) was only recently formally defined by Pearl (In: Causality, reason, and inference, Cambridge University Press, Cambridge UK 2000). This statistical methodology includes:

path analysis,
factor analysis (see also Chap. 23), and
linear regression (see Chap.1).

An SEM model looks like a complex regression model, but it is more. It extends the prior hypothesis of correlation to that of causality, and this is accomplished by a network of variables tested versus one another with standardized rather than unstandardized regression coefficients. The network computes the magnitudes of the standardized covariances in a multifactorial data file, their p-values of differences versus zero, and their correlation coefficients. It is also used to construct a DAG (directed acyclic graph), which is a probabilistic graphical model of nodes (the variables) and connecting arrows presenting the conditional dependencies of the nodes. We will assess, whether the Amos (analysis of moment structures) add-on module of SPSS statistical software, frequently used in econo-/sociometry but little used in medicine, is able to perform an SEM analysis of pharmacodynamic data. Can SEM modeling in Amos (analysis of moment structures) demonstrate direct and indirect effects of non-compliance and counseling on treatment efficacy and quality of life?

We will use the underneath example.

stool	counseling	noncompliance	qol
stools /month	counselings /month	drug noncompliances /month	quality of life score
24,00	8,00	25,00	69,00
30,00	13,00	30,00	110,00
25,00	15,00	25,00	78,00
35,00	10,00	31,00	103,00
39,00	9,00	36,00	103,00
30,00	10,00	33,00	102,00
27,00	8,00	22,00	76,00
14,00	5,00	18,00	75,00
39,00	13,00	14,00	99,00
42,00	15,00	30,00	107,00

The first 10 patients of the 35 patient data file is above. The entire data file is in extras.springer.com, and is entitled "chap12amos2.sav". It is previously used by the authors in Machine learning in medicine a complete overview, Chap. 49, Springer Heidelberg Germany, 2015. You may open the data file in your computer with SPSS installed.

We will use SEM modeling for estimating variances and covariances in these data.

$$\text{Variance} = \sum \left[(x - x_{mean})^2 \right].$$

It is a measure for the spread of the data of the variable x.

$$\text{Covariance} = \sum \left[(x_1 - x_{1mean}) \, (x_2 - x_{2mean}) \right].$$

It is a measure for the strength of association between the two variables x_1 and x_2. If the covariances are significantly larger than zero, this would mean that there is a significant association between them. For analysis your computer must be installed with Amos software program, which can be downloaded as a free trial download at www.jmp.com.

Command:
Analyze....click IBM SPSS Amos

The work area of Amos appears. The menu is in the second upper row. The toolbar is on the left. In the empty area on the right you can draw your networks.

click File....click Save as....Browse the folder you selected in your personal computer, and enter amos2....click Save.

In the first upper row the title amos2 has appeared, in the bottom rectangle left from the empty area the title amos2 has also appeared.

click Diagram....left click "Draw Observed" and drag to empty area....click the green rectangle and a colorless rectangle appears....left click it and a red rectangle appears....do this 3 more times and have the rectangles at different places

The underneath figure shows how your screen will look by now. There are four rectangle nodes for observed variables.

Next, we will have to enter the names of the variables.

Command:
right click in the left upper rectangle....click Object Properties....Variable name: type "noncompliance"....close dialog box....the name is now in the rectangle....do the same for the other three rectangles.

Next, arrows have to be added to the diagram.

Command:
click Diagram....click Draw Covariances.

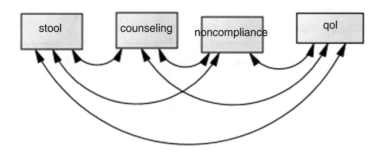

The above figure is now in the empty area. We will, subsequently, perform the analysis.

Command:

Analyze....Calculate Estimates....click File....click Save as....Browse for the folder of your choice and enter a name....click Save....click the new path diagram button that has appeared in the upper white rectangle left from the empty area.

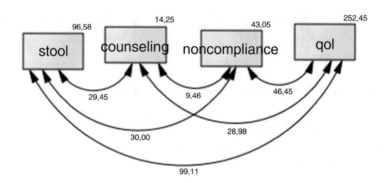

Unstandardized covariances of the variables are now shown in the graph, and variances of the variables are in the right upper corner of the nodes.

We will also view the text output.

Command:

click View....click Text output.

Estimates (Group number 1 - Default model)

Scalar Estimates (Group number 1 - Default model)

<u>Maximum Likelihood Estimates</u>

Covariances: (Group number 1 - Default model)

			Estimate	S.E.	C.R.	P	Label
stool	<-->	counseling	29,449	8,125	3,624	***	
counseling	<-->	noncompliance	9,461	4,549	2,080	,038	
noncompliance	<-->	qol	46,454	19,574	2,373	,018	
stool	<-->	noncompliance	30,003	12,197	2,460	,014	
counseling	<-->	qol	28,980	11,428	2,536	,011	
stool	<-->	qol	99,109	31,718	3,125	,002	

Variances: (Group number 1 - Default model)

	Estimate	S.E.	C.R.	P	Label
stool	96,576	23,423	4,123	***	
counseling	14,261	3,459	4,123	***	
noncompliance	43,050	10,441	4,123	***	
qol	252,462	61,231	4,123	***	

The above table shows the same values as the graph did, but p-values are added to the covariances. All of them, except stool versus counseling, were statistically significant with p-values from 0.002 to 0.038, meaning, that all of these variables were closer associated with one another than could happen by chance.

Unstandardized covariances of variables with different units are not appropriate, and, therefore, Amos also produces standardized values (= unstandardized divided by their own standard errors).

click View....click Analysis Properties....click Output tab....mark Standardized estimates....close dialog box....choose Analyze....click Calculate Estimates....click the path diagram button....click standardized estimates in third white rectangle left from the empty area.

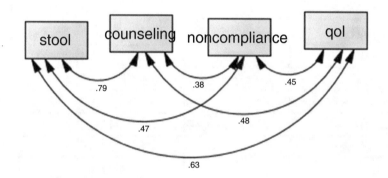

The standardized covariances are given.

Finally, we will view the standardized results as table.

click View....click Text Output.

The output now shows separately all of correlation coefficients between the variables.

A correlation coefficient < 0.25 indicates a poor, > 0.25 a reasonable, >0.50 a strong correlation.

Variables

Name	Label	Observed	Variance Estimate	SE
stool		☑	96,58	23,42
counseling		☑	14,26	3,46
noncompliance		☑	43,05	10,44
qol		☑	252,46	61,23

Regression weights

Dependent	Independent	Estimate	SE	Standardized

Covariances

Variable 1	Variable 2	Estimate	SE	Correlation
stool	counseling	29,45	8,13	,79
counseling	noncompliance	9,46	4,55	,38
noncompliance	qol	46,45	19,57	,45
stool	noncompliance	30,00	12,20	,47
counseling	qol	28,98	11,43	,48
stool	qol	99,11	31,72	,63

Unfortunately, overall statistics can not be computed, and, instead, overall goodness of fit tests are given. They do not prove, whether a model is significantly different from a null-hypothesis, but they do explain, whether it provides better or worse information, than any other DAG model obtained from the same data.

4   Bayesian Networks

Bayesian networks is like Structural equation modeling, but it goes on step further. It includes not only variables that in the multiple linear regressions are statistically significant, but also those that are not so. As an example, two meta-analysis studies of Lazarou (JAMA 1998; 279: 1200–5) are given about papers that assessed adverse drug reactions in patients already admitted to hospital. They included 39 papers reporting percentages of hospitalized patients with adverse drug reactions (Var = variable).

Var 0001 = year of publication,
Var 0002 = investigator (1 = internist, 2 = otherwise),

Var 0003 = sample size of study,
Var 0004 = percentage of hospitalized patients for adverse drug reactions.

The results from the first 10 papers are underneath. The entire data file is in extras.
springer.com, and is entitled "chap12bayesian.sav". It is previously used by the
authors in Modern meta-analysis, Chap. 12, Springer Heidelberg Germany, 2017.
Start by opening the data file in your computer with SPSS installed.

year of study	investigator type	study size	% admissions due to adverse effects
1995,00	2,00	379,00	5,30
1995,00	2,00	4031,00	4,40
1994,00	1,00	1024,00	10,30
1993,00	2,00	420,00	3,60
1981,00	1,00	815,00	14,80
1979,00	2,00	1669,00	16,80
1977,00	2,00	152,00	7,20
1977,00	1,00	334,00	10,20
1973,00	1,00	11526,00	22,50
1973,00	2,00	658,00	12,20

First, a traditional multiple linear regression with % admissions as outcome and
the other three variables as predictors will be performed.

Coefficients[a]

Model		Unstandardized Coefficients		Standardized Coefficients	t	Sig.
		B	Std. Error	Beta		
1	(Constant)	-95,518	101,603		-,940	,354
	year study	,054	,052	,180	1,048	,302
	investigator	-3,247	2,040	-,275	-1,591	,121
	study size	,000	,000	,057	,346	,732

a. Dependent Variable: %adverse effects

The above output table shows that none of the predictors of % severe (percentage
of patients in hospital with adverse drug reactions) were statistically significant, a
pretty disappointing result.

Next, we will perform Bayesian network assessments, bottom-up reasoning, and
we will use for the purpose the Knime (Konstanz information miner) and Weka
(Waikato knowledge environment analysis) software programs. In order for readers
to perform their own analyses examples of step by step analyses are given in
Machine learning in medicine a complete overview, the Chaps. 7, 70, 71,
74, Springer Heidelberg Germany, 2015, from the same authors.

The underneath directed acyclic graph, otherwise called Bayesian network, suggests, that the investigator type predicts the percentage of patients hospitalized for adverse drug reactions.

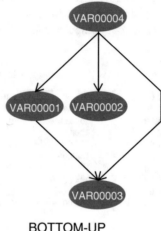

BOTTOM-UP

The underneath network based on the same data as those of the above network is different.

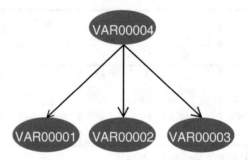

BOTTOM-UP

Which of the two networks is better? Akaike information index is often used for assessment. It is a goodness of fit test, calculated from the difference between the numbers of parameters minus their likelihood score. The smaller the Akaike information index, the better the fit of the model.

The Akaike information index for the two models are

$-6652.384..$ for the former

and

−37.662... for the latter.

This would mean, that the first of the two models provided a much better fit for the data given, than the latter one did.

5   Conclusion

Path analysis, structural equation modeling, and Bayesian networks are able to account not only direct but also indirect relationships across a network of variables. Unfortunately, p-values are not provided, but Akaike information indexes can be used to assess and compare the goodness of data fit of one network model with others. In this chapter the results of traditional regression analyses were compared with those of the novel networks. It is shown, that worthwhile additional information is obtained by the network methodology.

In the pas few years the network methodology has also been used as a method for synthesizing information from networks of trials addressing the same question but involving different interventions (Mills et al, JAMA 2012; 308: 1246–53, and Cipriani et al, Ann Intern Med 2013; 159: 130–7). The key assumption with any type of network meta-analysis is, of course, the exchangeability assumption: patient and study characteristics of studies in the meta-analysis must be similar enough to be comparable. The exchangeability assumption can sometimes be tested, but this is virtually always hard to do. E.g., a trial with three treatment arms A versus B versus C, should have the same effects of A versus C and B versus C as separate trials of two treatment arms should have. Currently, Extended Bucher networks (Song et al, BMJ 2011; 343: d4909) and Lumley networks (Stat Med 2002; 21: 2313–24) can cover for more complex networks with contrast coefficients adding up to one (see also the Chap. 9 of the current edition for a review of contrast coefficient meta-analysis).

Reference

To readers requesting more background, theoretical and mathematical information of computations given, several textbooks complementary to the current production and written by the same authors are available: Statistics applied to clinical studies 5th edition, 2012, Machine learning in medicine a complete overview, 2015, SPSS for starters and 2nd levelers 2nd edition, 2015, Clinical data analysis on a pocket calculator 2nd edition, 2016, Understanding clinical data analysis, 2017, all of them edited by Springer Heidelberg Germany.

Chapter 13
Multivariate Analysis of Variance and Canonical Regression

Turning Multiple Dependent and Independent Variables into Something Understandable

Abstract Multivariate analysis of variance (Manova) is the standard method for the analysis of multivariate data. It is, however, limited due to number of variables it can handle and due to the flaw of positive correlations between variables minimizing the statistical power of the models. Also overall statistics are lacking. Canonical analysis is wonderful, because it can handle many more variables than Manova, accounts for the relative importance of the separate variables and their interactions, and provides overall statistics. Unlike other methods for combining the effects of multiple variables like factor analysis/partial least squares, canonical analysis is scientifically entirely rigorous. Data examples and computations are given.

Keywords Multivariate analysis of variance · Canonical regression

1 Introduction, History, and Background

In clinical research the outcome is often measured with multiple variables all of whom constitute a separate aspect or dimension of the question. For example, the expressions of a cluster of genes can be used as a functional unit to predict the efficacy of cytostatic treatment, also repeated measurements can be used as endpoint in randomized longitudinal trials, and multi-item personal scores can be used for the evaluation of antidepressants. Many more examples can be given. Important methods for analysis are summarized.

1. Multivariate analysis of variance (Manova, see also the Chap. 8).
2. Latent variables methods, like factor analysis, principal components,
   partial least squares, discriminant analysis.
3. Canonical regression.

Electronic Supplementary Material The online version of this chapter (https://doi.org/10.1007/978-3-030-61394-5_13) contains supplementary material, which is available to authorized users. The videos can be accessed by scanning the related images with the SN More Media App.

1. Multivariate analysis of variance (Manova) is the standard method for the analysis of multivariate data. Just like Anova it is based on sums of squares (see Chap. 7), and, in addition, it computes SSCP matrices (sums of squares and cross products matrices). A problem with Manova is, that it rapidly loses statistical power with increasing numbers of variables, and that computer commands may not be executed due to numerical problems with higher order calculations among components. Also, clinically, we are often more interested in the combined effects of the clusters of variables than in the separate effects of the different variables. As a simple solution composite variables can be used as add-up sums of separate variables, but add-up sums do not account the relative importance of the separate variables, their interactions, and differences in units.

   Multivariate analysis of variance (MANOVA) can be thought of as a multivariate generalization of analysis of variance (ANOVA), in which more than one dependent variable is used to evaluate group differences. Samuel Wilks, American mathematician from Princeton Iowa played an important role in the development of Manova. In 1938 he invented an asymptotic distribution of the log likelihood ratio statistic suitable for estimating confidence intervals of maximal likelihoods, and named it the lambda statistic. Subsequently, other MANOVA test criteria were developed, although all share with lambda the difficulty of unusually complex sampling distributions (for example by Pillai, Hotelling, Roy). Instead of using lambda directly, either an approximate chi-square or F ratio transformation of lambda were often applied. Important contribution were given by Ted Anderson (1918–2016) from Stanford University who applied the above principles successfully for analyzing a great variety of statistics and economy data.

2. Latent variables methodology will be explained in the Chap. 23, but here a brief review is already given. Many factors in life are complex and difficult to measure directly. Charles Spearman, a London UK psychometrician in the 40s, is famous for his search for a method to measure intelligence with the help of latent variables methodology. He identified two or more unmeasured factors to explain a much larger number of factors. Latent variables methodology is computationally less complex than Manova. The principles are shown in the underneath graph.

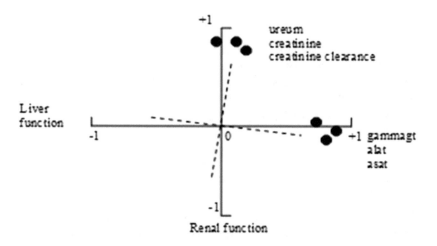

Of six measured serological tests used as variable, three are highly linearly correlated with one unmeasured novel variable, and three more measured variables are highly correlated with one more unmeasured novel variable. The two unmeasured variables are chosen such, that they provide the best correlation with the measured variables, and at the same time are orthogonal with one another. Using the novel unmeasured variables as a substitute for the 6 measured variables should provide a far more simple statistical model for answering scientific research questions. We should add, that the latent variables methodologies are, of course, rather subjective, and the construct of the latent variables are, of course, dependent on subjective decisions to cluster some of the measured variables and removing others. More information will be given in the Chap. 23.

3. Canonical regression, just like Manova, is scientifically more rigorous, than latent variables methods, leaving little room for subjective decisions. It was invented by Harold Hotelling (1895–1973), a professor of statistics in the early seventies at Chapel Hill University North Carolina, who used the term canonical, stemming from the Hebrew, and meaning being an important part or element of something, just like a canunic is in the Catholic church. It works with canonical weights, rather than subjectively gathered latent variables. Canonical weights are like regression coefficients in linear regression. In its simplest version they are the best fit a and b values (intercepts and regression coefficients) of the regression equations (CV = canonical variable) in the underneath mathematical model, used to provide a maximal correlation between two orthogonal canonical variates (CVs):

$$CVx = a_1 x_1 + a_2 x_2 + a_3 x_3 + a_4 x_4 \ldots\ldots$$

$$CVy = b_1\, y_1 + b_2\, y_2 + b_3\, y_3 + b_4\, y_4 \ldots \ldots$$

This means, that they provide the best fit a and b values of the dependent and independent variables in the data model.

In this chapter, first, a data analysis with Manova will be given. Then, a canonical analysis will be explained using the same data example. Latent variables will again be reviewed in a separate chapter, the Chap. 23.

2   Multivariate Analysis of Variance (Manova)

A 250 patients' data-file was supposed to include 27 variables consistent of both patients' microarray gene expression levels and their drug efficacy scores. All of the variables were standardized by scoring them on 11 points linear scales (0–10). The following genes were highly expressed: the genes (Gs) 1–4, 16–19, and 24–27. Four variables were supposed to represent drug efficacy scores and were clustered as the outcome variables (Os) 1–4.

G1	G2	G3	G4	G16	G17	G18	G19	G24	G25	G26	G27	O1	O2	O3	O4
8	8	9	5	7	10	5	6	9	9	6	6	6	7	6	7
9	9	10	9	8	8	7	8	8	9	8	8	8	7	8	7
9	8	8	8	8	9	7	8	9	8	9	9	9	8	8	8
8	9	8	9	6	7	6	4	6	6	5	5	7	7	7	6
10	10	8	10	9	10	10	8	8	9	9	9	8	8	8	7
7	8	8	8	8	7	6	5	7	8	8	7	7	6	6	7
5	5	5	5	5	6	4	5	5	6	6	5	6	5	6	4
9	9	9	9	8	8	8	8	9	8	3	8	8	8	8	8
9	8	9	8	9	8	7	7	7	7	5	8	8	7	6	6
10	10	10	10	10	10	10	10	10	8	8	10	10	10	9	10
2	2	8	5	7	8	8	8	9	3	9	8	7	7	7	6
7	8	8	7	8	6	6	7	8	8	8	7	8	7	8	8
8	9	9	8	10	8	8	7	8	8	9	9	7	7	8	8

The data from the first 13 patients are shown above. The entire data file entitled "chap13manova" is in extras.springer.com. It is previously used by the authors in Machine learning in medicine a complete overview, Chap. 22, Springer Heidelberg Germany, 2016. First, Manova was performed with the four drug efficacy scores as outcome variables and the twelve gene expression levels as covariates. Start by opening the data file in your computer with SPSS statistical software installed. We can now use the Menu commands.

Command:
click Analyze....click General Linear Model....click Multivariate....Dependent Variables: enter the four drug efficacy scores....Covariates: enter the 12 genes.... OK.

Multivariate Tests[a]

Effect		Value	F	Hypothesis df	Error df	Sig.
Intercept	Pillai's Trace	,043	2,657[b]	4,000	234,000	,034
	Wilks' Lambda	,957	2,657[b]	4,000	234,000	,034
	Hotelling's Trace	,045	2,657[b]	4,000	234,000	,034
	Roy's Largest Root	,045	2,657[b]	4,000	234,000	,034
geneone	Pillai's Trace	,006	,362[b]	4,000	234,000	,835
	Wilks' Lambda	,994	,362[b]	4,000	234,000	,835
	Hotelling's Trace	,006	,362[b]	4,000	234,000	,835
	Roy's Largest Root	,006	,362[b]	4,000	234,000	,835
genetwo	Pillai's Trace	,027	1,595[b]	4,000	234,000	,176
	Wilks' Lambda	,973	1,595[b]	4,000	234,000	,176
	Hotelling's Trace	,027	1,595[b]	4,000	234,000	,176
	Roy's Largest Root	,027	1,595[b]	4,000	234,000	,176
genethree	Pillai's Trace	,042	2,584[b]	4,000	234,000	,038
	Wilks' Lambda	,958	2,584[b]	4,000	234,000	,038
	Hotelling's Trace	,044	2,584[b]	4,000	234,000	,038
	Roy's Largest Root	,044	2,584[b]	4,000	234,000	,038
genefour	Pillai's Trace	,013	,744[b]	4,000	234,000	,563
	Wilks' Lambda	,987	,744[b]	4,000	234,000	,563
	Hotelling's Trace	,013	,744[b]	4,000	234,000	,563
	Roy's Largest Root	,013	,744[b]	4,000	234,000	,563
genesixteen	Pillai's Trace	,109	7,192[b]	4,000	234,000	,000
	Wilks' Lambda	,891	7,192[b]	4,000	234,000	,000
	Hotelling's Trace	,123	7,192[b]	4,000	234,000	,000
	Roy's Largest Root	,123	7,192[b]	4,000	234,000	,000
geneseventeen	Pillai's Trace	,080	5,118[b]	4,000	234,000	,001
	Wilks' Lambda	,920	5,118[b]	4,000	234,000	,001
	Hotelling's Trace	,087	5,118[b]	4,000	234,000	,001
	Roy's Largest Root	,087	5,118[b]	4,000	234,000	,001
geneeighteen	Pillai's Trace	,023	1,393[b]	4,000	234,000	,237
	Wilks' Lambda	,977	1,393[b]	4,000	234,000	,237
	Hotelling's Trace	,024	1,393[b]	4,000	234,000	,237
	Roy's Largest Root	,024	1,393[b]	4,000	234,000	,237
genenineteen	Pillai's Trace	,092	5,938[b]	4,000	234,000	,000
	Wilks' Lambda	,908	5,938[b]	4,000	234,000	,000
	Hotelling's Trace	,102	5,938[b]	4,000	234,000	,000
	Roy's Largest Root	,102	5,938[b]	4,000	234,000	,000
genetwentyfour	Pillai's Trace	,045	2,745[b]	4,000	234,000	,029
	Wilks' Lambda	,955	2,745[b]	4,000	234,000	,029
	Hotelling's Trace	,047	2,745[b]	4,000	234,000	,029
	Roy's Largest Root	,047	2,745[b]	4,000	234,000	,029
genetwentyfive	Pillai's Trace	,017	1,037[b]	4,000	234,000	,389
	Wilks' Lambda	,983	1,037[b]	4,000	234,000	,389
	Hotelling's Trace	,018	1,037[b]	4,000	234,000	,389
	Roy's Largest Root	,018	1,037[b]	4,000	234,000	,389
genetwentysix	Pillai's Trace	,027	1,602[b]	4,000	234,000	,174
	Wilks' Lambda	,973	1,602[b]	4,000	234,000	,174
	Hotelling's Trace	,027	1,602[b]	4,000	234,000	,174
	Roy's Largest Root	,027	1,602[b]	4,000	234,000	,174
genetwentyseven	Pillai's Trace	,045	2,751[b]	4,000	234,000	,029
	Wilks' Lambda	,955	2,751[b]	4,000	234,000	,029
	Hotelling's Trace	,047	2,751[b]	4,000	234,000	,029
	Roy's Largest Root	,047	2,751[b]	4,000	234,000	,029

a. Design: Intercept + geneone + genetwo + genethree + genefour + genesixteen + geneseventeen + geneeighteen + genenineteen + genetwentyfour + genetwentyfive + genetwentysix + genetwentyseven

b. Exact statistic

The above table of multivariate tests is in the output. It shows, that Manova can be considered, just like analysis of variance (Anova, Chap. 7), as a regression model with intercepts and regression coefficients. Just like Anova, it is based on normal distributions, and the results as given indicate, that the model is adequate for the data. Generally, Pillai's method gives the best robustness. We can conclude that the genes 3, 16, 17, 19, 24, and 27 are significant predictors of four drug efficacy outcome scores adjusted for one another. However, we do not know the relative importance of one outcome variable versus the other. Separate Anovas would be needed for that purpose. Also, unlike Anova, Manova does not give overall p-values, but rather separate p-values for separate covariates. However, in the given example, the genes are considered a cluster of genes forming a single functional unit. Also, the outcome variables are considered a cluster presenting different dimensions or aspects of drug efficacy. And, so, we are, particularly, interested in the combined effect of the set of covariates on the set of outcomes, rather than we are in modeling the separate variables. Instead of Manova some latent variables model can be constructed of the data (Chap. 23), but latent variables are subjective, and a canonical regression is not, and may be a better choice.

3   Canonical Regression

Like Manova, canonical analysis is based on multiple linear regression, used to find the best fit correlation coefficients for your data. However, because it works with Wilks' statistic and beta distributions rather than Pillai's statistic and normal distributions, it is able to more easily calculate overall correlation coefficients between sets of variables. Yet, it also assesses, how a set of variables, as a whole is related to separate variables. Along this way, an overall canonical model can be further improved by removing unimportant variables. Canonical analysis may be arithmetically equivalent to factor-analysis/partial least squares analysis, but, conceptionally, it is very different. Unlike the latter, the former method does not produce new (latent) variables, but rather makes use of two sets of manifest variables. Also, unlike the latter, it complies with all of the requirements of traditional linear regression, and is, therefore, scientifically rigorous. A canonical analysis should start with a correlation matrix. Variables with large correlation coefficients must be removed from the model. If in canonical models the clusters of predictor and outcome variables have a significant relationship, then this finding can, just like with linear regression, be used for making predictions about individual patients. We will again use SPSS statistical software. The Menu does not offer canonical analysis, but the Syntax program does. Canonical analysis should start with a collinearity matrix. This is because the uncertainty of the canonical weights, being the main outcome of a canonical regression, are severely overestimated in case of collinearity. Variable versus variable correlation coefficients larger than 0.80 means that the models is collinear and that the collinear variables should be removed from the model. In order to assess collinearity, a correlation matrix must be first constructed. We will use the data example from the previous section once more.

Command:

click File....click New....click Syntax....the Syntax Editor dialog box is displayed....enter the following text: "correlations/variables="and subsequently enter all of the gene-names....click Run.

Correlations

		geneone	genetwo	genethree	genefour	genesixteen	geneseventeen	geneeighteen	genenineteen	genetwentyfour	genetwentyfive	genetwentysix	genetwentyseven
geneone	Pearson Correlation	1	,644	,543	,504	,327	,349	,316	,396	,349	,343	,336	,283
	Sig. (2-tailed)		,000	,000	,000	,000	,000	,000	,000	,000	,000	,000	,000
	N	250	250	250	250	250	250	250	250	250	250	250	250
genetwo	Pearson Correlation	,644	1	,776	,449	,554	,437	,418	,539	,497	,359	,509	,326
	Sig. (2-tailed)	,000		,000	,000	,000	,000	,000	,000	,000	,000	,000	,000
	N	250	250	250	250	250	250	250	250	250	250	250	250
genethree	Pearson Correlation	,543	,776	1	,462	,579	,491	,477	,590	,524	,375	,521	,288
	Sig. (2-tailed)	,000	,000		,000	,000	,000	,000	,000	,000	,000	,000	,000
	N	250	250	250	250	250	250	250	250	250	250	250	250
genefour	Pearson Correlation	,504	,449	,462	1	,250	,384	,420	,329	,285	,337	,243	,375
	Sig. (2-tailed)	,000	,000	,000		,000	,000	,000	,000	,000	,000	,000	,000
	N	250	250	250	250	250	250	250	250	250	250	250	250
genesixteen	Pearson Correlation	,327	,554	,579	,250	1	,634	,460	,636	,567	,336	,580	,284
	Sig. (2-tailed)	,000	,000	,000	,000		,000	,000	,000	,000	,000	,000	,000
	N	250	250	250	250	250	250	250	250	250	250	250	250
geneseventeen	Pearson Correlation	,349	,437	,491	,384	,634	1	,629	,496	,543	,343	,497	,431
	Sig. (2-tailed)	,000	,000	,000	,000	,000		,000	,000	,000	,000	,000	,000
	N	250	250	250	250	250	250	250	250	250	250	250	250
geneeighteen	Pearson Correlation	,316	,418	,477	,420	,460	,629	1	,504	,442	,278	,416	,415
	Sig. (2-tailed)	,000	,000	,000	,000	,000	,000		,000	,000	,000	,000	,000
	N	250	250	250	250	250	250	250	250	250	250	250	250
genenineteen	Pearson Correlation	,396	,539	,590	,329	,636	,496	,504	1	,565	,359	,584	,345
	Sig. (2-tailed)	,000	,000	,000	,000	,000	,000	,000		,000	,000	,000	,000
	N	250	250	250	250	250	250	250	250	250	250	250	250
genetwentyfour	Pearson Correlation	,349	,497	,524	,285	,567	,543	,442	,565	1	,490	,732	,532
	Sig. (2-tailed)	,000	,000	,000	,000	,000	,000	,000	,000		,000	,000	,000
	N	250	250	250	250	250	250	250	250	250	250	250	250
genetwentyfive	Pearson Correlation	,343	,359	,375	,337	,336	,343	,278	,359	,490	1	,512	,445
	Sig. (2-tailed)	,000	,000	,000	,000	,000	,000	,000	,000	,000		,000	,000
	N	250	250	250	250	250	250	250	250	250	250	250	250
genetwentysix	Pearson Correlation	,336	,509	,521	,243	,580	,497	,416	,584	,732	,512	1	,477
	Sig. (2-tailed)	,000	,000	,000	,000	,000	,000	,000	,000	,000	,000		,000
	N	250	250	250	250	250	250	250	250	250	250	250	250
genetwentyseven	Pearson Correlation	,283	,326	,288	,375	,284	,431	,415	,345	,532	,445	,477	1
	Sig. (2-tailed)	,000	,000	,000	,000	,000	,000	,000	,000	,000	,000	,000	
	N	250	250	250	250	250	250	250	250	250	250	250	250

The above collinearity matrix comes up, and shows that none of the correlation coefficients were larger than 0.8, and, so, high collinearity was not in the data.

Unlike Manova, canonical regression is able to calculate overall correlation coefficients between sets of variables, and, in addition, it can assess, how sets of variables as a whole are related to separate variables. In order to assess the overall effect of clusters of genes on the cluster of drug efficacy scores, a canonical regression will be performed. Canonical regression is like collinearity matrices in SPSS' Syntax program.

Command:
click File....click New....click Syntax....the Syntax Editor dialog box is displayed....enter the following text: "manova" and subsequently enter all of the outcome variables....enter the text "WITH"....then enter all of the gene-names.... then enter the following text: /discrim all alpha(1)/print=sig(eigen dim).... click Run.

Numbers variables (covariates versus outcome variables) are in the output.

Numbers variables (covariates versus outcome variables) are in the output.

	Canon cor	Sq cor	Wilks L	F	Hypoth df	Error df	p
12 v 4	0.87252	0.7613	0.19968	9.7773	48.0	903.4	0.0001
7 v 4	0.87054	0.7578	0.21776	16.227	28.0	863.2	0.0001
7 v 3	0.87009	0.7571	0.22043	22.767	21.0	689.0	0.0001

The above table is given (cor = correlation coefficient, sq = squared, L = lambda, hypoth = hypothesis, df = degree of freedom, p = p-value, v = versus). The upper row, shows the result of the statistical analysis. The correlation coefficient between the 12 predictor and 4 outcome variables equals 0.87252. A squared correlation coefficient of 0.7613 means, that 76% of the variability in the outcome variables is explained by the 12 covariates. The cluster of predictors is a very significant predictor of the cluster of outcomes, and can be used for making predictions about individual patients with similar gene profiles. Repeated testing after the removal of separate variables gives an idea about relatively unimportant contributors as estimated by their coefficients, which are kind of canonical b-values (regression coefficients). The larger they are, the more important they are.

Canon Cor

Model	12 v 4	7 v 4	7 v 3
Outcome 1	-0.24620	-0.24603	0.25007
Outcome 2	-0.20355	-0.19683	0.20679
Outcome 3	-0.02113	-0.02532	
Outcome 4	-0.07993	-0.08448	0.09037
Gene 1	0.01177		
Gene 2	-0.01727		
Gene 3	-0.05964	-0.08344	0.08489
Gene 4	-0.02865		
Gene 16	-0.14094	-0.13883	0.13755
Gene 17	-0.12897	-0.14950	0.14845
Gene 18	-0.03276		
Gene 19	-0.10626	-0.11342	0.11296
Gene 24	-0.07148	-0.07024	0.07145
Gene 25	-0.00164		
Gene 26	-0.05443	-0.05326	0.05354
Gene 27	0.05589	0.04506	-0.04527

The above table gives an overview of the canonical coefficients, otherwise called canonical weights (the b-values of the canonical regression). In the left column are the canonical coefficients, otherwise called canonical weights (the multiple b-values of canonical regression), (Canon Cor = canonical correlation coefficient, v = versus, Model = analysis model after removal of one or more variables). The outcome 3, and the genes 2, 4, 18 and 25 contributed little to the overall result. When restricting the model by removing the variables with canonical coefficients smaller than 0.05 or larger than -0.05 (the middle and right columns of the table), the results were largely unchanged. And so were the results of the overall tests (the 2nd and 3rd rows of the above numbers variables table). Seven versus three variables produced virtually the same correlation coefficient, but with much more power (lambda increased from 0.1997 to 0.2204, the F value from 9.7773 to 22.767, in spite of a considerable fall in the degrees of freedom. It, therefore, does make sense to try and remove the weaker variables from the model ultimately to be used. The weakest contributing covariates of the Manova were virtually identical to the weakest canonical predictors, suggesting that the two methods are closely related and one method confirms the results of the other.

A table of standardized Canon Cor (canonical correlation coefficients) is also given.

Model	12 v 4	7 v 4	7 v 3
Outcome 1	-0.49754	-0.49720	0.50535
Outcome 2	-0.40093	-0.38771	0.40731
Outcome 3	-0.03970	-0.04758	
Outcome 4	-0.15649	-0.16539	0.17693
Gene 1	0.02003		
Gene 2	-0.03211		
Gene 3	-0.10663	-0.14919	0.15179
Gene 4	-0.04363		
Gene 16	-0.30371	-0.29918	0.29642
Gene 17	-0.23337	-0.27053	0.26862
Gene 18	-0.06872		
Gene 19	-0.23696	-0.25294	0.25189
Gene 24	-0.18627	-0.18302	0.18618
Gene 25	-0.00335		
Gene 26	-0.14503	-0.14191	0.14267
Gene 27	0.12711	0.10248	-0.10229

The above table is similar to the previous one, but the canonical correlations have been standardized, id est, z-transformed, meaning, that they are put in an interval between −1 and +1 with zero being the lowest possible canonical correlation and − and + 1 being the largest negative and positive correlations.

4   Conclusion

Manova is limited due to number of variables it can handle and due to the flaw of positive correlations between variables minimizing the statistical power of the models. Also overall statistics are lacking. Canonical analysis is wonderful, because it can handle many more variables than Manova, accounts for the relative importance of the separate variables and their interactions, and provides overall statistics. Unlike other methods for combining the effects of multiple variables, like factor analysis/ partial least squares (Chap. 23), canonical analysis is scientifically entirely rigorous.

Canonical regression and optimal scaling are sometimes combined. This combination is highly relevant, but not yet available in SPSS statistical software, although optimal scaling is in SPSS statistical software in the univariate version (see also the Chap. 16). In the past few years it is increasingly applied under the title sparse canonical, otherwise called penalized canonical correlation analysis. It is available in PMA package in R statistical software. Waaijenborg and Zwinderman (Stat Applic in Gen Mol Biol 2008; 7:1, and Parhomenko et al 2007; 1: s 19) contributed to its development, and showed that, with shrinking factors (λ-factors) included, regression coefficients reduced from b to $b/(1+\lambda)$, enabling dimension reductions in data models with very many variables, and that this method will be particularly suitable, if you are searching for a limited number of strong predictors (see also Machine learning in medical research 2nd edition, 2020, Chap 53, Springer Heidelberg Germany, from the same authors as the current edition). Canonical are somewhat like latent variables, and serve as a data dimension reduction method, helpful to avoid overdispersion and provide better sensitivity of testing. Optimal scaling and regularization furthermore optimize sensitivity of testing.

Reference

To readers requesting more background, theoretical and mathematical information of computations given, several textbooks complementary to the current production and written by the same authors are available: Statistics applied to clinical studies 5th edition, 2012, Machine learning in medicine a complete overview, 2015, SPSS for starters and 2nd levelers 2nd edition, 2015, Clinical data analysis on a pocket calculator 2nd edition, 2016, Understanding clinical data analysis, 2017, all of them edited by Springer Heidelberg Germany.

Chapter 14
More on Poisson Regressions

Poisson Regressions with Event Outcomes per Person or per Population and per Period of Time

Abstract In the Chap. 4, binary Poisson regressions were assessed of parallel groups with a binary outcome. In the Chap. 5, Poisson regressions were used for data with polytomous outcomes. This chapter will address additional models applying Poisson distributions. For rate analyses Poisson regression is very sensitive, and, generally, better so than standard linear regression. Linear regression measures events per population (or person), but does not explicitly include time as a covariate, although, implicitly, it is often assumed, albeit not plainly expressed. Poisson regression cannot only be used for counted events per person per period of time, but also for numbers of yes/no events per population per period of time. It is, then, similar to logistic regression, but different from it, in that it uses a log instead of logit (log odds) transformed dependent variable. It is more adequate and often provides better statistics than logistic regression does, because, again, time is explicitly included. In observational research event rates are often very much age and sex dependent and a model routinely adjusting these confounders are welcome. Examples and the analysis in SPSS statistical software is given, including intercept only Poisson regressions and loglinear Poisson models for incident rates with varying incident risks.

Keywords Poisson distribution · Poisson regressions · Event outcomes per person per period of time · Event outcomes per population and per period of time · Intercept only Poisson regressions · Loglinear models for incident rates with varying incident risks

---

Electronic Supplementary Material The online version of this chapter (https://doi.org/10.1007/978-3-030-61394-5_14) contains supplementary material, which is available to authorized users. The videos can be accessed by scanning the related images with the SN More Media App.

1   Introduction, History, and Background

The French mathematician, Simeon Denis Poisson (1781–1840) introduced the frequency distribution, bearing his name, in 1837, and published it together with his probability theory in his work "Recherches sur la probabilité des jugements en matière criminelle et en matière civile". The work theorized about the number of wrongful convictions in a given country by focusing on certain random variables that count, among other things, the number of discrete occurrences (sometimes called "events" or "arrivals") that take place during a time - interval of given length. The result had already been given in 1711 by De Moivre (Saumur France) in the latin paper "De Mensura Sortis; de Probabilitate Eventuum in Ludis a Casu Fortuito Pendentibus". This has prompted some authors to argue, that the Poisson distribution should bear the name of De Moivre. In 1860, Newcomb (Washington DC) fitted the Poisson distribution to the number of stars found in a unit of space. A further application of this distribution was made by Bortkiewicz (Petersburg Russia) in 1898, when he was given the task of investigating the number of soldiers in the Prussian army killed accidentally by horse kicks; this experiment introduced the Poisson distribution to the field of "reliability engineering".

In this edition the Poisson distribution has already been addressed several times. In the Chap. 4, binary Poisson regressions were compared to negative binomial regressions as two methods for assessing the effect of two parallel group treatment modalities on a binary outcome. In the Chap. 5, Poisson and negative binomial regressions were compared for the assessment the effect of various predictors on polytomous outcomes using multivariate regressions with dummy outcome variables.

This chapter will address the Poisson regression (after Poisson 1781–1840 Paris) for the analysis of counted event rates. This is different from the *linear regression*, because it uses a log transformed dependent variable. For rates, defined as numbers of events per person with amounts of observation time as a weighting factor, Poisson regression is very sensitive, and, generally, better so than standard linear regression. Linear regression measures events per population (or person), but does not explicitly include time as a covariate, although, implicitly, it is often assumed, albeit not plainly expressed. In contrast, Poisson regression should explicitly include time.

Poisson regression can be used for counted events per person per period of time, and, in addition, for numbers of yes/no events per population per period of time. It is, then, similar to *logistic regression*, but different from it, in that it uses a log instead of logit (log odds) transformed dependent variable. It is more adequate and often provides better statistics than logistic regression does, because, again, time is explicitly included.

Poisson regressions can be used for assessing the effect of various predictors on frequency counts of morbidities classified into multiple cells. In addition, Poisson regression can be used for assessing incident rates with varying incident risks.

In summary, Poisson distributions serve many statistical models.

1. Assessing the effect of two parallel group treatment modalities on a binary outcome.
2. Assessing the effect of various predictors on polytomous outcomes using multivariate regressions with dummy outcome variables.
3. Analysis of counted event rates defined as numbers of events per person or per population per period of time.
4. Assessment of the numbers of yes/no events per population per period of time.
5. Assessing the effect of various predictors on frequency counts of morbidities classified into multiple cells.
6. Assessing incidence rates with varying incident risks.

1 and 2 have been reviewed in the Chaps. 4 and 5, the remainder will be reviewed in the current chapter.

2 Poisson Regression with Event Outcomes per Person per Period of Time

As an example 50 patients were followed for numbers of paroxysmal atrial fibrillations (PAFs) while on a parallel-group treatment with two different treatments.

The scientific questions were: do psychological and social factor scores and different treatments affect the numbers of PAFs per person per period of time. The data file is in extras.springer.com and is entitled "chap14poissonoutcomerates". It is previously used by the authors in SPSS for starters and 2nd levelers, Chap. 21, Springer Heidelberg Germany, 2015. The first 10 patients are in the table underneath.

PAFs	Treat	Psych	Soc	Days
4	1	56,99	42,45	73
4	1	37,09	46,82	73
2	0	32,28	43,57	76
3	0	29,06	43,57	74
3	0	6,75	27,25	73
13	0	61,65	48,41	62
11	0	56,99	40,74	66
7	1	10,39	15,36	72
10	1	50,53	52,12	63

Treat = treatment modality, Psych = psychological score, Soc = Social score, PAFs = counted numbers of paroxysmal atrial fibrillations per person.

SPSS statistical software will be used for analysis. Start with opening the data file in your computer that has SPSS installed. Then command.

Command:

Analyze....Regression....Linear....Dependent Variable: episodes of paroxysmal atrial fibrillation....Independent: treatment modality, psychological score, social score, days of observation....click OK.

Coefficients[a]

Model		Unstandardized Coefficients		Standardized Coefficients	t	Sig.
		B	Std. Error	Beta		
1	(Constant)	49,059	5,447		9,006	,000
	treat	-2,914	1,385	-,204	-2,105	,041
	psych	,014	,052	,036	,273	,786
	soc	-,073	,058	-,169	-1,266	,212
	days	-,557	,074	-,715	-7,535	,000

a. Dependent Variable: paf

The above table is in the output sheets and shows, that treatment modality is weakly significant, and psychological and social score are not. Furthermore, days of observation is very significant. However, it is not appropriate to include this variable if your outcome is the numbers of events per person per time unit. Therefore, we will perform a linear regression, and adjust the outcome variable for the differences in days of observation using weighted least square (WLS) regression.

Command: Analyze....Regression....Linear....Dependent: episodes of paroxysmal atrial fibrillation....Independent: treatment modality, psychological score, social score ....WLS Weight: days of observation....click OK.

Coefficients[a,b]

Model		Unstandardized Coefficients		Standardized Coefficients	t	Sig.
		B	Std. Error	Beta		
1	(Constant)	10,033	2,862		3,506	,001
	treat	-3,502	1,867	-,269	-1,876	,067
	psych	,033	,069	,093	,472	,639
	soc	-,093	,078	-,237	-1,194	,238

a. Dependent Variable: paf

b. Weighted Least Squares Regression - Weighted by days

The above table is in the output sheets of your computer. It shows the results. A largely similar pattern is observed, but treatment modality is no more statistically significant. We will now perform a Poisson regression which is more appropriate for events per person per period of time.

For analysis the module Generalized Linear Models is required. It consists of two submodules: Generalized Linear Models and Generalized Estimation Models. The first submodule covers many statistical models like gamma regression (Chap. 19), Tweedie regression (Chap. 19), Poisson regression (the current chapter and the Chaps. 4 and 5), and the analysis of paired outcomes with predictors. The second submodule is for analyzing binary outcomes. For the current analysis the statistical model Poisson regression in the module Generalized Linear Models is required.

Command:
Analyze....Generalized Linear Models....mark: Custom....Distribution: Poisson .....Link function: Log....Response: Dependent variable: numbers of episodes of PAF....Scale Weight Variable: days of observation....Predictors: Main Effect: treatment modality....Covariates: psychological score, social score....Model: main effects: treatment modality, psychological score, social score....Estimation: mark Model-based Estimation....click OK.

Parameter Estimates

Parameter	B	Std. Error	95% Wald Confidence Interval		Hypothesis Test		
			Lower	Upper	Wald Chi-Square	df	Sig.
(Intercept)	1,868	,0206	1,828	1,909	8256,274	1	,000
[treat=0]	,667	,0153	,637	,697	1897,429	1	,000
[treat=1]	0[a]	.	.	.	.	.	.
psych	,006	,0006	,005	,008	120,966	1	,000
soc	-,019	,0006	-,020	-,017	830,264	1	,000
(Scale)	1[b]						

Dependent Variable: paf
Model: (Intercept), treat, psych, soc

a. Set to zero because this parameter is redundant.

b. Fixed at the displayed value.

The above table is given in the output. All of a sudden, all of the predictors including treatment modality, psychological and social score are very significant predictors of the numbers of PAF events per person per period of time. Poisson regression is, obviously, much more precise to test the effect of the various predictors on this output.

3   Poisson Regression with Yes / No Event Outcomes per Population per Period of Time

Poisson regression cannot only be used for counted events per person per period of time, but also for numbers of yes/no events per population per period of time. It is, then, similar to logistic regression, but also different from it, in that it uses a log instead of logit (log odds) transformed dependent variable. It is more adequate and often provides better statistics than logistic regression does, because, again, unlike with logistic regression, time is explicitly included. An example of the effect of two parallel-group treatment modalities on the presence or not of a severe cardiac arrhythmia (torsade de pointes) is used. The data file is in extras.springer.com and is entitled "chap14poissonbinary.sav". It is previously used by the authors in SPSS for starters and 2nd levelers, Springer Heidelberg Germany, Chap. 47. The first 10 patients are in the underneath table.

treat	presence of torsade de pointes.
,00	1,00
,00	1,00
,00	1,00
,00	1,00
,00	1,00
,00	1,00
,00	1,00
,00	1,00
,00	1,00
,00	1,00

SPSS statistical software will be used for analysis. Start with opening the data file in your computer that has SPSS installed. We will start with a traditional binary logistic regression analysis. Then command.

Command:
Analyze....Regression....Binary Logistic....Dependent: torsade....Covariates: treatment....click OK.

Variables in the Equation

	B	S.E.	Wald	df	Sig.	Exp(B)
Step 1[a] VAR00001	1,224	,626	3,819	1	,051	3,400
Constant	-,125	,354	,125	1	,724	,882

a. Variable(s) entered on step 1: VAR00001.

The above table shows that the treatment is not statistically significant. A Poisson regression will be performed subsequently. For this analysis the module Generalized Linear Models is required. It consists of two submodules: Generalized Linear Models and Generalized Estimation Models.

Command:
Analyze....Generalized Linear Models....Generalized Linear Models ....mark Custom....Distribution: Poisson ....Link Function: Log....Response: Dependent Variable: torsade.... Predictors: Factors: treat....click Model....click Main Effect: enter "treat.....click Estimation: mark Robust Tests....click OK.

Parameter Estimates

Parameter	B	Std. Error	95% Wald Confidence Interval		Hypothesis Test		
			Lower	Upper	Wald Chi-Square	df	Sig.
(Intercept)	-,288	,1291	-,541	-,035	4,966	1	,026
[VAR00001=,00]	-,470	,2282	-,917	-,023	4,241	1	,039
[VAR00001=1,00]	0ª	.	.	.	.	.	.
(Scale)	1ᵇ						

Dependent Variable: torsade
Model: (Intercept), VAR00001

a. Set to zero because this parameter is redundant.

b. Fixed at the displayed value.

The above table is in the output. It shows the results of the Poisson regression for binary outcomes. The predictor treatment modality is statistically significant at p = 0.039. According to the Poisson model the treatment modality is a significant predictor of torsades de pointes.

4 Poisson Regressions Routinely Adjusting Age and Sex Dependence, Intercept-Only Models

In observational research event rates are often very much age and sex dependent and a model routinely adjusting these confounders is welcome. An example will be given from Kirkwood and Sterne (Standardization, in: Medical Statistics, Chap. 25, Blackwell Science, Oxford UK 2003) studied the age and sex adjusted mortality rate of onchocerciasis patients already blind and tested it versus non blind patients using Poisson regression.

Age	non-blind			blind			expected	
Males	pyars	deaths	deaths/ pyars x1000	pyars	deaths	deaths/ /pyars x1000	ratio blind/non-blind	
30-39	2400	19	7.9	120	3	25.0	0.95	
40-49	1590	21	13.2	171	7	40.9	2.26	
50-59	1120	20	17.9	244	13	53.3	4.36	
60-	610	20	32.8	237	24	101.3	7.77	
Females								
30-39	3100	23	7.4	84	2	23.8	0.62	
40-49	1610	22	13.7	69	3	43.5	0.94	
50-59	930	16	17.2	168	8	47.6	2.89	
60.	270	8	29.6	93	9	96.8	2.76	

pyars = numbers of person-years
deaths = number of deaths
expected ratio blind/non-blind = (blind pyars / non-blind pyars) x (non-blind deaths)

A regression with (blind deaths / pyars) x 1000 as outcome and expected ratio blind / non-blind as scale weighted variable should provide an age and sex adjusted mortality rate in the subgroup of blind persons. The data file is entitled "chap14poissonstandardized" and is in www.springer.com. For analysis SPSS Generalized Linear Models is used again. Start by opening the data file in your computer with SPSS mounted.

Command:
Analyze....Generalized Linear Models....Generalized Linear Models ....click Type of Model....mark Poisson loglinear....click Response....Dependent Variable: enter blinddeaths....Scale Weight Variable: enter expratioblinddeaths....A message comes up: No effects have been specified, an intercept-only model will be fit. Do you want to fit an intercept-only model?....click Yes.

The regression coefficient is given in the output sheets: B = 3.219. This means that B = ln outcome (ln = natural logarithm). Outcome = antiln 3.219 = 25.003. The age and sex adjusted mortality rate of the blind persons equals 25.003.

Parameter Estimates

Parameter	B	Std. Error	95% Wald Confidence Interval		Hypothesis Test		
			Lower	Upper	Wald Chi-Square	df	Sig.
(Intercept)	2,633	,0564	2,523	2,744	2176,122	1	,000
(Scale)	1[a]						

Dependent Variable: blinddeath
Model: (Intercept)

a. Fixed at the displayed value.

The 95 % confidence interval of this mortality rate can be computed:

The regression coefficient B is log transformed, and anti logging is required for making predictions about the age and sex adjusted mortality rates.

B = 2.633 with 95% confidence 2.523 to 2,744.

Antiln B = 13.92 with 95% confidence 12.466 to 15.549.

In this data file Poisson regression with an intercept only model is helpful to predict age and sex matched blind deaths with their 95% confidence intervals from case control data.

5 Loglinear Models for Assessing Incident Rates with Varying Incident Risks

Data files that assess the effect of various predictors on frequency counts of morbidities / mortalities can be classified into multiple cells with varying incident risks (like, e.g., the incident risk of infarction). The underneath table gives an example.

In patients at risk of infarction with little soft drink consumption, and consumption of wine and other alcoholic beverages, the incident risk of infarction equals 240/930 = 24.2 %, in those with lots of soft drinks, no wine, and no alcohol otherwise it is 285/1043 = 27.3 %.

soft drink (1 = little)	wine (0 = no)	alc beverages (0 = no)	infarcts number	population number
1,00	1,00	1,00	240	993
1,00	1,00	,00	237	998
2,00	1,00	1,00	236	1016
2,00	1,00	,00	236	1011
3,00	1,00	1,00	221	1004
3,00	1,00	,00	221	1003
1,00	,00	1,00	270	939
1,00	,00	,00	269	940
2,00	,00	1,00	274	979
2,00	,00	,00	273	966
3,00	,00	1,00	284	1041
3,00	,00	,00	285	1043

The general loglinear model using Poisson distributions (see also Statistics applied to clinical studies 5th edition, Chap.23, Poisson regression, pp 267–275, Springer Heidelberg Germany, 2012, from the same authors) is an appropriate method for statistical testing. This section of the current chapter is to assess this methodology, frequently used by banks and insurance companies but little by clinicians so far.

The example in the above table will be applied. We wish to investigate the effect of soft drink, wine, and other alcoholic beverages on the risk of infarction. The SPSS data file is in extras.springer.com, and is entitled "chap14loglinear". It is previously used by the authors in Machine learning in medicine a complete overview, 2015. Start by opening the file in your computer.

Command:
Analyze....Loglinear ....General Loglinear Analysis....Factor(s): enter softdrink, wine, other alc beverages....click "Data" in the upper text row of your screen.... click Weigh Cases....mark Weight cases by....Frequency Variable: enter "infarcts".... click OK....return to General Loglinear Analysis....Cell structure: enter "population".... Options ....mark Estimates....click Continue....Distribution of Cell Counts: mark Poisson....click OK.

The underneath pretty dull table gives some wonderful information. The risk estimates of the soft drink classes 1 and 2 are not significantly different from zero. These classes have, thus, no greater risk of infarction than class 3. However, the regression coefficient of no wine is greater than zero at p = 0.016. No wine drinkers have a significantly greater risk of infarction than the wine drinkers have. No "other alcoholic beverages" did not protect from infarction better than the consumption of it. The three predictors did not display any interaction effects. This result would be in agreement with the famous French paradox. Data files that assess the effect of various predictors on frequency counts of morbidities / mortalities can be classified into multiple cells with varying incident risks (like ,e.g., the incident risk of infarction). The general loglinear model using Poisson distributions is an appropriate method for statistical testing. It can identify subgroups with significantly larger incident risks than other subgroups.

Parameter Estimates[b,c]

Parameter	Estimate	Std. Error	Z	Sig.	95% Confidence Interval	
					Lower Bound	Upper Bound
Constant	-1,513	,067	-22,496	,000	-1,645	-1,381
[softdrink = 1,00]	,095	,093	1,021	,307	-,088	,278
[softdrink = 2,00]	,053	,094	,569	,569	-,130	,237
[softdrink = 3,00]	0[a]	.	.	.	.	.
[wine = ,00]	,215	,090	2,403	,016	,040	,391
[wine = 1,00]	0[a]	.	.	.	.	.
[alcbeverages = ,00]	,003	,095	,029	,977	-,184	,189
[alcbeverages = 1,00]	0[a]	.	.	.	.	.
[softdrink = 1,00] * [wine = ,00]	-,043	,126	-,345	,730	-,291	,204
[softdrink = 1,00] * [wine = 1,00]	0[a]	.	.	.	.	.
[softdrink = 2,00] * [wine = ,00]	-,026	,126	-,209	,834	-,274	,221
[softdrink = 2,00] * [wine = 1,00]	0[a]	.	.	.	.	.
[softdrink = 3,00] * [wine = ,00]	0[a]	.	.	.	.	.
[softdrink = 3,00] * [wine = 1,00]	0[a]	.	.	.	.	.
[softdrink = 1,00] * [alcbeverages = ,00]	-,021	,132	-,161	,872	-,280	,237
[softdrink = 1,00] * [alcbeverages = 1,00]	0[a]	.	.	.	.	.
[softdrink = 2,00] * [alcbeverages = ,00]	,003	,132	,024	,981	-,256	,262
[softdrink = 2,00] * [alcbeverages = 1,00]	0[a]	.	.	.	.	.
[softdrink = 3,00] * [alcbeverages = ,00]	0[a]	.	.	.	.	.
[softdrink = 3,00] * [alcbeverages = 1,00]	0[a]	.	.	.	.	.
[wine = ,00] * [alcbeverages = ,00]	-,002	,127	-,018	,986	-,251	,246
[wine = ,00] * [alcbeverages = 1,00]	0[a]	.	.	.	.	.
[wine = 1,00] * [alcbeverages = ,00]	0[a]	.	.	.	.	.
[wine = 1,00] * [alcbeverages = 1,00]	0[a]	.	.	.	.	.
[softdrink = 1,00] * [wine = ,00] * [alcbeverages = ,00]	,016	,178	,089	,929	-,334	,366
[softdrink = 1,00] * [wine = ,00] * [alcbeverages = 1,00]	0[a]	.	.	.	.	.
[softdrink = 1,00] * [wine = 1,00] * [alcbeverages = ,00]	0[a]	.	.	.	.	.
[softdrink = 1,00] * [wine = 1,00] * [alcbeverages = 1,00]	0[a]	.	.	.	.	.
[softdrink = 2,00] * [wine = ,00] * [alcbeverages = ,00]	,006	,178	,036	,971	-,343	,356
[softdrink = 2,00] * [wine = ,00] * [alcbeverages = 1,00]	0[a]	.	.	.	.	.
[softdrink = 2,00] * [wine = 1,00] * [alcbeverages = ,00]	0[a]	.	.	.	.	.
[softdrink = 2,00] * [wine = 1,00] * [alcbeverages = 1,00]	0[a]	.	.	.	.	.
[softdrink = 3,00] * [wine = ,00] * [alcbeverages = ,00]	0[a]	.	.	.	.	.
[softdrink = 3,00] * [wine = ,00] * [alcbeverages = 1,00]	0[a]	.	.	.	.	.
[softdrink = 3,00] * [wine = 1,00] * [alcbeverages = ,00]	0[a]	.	.	.	.	.
[softdrink = 3,00] * [wine = 1,00] * [alcbeverages = 1,00]	0[a]	.	.	.	.	.

a. This parameter is set to zero because it is redundant.
b. Model: Poisson
c. Design: Constant + softdrink + wine + alcbeverages + softdrink * wine + softdrink * alcbeverages + wine * alcbeverages + softdrink * wine * alcbeverages

6   Conclusion

For rates, defined as numbers of events per person per period of time, Poisson regression is very sensitive, and, generally, better so than standard linear regression. Linear regression measures events per population (or person), but does not explicitly include time as a covariate, although, implicitly, it is often assumed, albeit not plainly expressed. Poisson regression cannot only be used for counted events per person per period of time, but also for numbers of yes/no events per population per period of time. It is, then, similar to logistic regression, but different from it, in that it uses a log instead of logit (log odds) transformed dependent variable. It is more adequate and often provides better statistics than logistic regression does, because, again, time is explicitly included. In observational research event rates are often very much age and sex dependent and a model routinely adjusting these confounders are welcome. Examples and the analysis in SPSS statistical software is given, including intercept only Poisson regressions and loglinear models for incident rates with varying incident risks.

Reference

To readers requesting more background, theoretical and mathematical information of computations given, several textbooks complementary to the current production and written by the same authors are available: Statistics applied to clinical studies 5th edition, 2012, Machine learning in medicine a complete overview, 2015, SPSS for starters and 2nd levelers 2nd edition, 2015, Clinical data analysis on a pocket calculator 2nd edition, 2016, Understanding clinical data analysis, 2017, all of them edited by Springer Heidelberg Germany.

Chapter 15
Regression Trend Testing

Providing Better Statistics with Fewer Degrees of Freedom

Abstract Current clinical trials often involve more than two treatments or treatment modalities. In such situations small differences in efficacies are to be expected, and we need, particularly, sensitive tests. A standard approach to the analysis of such data is multiple groups analysis of variance (Anova) and multiple groups chi-square tests, but a more sensitive approach may be a trend-analysis. A trend means an association between the order of treatment and the magnitude of response. Examples are given in this chapter. Trend tests provide markedly better sensitivity for demonstrating incremental effects from incremental treatment dosages, than traditional statistical tests do. We should add that, within the context of a clinical trial, demonstrating trends, may provide better evidence of causal treatment effects than simple comparisons of treatment modalities may do.

Keywords Regression trends · Linear trend testing of continuous data · Linear trend testing of discrete data

1 Introduction, History, and Background

Current clinical trials often involve more than two treatments or treatment modalities, e.g., dose-response and dose-finding trials, studies comparing multiple drugs from one class with different potencies, or different formulas from one drug with various bio-availabilities and other pharmacokinetic properties. In such situations small differences in efficacies are to be expected, and we need, particularly, sensitive tests. A standard approach to the analysis of such data is multiple groups analysis of variance (Anova) and multiple groups chi-square tests, but a more sensitive, although, so far, little used, approach may be a trend-analysis. A trend means an association between the order of treatment and the magnitude of response.

Electronic Supplementary Material The online version of this chapter (https://doi.org/10.1007/978-3-030-61394-5_15) contains supplementary material, which is available to authorized users. The videos can be accessed by scanning the related images with the SN More Media App.

257

Trend analysis is the widespread practice of collecting information and attempting to spot a pattern. In economy, the term "trend analysis" has more formally obtained the meaning of an upward direction leading to increased profit for the investor. Although trend analysis is often used to predict future events, it could be used to estimate uncertain events in the past, such as how many ancient kings probably ruled between two dates, based on data such as the average years which other known kings reigned. In project management, trend analysis is a mathematical technique that uses historical results to predict future outcome. This is achieved by tracking variances in cost and schedule performance. In this context, it is a project management quality control tool. In statistics, trend analysis often refers to techniques for extracting an underlying pattern of behavior in a time series which would otherwise be partly or nearly completely hidden by noise. If the trend could be assumed to be linear, then trend analysis can be undertaken within a formal regression analysis, described as trend estimation. If the trends had shapes other than linear, trend testing can be performed by non – parametric methods, using, e.g., the Kendall rank correlation coefficient. Particularly valuable is the Mann-Kendall trend test, because it covers monotonous up – and downwards trends in a nonparametric way. Henry Mann, a mathematician from Vienna published in 1945 a paper entitled "Nonparametric against trend testing". Maurice Kendall (1907–1983 from Redhill UK) invented the Kendall rank correlation coefficient (which he named Tau). Maurice Kendall must not be mixed up with David Kendall (1918–2007, a statistician involved in the queuing theory, from Cambridge UK).

This edition will only address linear trend tests of continuous as well as binary outcome data. Non-parametric trend testing is, however, quite straightforward, and particularly adequate for testing ordinal outcome data (see, e.g., the Chap. 9 of Clinical data analysis on a pocket calculator, 2nd edition, ed Springer Heidelberg Germany, 2016, from the same authors).

In the current edition, however, non – parametric trend testing was not applied for analyzing ordinal data. A complementary log – log link function was applied instead. The authors, thereby, followed SPSS' statistical software's choice (see the Chap. 5).

We should add, that, for testing and visualization of nonlinear trends, "smoothing methodologies" is a more modern alternative possibility. Particularly, Ramsay's "functional data analysis" provides a better sensitivity to make predictions from complex data, while integrating various methodologies (see the Chaps 25 and 26).

2   Linear Trend Testing of Continuous Data

With linear trend testing of continuous data the outcome variable is continuous, the predictor variable is categorical, and can be measured either as nominal (just like names) or as ordinal variable (a stepping pattern not necessarily with equal intervals). In the Variable View of SPSS the command "Measure" may, therefore, be changed into nominal or ordinal, but, since we assume an incremental function, the default measure scale is OK as well.

As an example in a parallel-groups study the effects of three incremental dosages of antihypertensive treatments will be assessed. The mean reduction of mean blood pressure per group is tested.

outcome (mean blood pressure, mm Hg)	treatment group
113,00	1,00
131,00	1,00
112,00	1,00
132,00	1,00
114,00	1,00
130,00	1,00
115,00	1,00
129,00	1,00
122,00	1,00
118,00	2,00

The entire data file is in extras.springer.com, and is entitled "chap15trendcontinuous". It is previously used by the authors in SPSS for starters and 2nd levelers, Chap. 15, Springer Heidelberg Germany, 2016. We will, first, perform a one way analysis of variance (Anova) to see, if there are any significant differences in the data. If not, we will perform a trend test using simple linear regression. For analysis the statistical model One Way Anova in the module Compare Means is used. Open the file in your computer mounted with SPSS.

Command:
Analyze....Compare Means....One-Way ANOVA....Dependent List: blood pressure Factor: treatment...click OK.

ANOVA

VAR00002

	Sum of Squares	df	Mean Square	F	Sig.
Between Groups	246,667	2	123,333	2,035	,150
Within Groups	1636,000	27	60,593		
Total	1882,667	29			

The above table shows that there is no significant difference in efficacy between the treatment dosages, and so, sadly, this is a negative study. However, a trend test having just 1 degree of freedom has more sensitivity than a usual one way Anova, and it could, therefore, be statistically significant even so. For analysis the model Linear in the module Regression is required.

Command:

Analyze....Regression....Linear....Dependent: blood   pressure....Independent(s): treatment....click OK.

ANOVA[b]

Model		Sum of Squares	df	Mean Square	F	Sig.
1	Regression	245,000	1	245,000	4,189	,050[a]
	Residual	1637,667	28	58,488		
	Total	1882,667	29			

a. Predictors: (Constant), VAR00001

b. Dependent Variable: VAR00002

Coefficients[a]

Model		Unstandardized Coefficients		Standardized Coefficients	t	Sig.
		B	Std. Error	Beta		
1	(Constant)	125,333	3,694		33,927	,000
	treatment	-3,500	1,710	-,361	-2,047	,050

a. Dependent Variable: blood pressure

The above tables in the output sheets show, that treatment dosage is a significant predictor of treatment response at a p-value of 0,05. There is, thus, a significantly incremental response with incremental dosages. Trend tests provide, obviously, markedly better sensitivity for demonstrating incremental effects from incremental treatment dosages, than traditional statistical tests do. One way ANOVA using 2 degrees of freedom was not statistically significant (p = 0,150) in the example given, while a linear regression with 1 degree of freedom was significant at p = 0,05.

3   Linear Trend Testing of Discrete Data

With discrete data linear regression is impossible, but a chi-square test for trends can be performed. As an example, in a 106 parallel-groups study the effects of three incremental dosages of an antihypertensive drug were assessed. The proportion of responders in each of the three groups was used as outcome measure. The first 10 patients are in the above table. The entire data file is in extras.springer.com, and is entitled "chap15trendbinary". It has been previously used by the authors in SPSS for starters and 2nd levelers, Chap. 40, Springer Heidelberg Germany, 2016. Open the data file in your computer installed with SPSS.

responder	treatment
1,00	1,00
1,00	1,00
1,00	1,00
1,00	1,00
1,00	1,00
1,00	1,00
1,00	1,00
1,00	1,00
1,00	1,00
1,00	2,00

responder: normotension 1, hypertension 0
treatment: incremental treatment dosages 1-3

A multiple groups chi-square test will be performed. For analysis the statistical model Crosstabs in the module Descriptive Statistics is used.

Command:
Analyze....Descriptive   Statistics....Crosstabs....Row(s):   responder....Column(s): treatment....Statistics....Chi-Square Test....click OK.

Chi-Square Tests

	Value	df	Asymp. Sig. (2-sided)
Pearson Chi-Square	3,872[a]	2	,144

The above table shows, that, indeed, the Pearson chi-square value for multiple groups testing is not significant with a chi-square value of 3,872 and a p-value of

0,144, and we have to conclude that there is, thus, no significant difference between the odds of responding to the three dosages.

Subsequently, a chi-square test for trends can be executed, a test, that, essentially, assesses, whether the above odds of responding (number of responder / numbers of non-responders per treatment group) increase significantly. The "linear-by-linear association" from the same table is appropriate for the purpose. It has approximately the same chi-square value, but it has only 1 degree of freedom, and, therefore, it reaches statistical significance with a p-value of 0,050. There is, thus, a significant incremental trend of responding with incremental dosages.

Linear-by-Linear Association	3,829	1	,050
N of Valid Cases	106		

a. 0 cells (,0%) have expected count less than 5.
    The minimum expected count is 11,56.

The trend in this example can also be tested using logistic regression with responding as outcome variable and treatment as independent variable (enter the latter as covariate, not as categorical variable).

4   Conclusion

Trend tests provide markedly better sensitivity for demonstrating incremental effects from incremental treatment dosages, than traditional statistical tests do. Current clinical trials often involve more than two treatments or treatment modalities, e.g., dose-response and dose-finding trials, studies comparing multiple drugs from one class with different potencies, or different formulas from one drug with various bio-availabilities and other pharmacokinetic properties. In such situations small differences in efficacies are to be expected and we need, particularly, sensitive tests. A standard approach to the analysis of such data is multiple groups analysis of variance (Anova) and multiple groups chi-square tests, but a more sensitive, although so far little used, approach may be a trend-analysis. A trend means an association between the order of treatment and the magnitude of response. We should add that, within the context of a clinical trial, demonstrating trends, may provide better evidence of causal treatment effects than simple comparisons of treatment modalities may do.

Reference

To readers requesting more background, theoretical and mathematical information of computations given, several textbooks complementary to the current production and written by the same authors are available: Statistics applied to clinical studies 5th edition, 2012, Machine learning in medicine a complete overview, 2015, SPSS for starters and 2nd levelers 2nd edition, 2015, Clinical data analysis on a pocket calculator 2nd edition, 2016, Understanding clinical data analysis, 2017, all of them edited by Springer Heidelberg Germany.

Chapter 16
Optimal Scaling and Automatic Linear Regression

Better Power of Testing if Continuous Predictor Variables Are Parametrically Turned into Discretized Ones

Abstract Optimal scaling is a method designed to optimize the statistical power of the relationship between the predictor and outcome variables. It makes use of processes like discretization (converting continuous variables into discretized values), and regularization (correcting discretized variables for overfitting, otherwise called overdispersion). The current chapter gives examples and shows, that in order to fully benefit from optimal scaling a regularization procedure is important. This chapter also addresses automatic linear regression in SPSS. It provides much better statistics of regression data than traditional multiple linear regression does. Optimal scaling is a major contributor to the benefits of the automatic linear regression module in SPSS statistical software. We conclude, that optimal scaling using discretization, is a method for an improved analysis of clinical trials, where the consecutive levels of the variables are unequal. In order to fully benefit from optimal scaling, a regularization procedure for the purpose of correcting overdispersion is desirable.

Keywords Optimal scaling · Automatic linear regression · Turning continuous predictors into discretized ones · Regularization

1 Introduction, History, Background

In clinical trials the research question is often measured with multiple variables. For example, the expressions of a number of genes can be used to predict the efficacy of cytostatic treatment, repeated measurements can be used in randomized longitudinal trials, and multi-item personal scores can be used for the evaluation of antidepressants. Many more examples can be given. Multiple linear regression analysis is often used for analyzing the effect of predictors on outcome variables. The underneath

Electronic Supplementary Material The online version of this chapter (https://doi.org/10.1007/978-3-030-61394-5_16) contains supplementary material, which is available to authorized users. The videos can be accessed by scanning the related images with the SN More Media App.

figure gives an example of a continuous predictor variable scored on an outcome scale of 0-10.

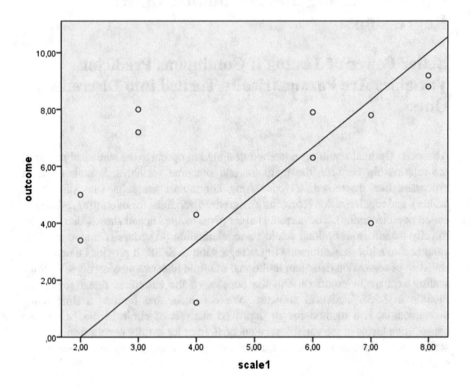

Patients with the predictor values 0, 1, 5, 9 and 10 are missing. Instead of a scale of integers between 0 and 10, other scales are possible, e.g. a scale of two or four scores. Any scale used is, of course, arbitrary and can be replaced with another one.

Scale 1: 0, 1, 2, 3, 4, 5, 6, 7, 8, 9, 10.

Scale 2: 1, 2 (1 = (0 to 5) ; 2 = (5 to 10).

Scale 3: 1, 2, 3, 4 (1 = (0 to 2.5) ; 2 = (2.5 to 5); 3 = (5 to 7.5) ; 4 = (7.5 to 10)

The underneath tables show, that linear regressions of each scale produced different regression coefficients, t-values, and p-values, one result better than the other. With the scales 2 and 3 a gradual improvement of the t-values and p-values is observed. Optimal scaling is a method designed to optimize the statistical power of the relationship between the predictor and outcome variables.

Coefficients[a]

		Unstandardized Coefficients		Standardized Coefficients		
Model		B	Std. Error	Beta	t	Sig.
1	(Constant)	3,351	1,647		2,034	,069
	scale1	,548	,302	,497	1,813	,100

a. Dependent Variable: outcome

Coefficients[a]

		Unstandardized Coefficients		Standardized Coefficients		
Model		B	Std. Error	Beta	t	Sig.
1	(Constant)	2,367	2,032		1,165	,271
	scale2	,497	,257	,521	1,932	,082

a. Dependent Variable: outcome

Coefficients[a]

		Unstandardized Coefficients		Standardized Coefficients		
Model		B	Std. Error	Beta	t	Sig.
1	(Constant)	2,217	1,647		1,346	,208
	scale3	,620	,246	,623	2,520	,030

a. Dependent Variable: outcome

Optimal scaling makes use of processes like discretization (converting continuous variables into discretized values), and regularization (correcting discretized variables for overfitting, otherwise called overdispersion). In order to transform continuous data into a discrete model, the quadratic approximation is convenient: $fx = fa + f^1a$ $(x-a)$ where f^1a is the first derivative of the function fa, and can be followed by the second, the third derivative, etc. The above function is called a Taylor series (Brook Taylor 1685-1731 from Cambridge UK), and the first part of it is called the quadratic approximation, otherwise called the delta method, and is helpful for finding the best fit cut-offs for discretization and regularization. The quadratic approximation is based on the principle, that the simplest model next to the linear is the quadratic model. Obviously, the magnitude of (any) function can be described by the first derivative of the same function (= slope of the function). This approach is helpful for assessing complex functions like those of standard errors, but also to find the best fit distance (discretization) between some x-value and an x-value close by, called an A-value, which is then used as the best fit scale for the data. In order to further improve the best fit scale for a variable, SPSS provides the possibility to cut a linear variable into two pieces (splines), combining two linear functions for modeling not entirely linear patterns (see the Chaps. 17 and 18). In addition to Taylor, Ludwig Hesse (1811-1874, from Koenigsberg Germany) was an important contributor to the above methodology.

To the implementation of optimal scaling into multivariate analysis, also a substantial role was played by Jan de Leeuw (1945-..., Leiden Netherlands). The 2016 book "Multivariate analysis with optimal scaling", edited by Bookdown.org, was first presented by De Leeuw in 1988, and, subsequently, implemented in his research (2010). Jan de Leeuw is now a distinguished professor emeritus of statistics, and founding chair of the Department of Statistics, University of California, Los Angeles.

Another important contributor to the implementation of optimal scaling in multivariate data is Aeilko H Zwinderman (58 years, and chair of statistics University of Amsterdam Netherlands, and co – author of the current work). He was a co – developer of the excellent big – data – methodology of sparse canonical correlation analysis (Machine learning in medicine-a complete overview, 2nd edition, 2020, Springer Heidelberg Germany, Chap. 53).

2   Optimal Scaling with Discretization and Regularization versus Traditional Linear Regression

The underneath table show the first 13 patient data of a 250 patient data file including 16 variables (Vars), G1 to G 27 were the predictors, highly expressed gene scores predicting clinical outcomes, O1 to O4 were outcome variables, efficacy of cytostatic treatments. An entire SPSS data file is in extras.springer.com, and is entitled "chap13manova". It is previously used in the Chap. 13.

G1	G2	G3	G4	G16	G17	G18	G19	G24	G25	G26	G27	O1	O2	O3	O4
8	8	9	5	7	10	5	6	9	9	6	6	6	7	6	7
9	9	10	9	8	8	7	8	8	9	8	8	8	7	8	7
9	8	8	8	8	9	7	8	9	8	9	9	9	8	8	8
8	9	8	9	6	7	6	4	6	6	5	5	7	7	7	6
10	10	8	10	9	10	10	8	8	9	9	9	8	8	8	7
7	8	8	8	8	7	6	5	7	8	8	7	7	6	6	7
5	5	5	5	5	6	4	5	5	6	6	5	6	5	6	4
9	9	9	9	8	8	8	8	9	8	3	8	8	8	8	8
9	8	9	8	9	8	7	7	7	7	5	8	8	7	6	6
10	10	10	10	10	10	10	10	10	8	8	10	10	10	9	10
2	2	8	5	7	8	8	8	9	3	9	8	7	7	7	6
7	8	8	7	8	6	6	7	8	8	8	7	8	7	8	8
8	9	9	8	10	8	8	7	8	8	9	9	7	7	8	8

Traditional linear regression in SPSS statistical software, installed in your computer, with the summary O1 to O4 as outcome and the gene scores as predictors was performed.

Command: Analyze....Regression....Linear....Dependent: enter Summary outcome.... Independent: enter G1 to G 27....click OK.

The output sheets show the results table given below.

Coefficients[a]

Model		Unstandardized Coefficients		Standardized Coefficients	t	Sig.
		B	Std. Error	Beta		
1	(Constant)	3.293	1.475		2.232	.027
	geneone	-.122	.189	-.030	-.646	.519
	genetwo	.287	.225	.078	1.276	.203
	genethree	.370	.228	.097	1.625	.105
	genefour	.063	.196	.014	.321	.748
	genesixteen	.764	.172	.241	4.450	.000
	geneseventeen	.835	.198	.221	4.220	.000
	geneeighteen	.088	.151	.027	.580	.563
	genenineteen	.576	.154	.188	3.751	.000
	genetwentyfour	.403	.146	.154	2.760	.006
	genetwentyfive	.028	.141	.008	.198	.843
	genetwentysix	.320	.142	.125	2.250	.025
	genetwentyseven	-.275	.133	-.092	-2.067	.040

a. Dependent Variable: summaryoutcome

The table shows that 6 of the 12 predictor variables were significant predictors of the summary outcome with p-values 0.04 or less. We will now investigate, whether optimal scaling methodology is capable of improving levels of statistical significance of this data example.

Command: Analyze....Regression....Optimal Scaling....Dependent Variable: Var 28 (Define Scale: mark spline ordinal 2.2)....Independent Variables: Var 1, 2, 3, 4, 16, 17, 18, 19, 24, 25, 26, 27 (all of them Define Scale: mark spline ordinal 2.2)....click OK.

In the output sheets the table below is observed.

Coefficients

	Standardized Coefficients		df	F	Sig.
	Beta	Bootstrap (1000) Estimate of Std. Error			
geneone	-.109	.110	2	.988	.374
genetwo	.193	.107	3	3.250	.023
genethree	-.092	.119	2	.591	.555
genefour	.113	.074	3	2.318	.077
genesixteen	.263	.087	4	9.065	.000
geneseventeen	.301	.114	2	6.935	.001
geneeighteen	.113	.136	1	.687	.408
genenineteen	.145	.067	1	4.727	.031
genetwentyfour	.220	.097	2	5.166	.007
genetwentyfive	-.039	.094	1	.170	.681
genetwentysix	.058	.107	2	.293	.746
genetwentyseven	-.127	.104	2	1.490	.228

Dependent Variable: summaryoutcome

There is no intercept anymore, and the t-tests have been replaced with F-tests. The optimally scaled model without either discretization or regularization shows similarly sized effects, compared to those of the traditional multiple linear regression model, although the number of statistical significant predictor variables fell from 6 to 5. Subsequently, an analysis including discretization will be performed.

Command: Analyze....Regression....Optimal Scaling....Dependent Variable: Var 28 (Define Scale: mark spline ordinal 2.2)....Independent Variables: Var 1, 2, 3, 4, 16, 17, 18, 19, 24, 25, 26, 27 (all of them Define Scale: mark spline ordinal 2.2)....Discretize: Method Grouping, Number categories 7)....click OK.

In the output sheets the underneath table is given.

Coefficients

| | Standardized Coefficients | | | | |
	Beta	Bootstrap (1000) Estimate of Std. Error	df	F	Sig.
geneone	-.109	.105	2	1.076	.343
genetwo	.193	.099	3	3.786	.011
genethree	-.092	.120	2	.587	.557
genefour	.113	.069	3	2.694	.047
genesixteen	.263	.086	4	9.285	.000
geneseventeen	.301	.112	2	7.161	.001
geneeighteen	.113	.138	1	.668	.415
genenineteen	.145	.064	1	5.185	.024
genetwentyfour	.220	.095	2	5.351	.005
genetwentyfive	-.039	.095	1	.168	.682
genetwentysix	.058	.112	2	.267	.766
genetwentyseven	-.127	.100	2	1.586	.207

Dependent Variable: summaryoutcome

Instead of 5 significant predictors, we now have 6 significant predictors. In order to fully benefit from optimal scaling a regularization procedure for the purpose of correcting overdispersion is desirable. We will assess, whether additional regularization can further optimize the statistical sensitivity of the optimal scaling model, we will use ridge regression for the purpose

Command: Analyze....Regression....Optimal Scaling....Dependent Variable: Var 28 (Define Scale: mark spline ordinal 2.2)....Independent Variables: Var 1, 2, 3, 4, 16, 17, 18, 19, 24, 25, 26, 27 (all of them Define Scale: mark spline ordinal 2.2)....Discretize: Method Grouping, Number categories 7)....Regularization .... mark: ridge regression....click OK.

Coefficients

	Standardized Coefficients		df	F	Sig.
	Beta	Bootstrap (1000) Estimate of Std. Error			
geneone	.028	.032	2	.805	.449
genetwo	.068	.020	3	11.089	.000
genethree	.053	.029	1	3.276	.072
genefour	.061	.020	3	9.505	.000
genesixteen	.146	.022	4	42.980	.000
geneseventeen	.150	.025	2	37.111	.000
geneeighteen	.103	.039	2	6.975	.001
genenineteen	.104	.020	2	26.348	.000
genetwentyfour	.120	.021	2	32.843	.000
genetwentyfive	.039	.037	2	1.069	.345
genetwentysix	.088	.024	2	13.212	.000
genetwentyseven	-.020	.046	1	.183	.670

Dependent Variable: summaryoutcome

this model is obviously superior to all of the other models. The numbers of generally very significant p-values rose to no less than 8. Instead of ridge regression, lasso or elastic net regression can be used as an alternative to the regularization. Lasso regression instead of ridge may improve the model by leaving out weak predictors, elastic net may do so with very large numbers of observations, but neither of them were, in the example given, as sensitive as ridge regression was. We conclude, that optimal scaling with discretization and regularization with ridge regression was capable of providing a statistical analysis of the data, that was considerably better sensitive, than traditional linear regression was.

3   Automatic Regression for Maximizing Relationships

Automatic linear regression is in the Statistics Base add-on module SPSS version 19 and up. X-variables are automatically transformed, in order to provide an improved data fit, and SPSS uses optimal scaling of not only continuous but also categorical predictors in addition to other methods for the purpose. A data example will be given. In a clinical crossover trial an old laxative is tested against a new one. Numbers of stools per month is the outcome. The old laxative and the patients' age are the predictor variables. Does automatic linear regression provide better statistics of these data than traditional multiple linear regression does.

Patno	newtreat	oldtreat	age categories
1,00	24,00	8,00	2,00
2,00	30,00	13,00	2,00
3,00	25,00	15,00	2,00
4,00	35,00	10,00	3,00
5,00	39,00	9,00	3,00
6,00	30,00	10,00	3,00
7,00	27,00	8,00	1,00
8,00	14,00	5,00	1,00
9,00	39,00	13,00	1,00
10,00	42,00	15,00	1,00

patno = patient number
newtreat = frequency of stools on a novel laxative
oldtreat = frequency of stools on an old laxative
agecategories = patients' age categories (1 = young, 2 = middle-age, 3 = old)

Only the first 10 patients of the 55 patients are shown above. The entire file is in extras.springer.com and is entitled "chap16automaticlinreg". It is previously used by the authors in Machine learning in medicine a complete overview, Chap.31, Springer Heidelberg Germany, 2015. Open the data file in your computer. We will first perform a traditional multiple linear regression.

Command:
Analyze....Regression....Linear....Dependent: enter newtreat....Independent: enter oldtreat and agecategories ....click OK.

Model Summary

Model	R	R Square	Adjusted R Square	Std. Error of the Estimate
1	,429[a]	,184	,133	9,28255

a. Predictors: (Constant), oldtreat, agecategories

ANOVA[a]

Model		Sum of Squares	df	Mean Square	F	Sig.
1	Regression	622,869	2	311,435	3,614	,038[b]
	Residual	2757,302	32	86,166		
	Total	3380,171	34			

a. Dependent Variable: newtreat

b. Predictors: (Constant), oldtreat, agecategories

Coefficients[a]

Model		Unstandardized Coefficients		Standardized Coefficients	t	Sig.
		B	Std. Error	Beta		
1	(Constant)	20,513	5,137		3,993	,000
	agecategories	3,908	2,329	,268	1,678	,103
	oldtreat	,135	,065	,331	2,070	,047

a. Dependent Variable: newtreat

The above tables are in the output, and we observe, that age is insignificant, while old treatment is borderline significant at $p = 0.047$. Subsequently, an automatic linear regression is performed.

Command:
click Transform....click Random Number Generators....click Set Starting Point ....click Fixed Value (2000000)....click OK....click Analyze....Regression.... Automatic Linear Regression....click Fields....newtreat drag to Target:.... patientno drag to Analysis Weight:....oldtreat and agecategories drag to Fields:....click Build Options....click Objectives....mark Create a standard model....click Basics....mark Automatically prepare data....click Model

Options. . . .mark Save predicted values to the dataset. . . .mark Export model. . . .File name: type "exportautomaticlinreg". . . .click Browse and save the export file in the appropriate folder of your computer. . . .click Run.

The underneath Automatic linear regression results show, that the two predictors agecategories and oldtreat have been transformed, respectively into merged categories and a variable without outliers.

Automatic Data Preparation

Target: newtreat

Field	Role	Actions Taken
(agecategories_transformed)	Predictor	Merge categories to maximize association with target
(oldtreat_transformed)	Predictor	Trim outliers

If the original field name is X, then the transformed field is displayed as (X_transformed). The original field is excluded from the analysis and the transformed field is included instead.

Both of the predictors are now statistically very significant with a correlation coefficient at $p < 0.0001$, and regression coefficients at p-values of respectively 0.001 and 0.007.

Coefficients

Target: newtreat

Model Term	Coefficient ▶	Sig.	Importance
Intercept	35,926	,000	
agecategories_transformed=0	-11,187	,001	0,609
agecategories_transformed=1	0,000[a]		0,609
oldtreat_transformed	0,209	,007	0,391

[a]This coefficient is set to zero because it is redundant.

Effects

Target: newtreat

Source	Sum of Squares	df	Mean Square	F	Sig.
Corrected Model ▶	1.289,960	2	644,980	9,874	,000
Residual	2.090,212	32	65,319		
Corrected Total	3.380,171	34			

Obviously, automatic linear regression provided much better statistics of these data than traditional multiple linear regression did. Optimal scaling is a major contributor to the benefits of the automatic linear regression module in SPSS statistical software.

4   Conclusion

Traditional linear regression has been demonstrated not to perform well in case of multiple independent variables. This is, because consecutive levels of variables are often unequal. Optimal scaling is a method adequate for such variables. It may, however, cause power loss due to overdispersion.

Also, a sharp increase of the t-values of some x-values is often observed, if other x-values are removed. This phenomenon is called instable regression coefficients, sometimes referred to as "bouncing betas" (Van der Kooij, Prediction accuracy and stability of regression, PhD thesis Univ Leiden Neth 2007), and arises, when predictors are correlated, or when a large number of predictors relative to the number of observations is present. Shrinking the regression coefficients has been demonstrated to be beneficial, not only to counterbalance overdispersion, but also to reduce this phenomenon of instability.

Limitations of the method should be mentioned. First, independence of the scale intervals on the magnitude of the outcome variable is an assumption, that may not be appropriate in some cases. Also, the spread in the data is sometimes inappropriately amplified with wide scale intervals. Finally, more than a single scale may be optimal.

The current chapter shows, that, in order to fully benefit from optimal scaling, a regularization procedure is desirable.

Automatic linear regression in SPSS provided much better statistics of these data, than traditional multiple linear regression did. Optimal scaling is a major contributor to the benefits of the automatic linear regression module in SPSS statistical software. We conclude.

1. Optimal scaling using discretization is a method for an improved analysis of clinical trials, where the consecutive levels of the variables are unequal.
2. In order to fully benefit from optimal scaling, a regularization procedure for the purpose of correcting overdispersion is desirable.

Reference

To readers requesting more background, theoretical and mathematical information of computations given, several textbooks complementary to the current production and written by the same authors are available: Statistics applied to clinical studies 5th edition, 2012, Machine learning in medicine a complete overview 2nd edition, 2020, SPSS for starters and 2nd levelers 2nd edition, 2015, Clinical data analysis on a pocket calculator 2nd edition, 2016, Understanding clinical data analysis, 2017, all of them edited by Springer Heidelberg Germany.

Chapter 17
Spline Regression Modeling

The Digital Clay of the Twenty First Century

Abstract Spline modeling is a mathematically refined modeling tool, that adequately fits complex data, even if they do not fit the traditional mathematical models. It is called the digital clay of the twenty-first century, although, so far, little used in clinical research spline modeling, but this is a matter of time. The usual model for the outcome analysis of clinical trials is the linear or log linear model. However, when the linear model is not significant, or when the data plots suggest nonlinearity, the quadratic and cubic models often produce a better fit. The current chapter shows, that spline modeling can produce curves that even better fit the outcome data of such clinical trials, than these traditional models do. Also, patterns in the data can be detected, that may go unobserved with traditional linear or curvilinear modeling. Spline modeling can adequately assess the relationships between an exposure and outcome variable in a clinical trial. It can detect patterns in a trial that are relevant, but go unobserved with simpler regression models. In clinical research spline modeling has great potential, given the presence of many nonlinear effects in this research field, and given its sophisticated mathematical refinement to fit any nonlinear effect in the most accurate way. Spline modeling should enable to improve making predictions from clinical research for the benefit of health decisions and health care. We hope that this brief introduction to spline modeling will stimulate clinical investigators to start using this wonderful method.

Keywords Spline regressions · Polynomial models

1   Introduction, History, and Background

Traditionally, nonlinear relationships like the smooth shapes of airplanes, boats, and motor cars were constructed from scale models using stretched thin wooden strips, otherwise called splines, producing smooth curves, assuming a minimum of strain in the materials used. In the past decades mechanical spline methods have been replaced with their mathematical counterparts. Already in 1964 this was introduced by Boeing (Ferguson, JACM 1964; 11:221–8) and General Motors (Birkhof, Proc General Motors Symposium 1964; 164–90). Mathematical spline modeling is, currently, widely applied even in fields like architecture, consumer product design, and even

T. J. Cleophas, A. H. Zwinderman, *Regression Analysis in Medical Research*,
https://doi.org/10.1007/978-3-030-61394-5_17

jewellery design, and it is sometimes called the "digital clay" of the twenty first century (Fifty-one spline models, www.tsplines.com). Before computers were used, numerical calculations were done by hand. Although piecewise-defined functions like the sign function or step function were used, polynomials were generally preferred, because they were easier to work with. Through the advent of computers, splines have gained importance. They were first used as a replacement for polynomials in interpolation, then as a tool to construct smooth and flexible shapes in computer graphics.

It is commonly accepted, that the first mathematical reference to splines is the 1946 paper by Schoenberg, 1913–1990, mathematician living in Galati Romania, and later on in Princeton NJ, which is probably the first place where the word "spline" was used in connection with smooth, piecewise polynomial approximations. The word "spline" was originally an East Anglian dialect word. In clinical research spline modeling better fitted survival data, than did the traditional Cox model, based on exponential patterns of deaths (Helland et al, Breast Cancer res and Treat 2016; 156: 249–57). However, otherwise, spline modeling is little applied in clinical research (Mulla, West Indian Med 2007; 56: 77–9). In this chapter a clinical trial assessing the effect of quantity of care on quality of care will be used as an example. Spline curves consistent of 4 or 5 cubic functions as explained underneath, will be applied. SPSS statistical software will be used for analysis. In this way this chapter will try and answer, whether spline curves from clinical research data can outperform traditional curves

in establishing relationships between exposure and outcome variables,
in detecting patterns in trial data, that are relevant, but remain unobserved with
     simpler regression models,
in the refinement to fit nonlinear effects, and
in their predictive property for the benefit of health decisions and health care.

2   Linear and the Simplest Nonlinear Models of the Polynomial Type

Clinical research is often involved in predicting an outcome from a predictor variable, and linear modeling is the commonest and simplest method for that purpose. The simplest except one is the quadratic relationship providing a symmetric curve, and the next simplest is the cubic model providing a sinus-like curve.

The equations are

Linear model	$y = a + bx$
Quadratic model	$y = a + bx^2$
Cubic model	$y = a + bx^3$.

The larger the regression coefficient b, the better the model fits the data. Instead of the terms linear, quadratic, and cubic, often, the terms first order, second order, and third order polynomial are applied.

An example is given of a clinical trial assessing the effect of quantity of care on quality of care. The predictor, quantity of care, is estimated as the number of interventions (endoscopies, small operations) per physician per day, the outcome variable, quality of care, is estimated as a quality of care score between 0 and 35 with 35 being the best possible outcome. The data are given underneath with younger (< 50 years) and older physicians (>50 years) as an additional variable.

Variable		
1	2	3
1,00	19,00	2,00
1,00	20,00	3,00
1,00	23,00	4,00
1,00	24,00	5,00
1,00	26,00	6,00
1,00	27,00	7,00
1,00	28,00	8,00
1,00	29,00	9,00
1,00	29,00	10,00
1,00	29,00	11,00
1,00	28,00	12,00
1,00	27,00	13,00
1,00	27,00	14,00
1,00	26,00	15,00
1,00	25,00	16,00
1,00	24,00	17,00
1,00	23,00	18,00
1,00	22,00	19,00
1,00	22,00	20,00
1,00	21,00	21,00
1,00	21,00	22,00
0,00	23,00	1,00
0,00	25,00	2,00
0,00	28,00	3,00
0,00	30,00	3,50
0,00	30,00	5,00
0,00	28,00	6,00
0,00	25,00	7,00
0,00	21,00	8,00
0,00	18,00	9,00
0,00	16,00	10,00
0,00	15,00	11,00
0,00	13,00	12,00
0,00	11,00	13,00
0,00	10,00	14,00
0,00	9,50	14,00
0,00	9,00	15,00
0,00	9,00	16,00
0,00	8,50	17,00
0,00	8,00	18,00
0,00	8,00	19,00
0,00	7,50	20,00

Variable 1 = age class of physicians (1 = younger, 0 = older)
Variable 2 = qol of care score
Variable 3 = quantity of care (daily numbers of medical interventions, like
          endoscopies and operations/ per physician)

Our first hypothesis was: the presence of a negative correlation between the quantity of care (numbers of interventions) and the quality of care score. The second hypothesis was: younger physicians perform better than older do. A linear model is, generally, successful to assess the first hypothesis, and a simple unpaired parametric or nonparametric test to assess the second hypothesis. Also a significant difference in steepness of the regression lines should tell us something about the differences in performance between the younger and older physicians. The above data can be entered in your computer with SPSS installed, simply by copy and paste commands.

If a linear model does not produce a significant result, or, if a non-linear model is anticipated, then curvilinear estimation should follow. SPSS statistical software is, thus, used to assess, whether curvilinear modeling produces a better fit than does the linear model.

Command:
Analyze. . .Regression. . .Curvilinear Estimation. . .Dependent: variable 2. . . .Variable: variable 3. . .Models: mark Linear, Quadratic, Cubic. . .mark Plot Models.

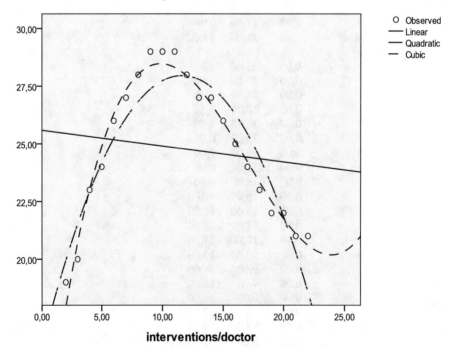

qual care score

interventions/doctor

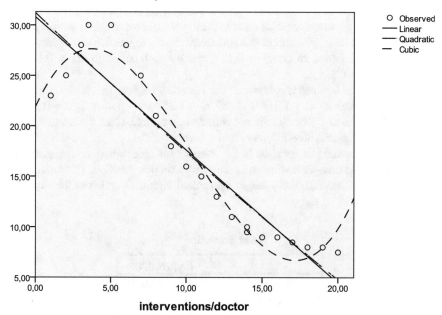

The above graphs are in the output sheets, and give the best fit regression lines/ curves.

Model Summary and Parameter Estimates

Dependent Variable:VAR00002

Equation	Model Summary					Parameter Estimates			
	R Square	F	df1	df2	Sig.	Constant	b1	b2	b3
Linear	,018	,353	1	19	,559	25,588	-,069		
Quadratic	,866	58,321	2	18	,000	16,259	2,017	-,087	
Cubic	,977	236,005	3	17	,000	10,679	4,195	-,301	,006

The independent variable is VAR00003.

Model Summary and Parameter Estimates

Dependent Variable:VAR00002

Equation	Model Summary					Parameter Estimates			
	R Square	F	df1	df2	Sig.	Constant	b1	b2	b3
Linear	,862	118,805	1	19	,000	30,800	-1,317		
Quadratic	,863	56,462	2	18	,000	31,213	-1,430	,005	
Cubic	,961	141,258	3	17	,000	21,895	3,314	-,540	,017

The independent variable is VAR00003.

Also the above tables are in the output sheets. They give the statistics of the best fit linear, quadratic and cubic model for the younger (upper table) and those of the older physician data (lower table). A statistically significant negative correlation in the older but not in the younger physicians (respectively regression coefficient (b) = −1.317 with SE (standard error) = 0.121, and b = −0.069 with SE = 0.116) is observed.

The difference in steepness between the linear regression lines was significant: difference in b-values = −1.317 + 0.069 = −1.248 with pooled SE = 0.168, p <0.001. However, this procedure is not entirely appropriate, since the b-value in the younger is not significantly different from a b-value of 0.

Instead, an unpaired comparison of the younger and older outcomes is performed: Mann-Whitney is non-parametric and a safe choice in this situation. The underneath table shows that, overall, the younger performed significantly better than did the older at p = 0.004.

Test Statistics[a]

	VAR00002
Mann-Whitney U	107,000
Wilcoxon W	338,000
Z	-2,859
Asymp. Sig. (2-tailed)	,004

a. Grouping Variable: VAR00001

According to the F-values and p-values the best fit is given by the cubic models. Also, the figures show, that, particularly, the cubic lines provide a pretty good fit for the data. They also show maximal outcome values, the top qualities of care produced in both groups, for the younger at approximately 10 interventions per day, for the older at approximately 5 interventions per day. However, the cubic lines are not perfect. In both groups the modeled top is considerably lower than the one expected from the data. Sometimes the distance between the best fit cubic line and the real data was pretty large, e.g., for the older physicians at 2,5,17 and 20 interventions. Also the cubic lines produce somewhat sinusoidal patterns, while there is no sinusoidal pattern in the data.

3   Spline Modeling

We will now use spline modeling in order to obtain a smooth spline curve of the data.

Spline modeling cuts the data into 4 or 5 intervals, and uses the best fit third order polynomial functions for each interval. In order to obtain a smooth spline curve, the junctions between two subsequent functions must have

(1) the same y value,
(2) the same slope,
(3) the same curvature.

All of these requirements are met if

(1) the two subsequent functions are equal at the junction,
(2) have the same first derivative at the junction,
(3) have the same second derivative at the junction.

There is a lot of matrix algebra involved, but a computer program can do the calculations for you, and readily provide you with the best fit spline curve. SPSS is used to produce the spline curves.

Command:
click Graphs...Chart Builder...Basic Elements...Choose Axes...y-x...Choose Elements Line...Element Properties...Spline...Close...click OK.

The underneath 1st and 3rd graphs are the polynomial graphs of respectively the data from the younger and older physicians. The 2nd and 4th graphs are the graphs observed in the output sheets of the spline analyses. They show the spline curves of our data

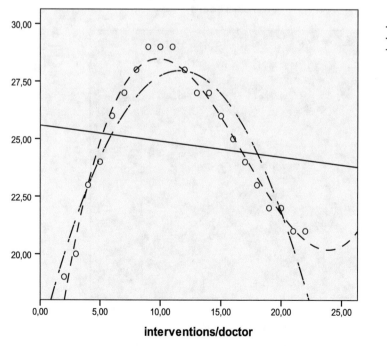

qual care score

interventions/doctor

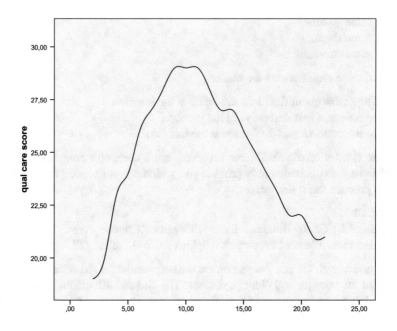

qual care score

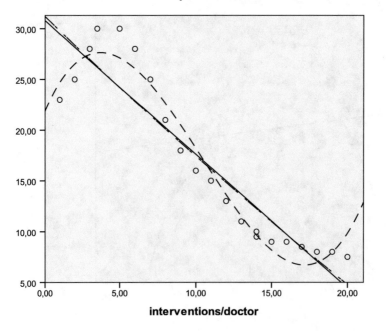

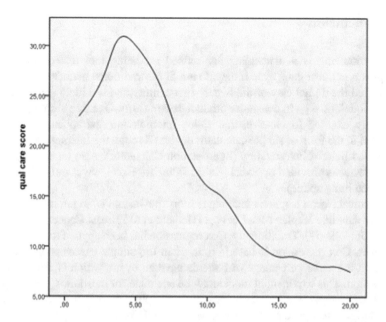

Spline curves, due to their sophisticated mathematical refinement, should provide a much better fit to complex nonlinear patterns, than the usual methods, used for that purpose.

However, with 4 or 5 junctions no simple b-value for assessing the degree of fit with the data can be calculated anymore. Instead, the adequacy of fit can be simply extrapolated from the curve. The underneath table compares the goodness of fit of the spline curve to that of the conventional curves.

Intervent per day	Predicted quality of care from cubic model I	Predicted quality of care from spline model II	Predicted quality of care from real data III	Difference III-I	Difference III-II
2	25	23	23	-2	0
5	27	30	30	+3	0
17	7	9	9	+2	0
21	10	8	8	-2	0

Intervent = interventions

The values extrapolated from the spline curve are exactly the same as those of the real values, unlike those of the conventional curves.

The two spline curves in our example outperformed the next best cubic curves, because

(1) unlike the cubic curves, they did not miss the top quality of care given in either subgroup,
(2) unlike the cubic curves, they, rightly, did not produce sinusoidal patterns,
(3) unlike the cubic curves, they provided a virtually 100% match of the original values.

4 Conclusion

Spline modeling is a mathematically refined modeling tool that adequately fits complex non linear data, even if they do not fit the traditional mathematical models. It is called the digital clay of the twenty-first century and is widely applied for non linear modeling, e.g., in consumer product design (Fifty-One free T-spline models). It is very exciting to observe, that spline interpolation /extrapolation is already provided in the form of simple calculator devices through the Internet for the benefit of a broad field of investigators (Interpolation Calculator). Also an excellent cubic spline device is provided by Excel (Cubic Spline for Excel, www.srs1software.com/download.htm# spline).

In clinical research spline modeling is little known, and is, so far, mainly used for survival data like Kaplan Meier curves (Helland et al , Breast Cancer res and Treat 2016; 156: 249–57). Traditionally, Cox regression has been applied for that purpose. However, Cox regression is based on an often too simple exponential model: per time unit the same percentage of patients have an event, which is a pretty strong assumption. This exponential model may be adequate for mosquitos, but less so for complex creatures like human beings. Indeed, spline modeling produced a much better fit for complex Kaplan-Meier curves than did Cox regression (Helland , Breast Cancer res and Treat 2016; 156: 249–57).

The usual model for the outcome analysis of clinical trials is the linear or log linear model. However, when the linear model is not significant, or when the data plots suggest nonlinearity, the quadratic and cubic models often produce a better fit. The current paper shows, that spline modeling can produce curves, that even better fit the outcome data of such clinical trials, than these traditional models do. Also patterns in the data can be detected, that may go unobserved with traditional linear or curvilinear modeling.

For nonlinear effects in clinical research also traditional models other than quadratic and cubic models are sometimes applied, like, e.g., the exponential models used in pharmacodynamic studies, and the rectangular hyperbolas from the Michaelis-Menten equations used in pharmacokinetic studies. There are some studies to show, that spline modeling is also feasible for pharmacodynamic (Mulla West Indian Med 2007; 56: 77–9) and pharmacokinetic modeling (Asyali, Turk J Eng & Com Sci 2010; 18: 1019–30) with similar, if not, better results, than those of the traditional models. However, data are scanty, and studies are just beginning.

In clinical research spline modeling has great potential given the presence of many nonlinear effects in this field of research, and given the sophisticated mathematical refinement of splines to fit any non linear effect in the most accurate way. In this way spline modeling should enable to improve making predictions from clinical research for the benefit of health decisions and health care.

In summary.

1. Spline modeling can adequately assess the relationships between exposure and an outcome variables in a clinical trial.
2. Spline modeling can detect patterns in a trial, that are relevant, but go unobserved with simpler regression models.

3. In clinical research spline modeling has great potential given the presence of many nonlinear effects in this research field, and given its sophisticated mathematical refinement to fit any non linear effect in the most accurate way.
4. Spline modeling should enable to improve making predictions from clinical research for the benefit of health decisions and health care.

We do hope, that this brief introduction to spline modeling will stimulate clinical investigators to start using this wonderful method.

Reference

To readers requesting more background, theoretical and mathematical information of computations given, several textbooks complementary to the current production and written by the same authors are available: Statistics applied to clinical studies 5th edition, 2012, Machine learning in medicine a complete overview, 2015, SPSS for starters and 2nd levelers 2nd edition, 2015, Clinical data analysis on a pocket calculator 2nd edition, 2016, Understanding clinical data analysis, 2017, all of them edited by Springer Heidelberg Germany.

Chapter 18
More on Nonlinear Regressions

Assessment of Nonlinear Effects in Clinical Research

Abstract Many tools are available for developing nonlinear models for character-izing data sets and making predictions from them. This edition has already reviewed many of them, for example, logistic and Cox regressions (Chap. 2), nonlinear mixed effects models (Chap. 10), Markov (Chap. 11) and Poisson regressions (Chap. 14), Spline regressions (Chap. 17). This chapter will review additional nonlinear models, and will try and summarize the most important ones.

Logit and probit transformation can sometimes be used to mimic a linear model. Logistic regression, Cox regression, Poisson regression, and Markov modeling are examples of log transformations of the y-axes of two dimensional regression models.

Log transformations of the y-axes or both the x- and y-axes are the basis of Box Cox transformation equation, ACE (alternating conditional expectations) and AVAS (additive and variance stabilization for regression) packages. They consist of simple empirical methods often successful for linear remodeling of nonlinear data.

Data that look sinusoidal, can, generally, be successfully modeled using polyno-mial regression or Fourier analysis. For exponential patterns like plasma concentra-tion time relationships, exponential modeling with or without Laplace transformations is a possibility. Spline and Loess modeling are modern methods, particularly, suitable for smoothing data patterns, if the data plot leaves you with no idea of the relationship between the y- and x-values. Loess tends to skip outlier data, while spine modeling rather tends to include them. So, if you are planning to investigate the outliers, the spline is your tool. We have to add, that traditional nonlinear modeling produces p-values, and modern methods do not. However, given the poor fit of many traditional models, these p-values do not mean too much. Also, it is reassuring to observe, that both Loess and spline provide a better fit to nonlinear data than does traditional modeling.

Keywords Nonlinear regressions · Testing for linearity · Logit and probit transformations · Trials and error methods · Box Cox transformation · ACE/AVAS packages · Sinusoidal data with poly nomial regressions · Exponential modeling · Spline modeling · Loess modeling

© The Author(s), under exclusive license to Springer Nature Switzerland AG 2021   291
T. J. Cleophas, A. H. Zwinderman, *Regression Analysis in Medical Research*,
https://doi.org/10.1007/978-3-030-61394-5_18

1   Introduction, History, and Background

Novel models for the assessment of nonlinear data are being developed for the benefit of making better predictions from the data.

This chapter reviews traditional and modern models. We come to conclude.

1. Logit and probit transformations are often successfully used to mimic a linear model. Logistic regression, Cox regression, Poisson regression, and Markov modeling are examples of log transformations.
2. Either the x- or y-axis or both of them can be logarithmically transformed. Also Box Cox transformation equations and ACE (alternating conditional expectations) or AVAS (additive and variance stabilization for regression) packages are simple empirical methods often successful for linearly remodeling of non-linear data.
3. Data that are sinusoidal, can, generally, be successfully modeled using polynomial regression or Fourier analysis.
4. For exponential patterns like plasma concentration time relationships exponential modeling with or without Laplace transformations is a possibility.
5. Spline and Loess are computationally intensive modern methods, suitable for smoothing data patterns, if the data plot leaves you with no idea of the relationship between the y- and x-values. There are no statistical tests to assess the goodness of fit of these methods, but it is always better than that of traditional models.

Nonlinear regression models have already been addressed in previous chapters of this edition. For example, the Chap. 10 addressed multi-exponential models. The Chap. 19 will address kernel regressions appropriate for nonlinear radial basis neural network analyses. Another example is the nonlinear relationship of the smooth shapes of airplanes, boats, and motor cars constructed from scale models using stretched thin wooden strips, otherwise called splines, producing smooth curves, assuming a minimum of strain in the materials used (already addressed in the Chap. 17). With the advent of the computer it became possible to replace it with statistical modeling for the purpose: already in 1964 it was introduced by Boeing and General Motors. A computer program was used to calculate the best fit line/curve, which is the line/curve with the shortest distance to the data. More complex models were required, and they were often laborious so that even modern computers had difficulty to process them. Software packages make use of iterations: 5 or more regression curves are estimated ("guesstimated"), and the one with the best fit is chosen. With large data samples the calculation time can be hours or days, and modern software will automatically proceed to use Monte Carlo calculations (Cleophas and Zwinderman, Statistics applied to clinical studies 5th edition, 2012, 619–26) in order to reduce the calculation times. The basics of the superb diagnostic quality of spline analyses has already been explained in the Chap. 17. Nowadays, many nonlinear data patterns can be developed mathematically, and this paper reviews some more of them.

For example, local regression or local polynomial regression, also known as moving regression, is a generalization of moving average and polynomial regression. Its most common methods, initially developed for scatterplot smoothing, are Loess (locally estimated scatterplot smoothing) and Lowess (locally weighted scatterplot smoothing), both pronounced as written. They are two strongly related non-parametric regression methods, that combine multiple regression models in a k-nearest-neighbor-based meta-model. Outside of econometrics, Loess is known and commonly referred to as the Savitzky–Golay filter. However, the Savitzky–Golay filter was already proposed 15 years before Loess, and A Savitzky–Golay filter, sometimes contracted to Sgolay-filter. It is a kind of digital filter to be applied for data smoothing purposes and increasing precision without distorting signal tendencies.

Loess and Lowess, thus, build on "classical" methods, such as linear and nonlinear least squares regression. They address situations in which the classical procedures do not perform well or cannot be effectively applied without undue labor. Loess combines much of the simplicity of linear least squares regression with the flexibility of nonlinear regression. It does so by fitting simple models to localized subsets of the data to build up a function, that describes the deterministic part of the variation in the data, point by point. In fact, one of the chief attractions of this method is, that the data analyst is not required to specify a global function of any kind to fit a model to the data, but only to fit segments of the data.

The trade-off for these advantages is an increased amount of computations. Because it is so computationally intensive, Loess would have been practically impossible to use in the time, that least squares regression was developed. We should add that other modern methods for process modeling are similar to Loess in this respect. All of the methods have been consciously designed to use our current computational ability to the fullest possible advantage to achieve goals not easily achieved by traditional approaches.

A smooth curve through a set of data points obtained with this statistical technique is called a loess curve, particularly when each smoothed value is given by a weighted quadratic least squares regression over the span of values of the y axis scattergram criterion variable. When each smoothed value is given by a weighted linear least squares regression over the span, this is known as a lowess curve; however, some authorities treat lowess and loess as synonyms.

In this Chap. we will address:
Tests for linearity

Logit and Probit Transformations (see also Chap. 4 (section 4))
Trial and error methods
Box Cox transformations (Box, son in law of Fisher famous of the F-Test, 1919–2013, Madison Wisconsin, Cox 1924- Birmingham UK)
ACE (alternating conditional expectations) package
AVAS (additive and variance stabilization for regression) package
Sinusoidal data with Polynomial regressions (see also Chap. 6 (section 6))
Exponential modeling
Spline modeling (see also Chap. 17)
Loess modeling.

2   Testing for Linearity

A first step with any data analysis is, to assess the data pattern from a scatter plot. The three underneath graphs give examples of nonlinear data sets: (1) relationship between age and systolic blood pressure, (2) effects of mental stress on fore arm vascular resistance, (3) relationship between time after polychlorobiphenyl (PCB) exposure and PCB concentrations in lake fish.

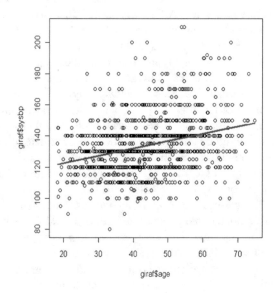

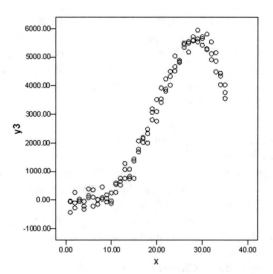

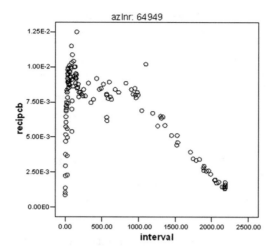

A considerable scatter is common, and it may be difficult to find the best fit model. Prior knowledge about patterns to be expected is helpful. Sometimes, a better fit of the data is obtained by drawing y versus x instead of the reverse. Residuals of y versus x with or without adjustments for other x-values are helpful for finding a recognizable data pattern. Statistically, we test for linearity by adding a non-linear term of x to the model, particularly, x squared or square root x, etc. If the squared correlation coefficient r^2 becomes larger by this action, then the pattern is, obviously, nonlinear. Statistical software like the curvilinear regression option in SPSS helps you identify the best fit model. The underneath graph and table give an example.

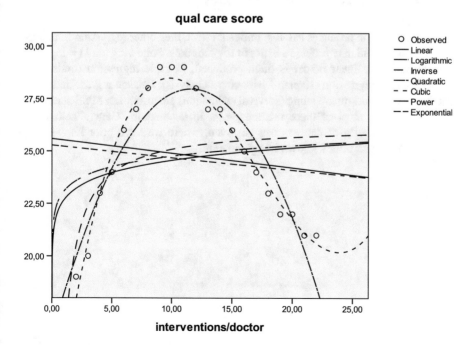

Model Summary and Parameter Estimates

Dependent Variable:qual care score

Equation	Model Summary					Parameter Estimates			
	R Square	F	df1	df2	Sig.	Constant	b1	b2	b3
Linear	,018	,353	1	19	,559	25,588	-,069		
Logarithmic	,024	,468	1	19	,502	23,086	,726		
Inverse	,168	3,829	1	19	,065	26,229	-11,448		
Quadratic	,866	58,321	2	18	,000	16,259	2,017	-,087	
Cubic	,977	236,005	3	17	,000	10,679	4,195	-,301	,006
Power	,032	,635	1	19	,435	22,667	,035		
Exponential	,013	,249	1	19	,624	25,281	-,002		

The independent variable is interventions/doctor.

Standard models of regression analyses: the effect of quantity of care (numbers of daily interventions, like endoscopies or small operations, per doctor) are assessed against quality of care scores. The best fit models for the data given in the graph are the quadratic and cubic models. In the next few sections various commonly used mathematical models are reviewed. The mathematical equations of these models are summarized in the appendix. They are helpful to make you understand the assumed nature of the relationships between the dependent and independent variables of the models used.

3   Logit and Probit Transformations

If linear regression produces a non-significant effect, then other regression functions can be chosen, and may provide a better fit for your data. Following logit ($=$ logistic) transformation a linear model is often produced. Logistic regression (odds ratio analysis), Cox regression (Kaplan-Meier curve analysis), Poisson regression (event rate analysis), Markov modeling (survival estimation) are examples. SPSS statistical software covers most of these methods, e.g., in its module "Generalized linear methods". Examples of datasets can be given, where we have prior knowledge, that they are linear after a known transformation.

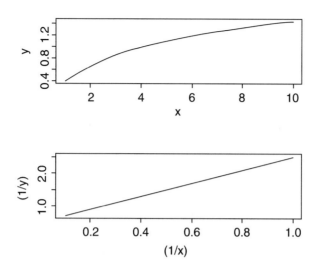

An example is given above of nonlinear relationship, that is linear after log transformation (Michaelis-Menten relationship between sucrose concentration on x-axis and invertase reaction rate on y-axis).

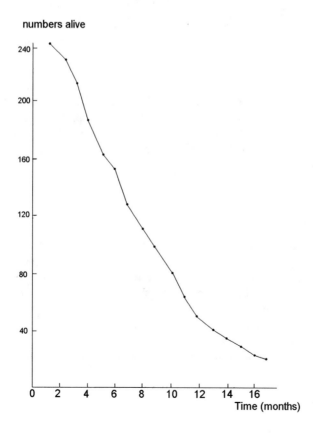

In numbers alive

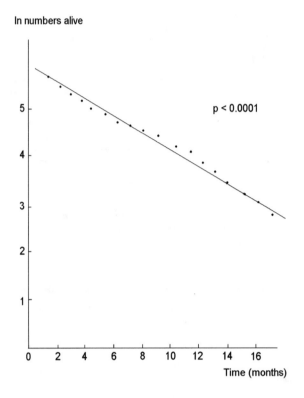

Above another example is given of a nonlinear relationship that is linear after logarithmic transformation (survival of 240 small cell carcinoma patients). As a particular caveat we should add here, that many examples can be given, but be aware. Most models in biomedicine have considerable residual scatter around the estimated regression line. For example, if the model applied is the following ($e =$ Euler's constant $= 2.7.....$, e = random variation):

$$y_i = \alpha \, e^{\beta x} + e_i,$$
then
$$\ln(y_i) \neq \ln(\alpha) + \beta x + e_i.$$

The smaller the e_i term is, the better fit is provided by the model. Another problem with logistic regression is that sometimes after iteration (= computer program for finding the largest log likelihood ratio for fitting the data) the results do not converse, i.e., a best log likelihood ratio is not established. This is due to (1) insufficient data size, (2) inadequate data, or (3) non-quadratic (linear) data patterns. An alternative for that purpose is probit modeling, which, generally, gives less iteration problems. The dependent variable of logistic regression (the log odds of responding) is closely related to log-probit (probit is the z-value corresponding to its area under curve value of the normal distribution). It can be shown that log-odds of responding $=$ logit $\approx (\pi/\sqrt{3})$ x probit.

4   "Trial and Error" Method, Box Cox Transformation, ACE /AVAS Packages

If logit or probit transformations do not work, then additional transformation techniques may be helpful. How do you find the best transformations? First, prior knowledge about the patterns to be expected is helpful. If this is not available, then the "trial and error" method can be recommended, particularly, logarithmically transforming either x- or y-axis or both of them.

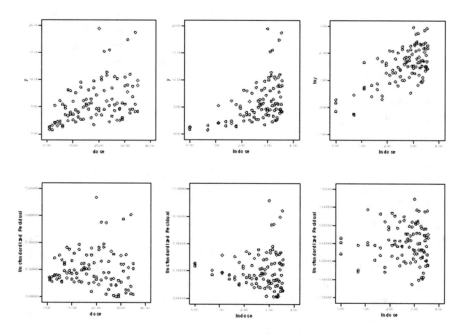

Trial and error methods can be used to find in the above graphs recognizable data patterns: relationship between isoproterenol dosages (on the x-axis) and relaxation of bronchial smooth muscle (on the y-axis).

log(y) vs x,
y vs log(x),
log(y) vs log(x).

The data fit of simple mathematical equations like the above ones can be successful and computed by heart (vs = versus). Box Cox transformation (http://itl.nist.gov), additive regression using ACE (alternating conditional expectations) and AVAS (additive and variance stabilization for regression). These packages describe modern non-parametric methods, otherwise closely related to the "trial and error" method, and can also be used for the purpose (http://pinard.gov). They are not in SPSS statistical software, but, instead, a free Box-Cox normality plot calculator is available on the Internet (www.essa.net). All of the methods in this section are largely empirical techniques to normalize non-normal data, that can, subsequently, be easily modeled, and they are available in virtually all modern software programs.

5 Sinusoidal Data with Polynomial Regressions

Clinical research is often involved in predicting an outcome from a predictor variable, and linear modeling is the commonest and simplest method for that purpose. The simplest except one is the quadratic relationship providing a symmetric curve, and the next simplest is the cubic model providing a sinus-like curve.

The equations are

Linear model $y = a + bx$
Quadratic model $y = a + bx^2$
Cubic model $y = a + bx^3$.

The larger the regression coefficient b, the better the model fits the data. Instead of the terms linear, quadratic, and cubic the terms first order, second order, and third order polynomial are applied.

If the data plot looks, obviously, sinusoidal, then higher order polynomial regression or Fourier analysis could be adequate (Cleophas and Zwinderman, Statistics applied to clinical studies 5th edition, 2012, 187–96). The latter of the two uses, instead of polynomials, regression equations like "$y = a + b.\cos x + c.\sin x$" (see also Statistics applied to clinical studies 5th edition, Chap. 16, Springer Heidelberg Germany, 2012, from the same authors). The equations are given in the appendix. The underneath graph gives an example of a polynomial model of the seventh order. It is used to describe ambulatory blood pressure measurements.

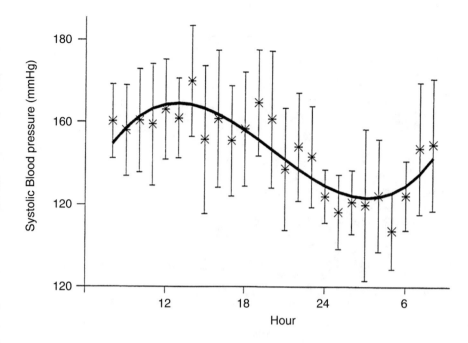

6   Exponential Modeling

For exponential-like patterns, like plasma concentration time relationships, exponential modeling is a possibility (Cleophas and Zwinderman, Statistics applied to clinical studies 5th edition, 2012, 213–5). Also multiple exponential modeling has become possible with the help of Laplace transformations. The nonlinear mixed effect exponential model (non-men model Sheiner, J Pharmacokin Pharmacodyn 1983; 11: 303–19) for pharmacokinetic studies is an example.

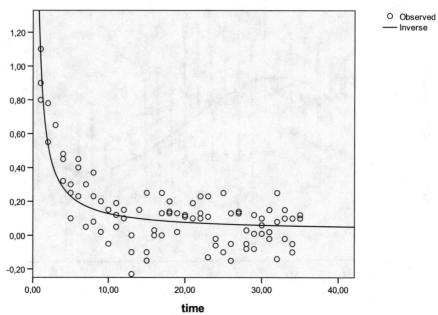

The above graph is an example of non-mem exponential model to describe plasma concentration-time relationship of zoledronic acid. The data plot shows, that the data spread is wide and, so, very accurate predictions can not be made in the given example. Nonetheless, the method is helpful to give an idea about some pharmacokinetic parameters like drug plasma half life and distribution volume.

7   Spline Modeling

If the above models do not adequately fit your data, you may use a method called spline modeling. It stems from the thin flexible wooden splines, formerly used by shipbuilders and car designers, to produce smooth shapes. Spline modeling will be, particularly, suitable for smoothing data patterns, if the data plot leaves you with no idea of the relationship between the y- and x-values. Basics of spline regressions is already described in the Chap. 17, using the effect of quantity of care on quality of care as example.

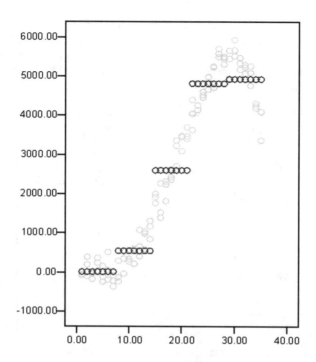

The above graph gives the data of another example, the effects of mental stress on fore arm vascular resistance. the data are, obviously nonlinear. Technically, the method of local smoothing, categorizing the x-values is used. It means that, if you have no idea about the shape of the relation between the y-values and the x-values of a two dimensional data plot, you may try and divide the x-values into 4 or 5 categories, where ϑ-values are the cut-offs of categories of x-values otherwise called the knots of the spline model.

- cat. 1: min $\leq$ x $< \theta_1$
- cat. 2: $\theta_1 \leq$ x $< \theta_2$
- ...
- cat. k: $\theta_{k-1} \leq$ x $<$ max.

Then, estimate y as the mean of all values within each category. Prerequisites and primary assumptions include

- the y-value is more or less constant within categories of the x-values,
- categories should have a decent number of observations,
- preferably, category boundaries should have some meaning.

The above graph shows, that a linear regression of five categories was possible, but the linear regression lines were not necessarily connected. The left graph underneath shows connected linear regressions, but the lines were only three. Instead of linear regression lines, a better fit for the data is provided by separate low-order polynomial regression lines (underneath right graph).

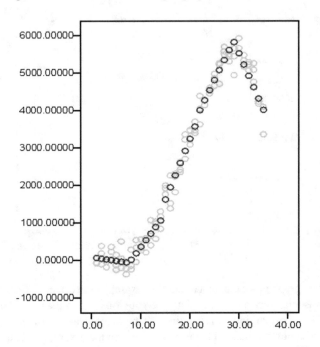

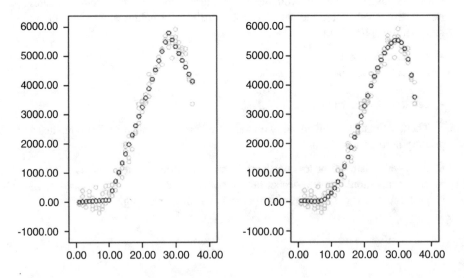

A polynomial curve with three intervals is connected at so-called knots. The knots are the x-values that connect one x-category with a subsequent one. Usually, cubic regressions, otherwise called third order polynomial regressions, are used. They have as simplest equation $y = a + bx^3$. Eventually, the separate lines are joined at the knots. With multiple intervals, this approach is called spline modeling. Spline modeling, thus, cuts the data into 4 or 5 intervals and uses the best fit third order polynomial functions for each interval.

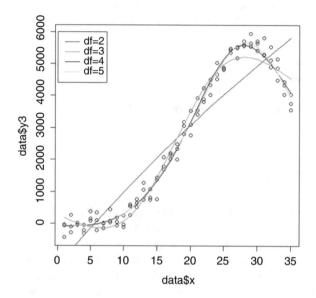

Above a spline regression of the same data is obtained with increasing numbers of knots.

In order to obtain a smooth spline curve, the junctions between two subsequent functions must have

(1) the same y value,
(2) the same slope,
(3) the same curvature.

All of these requirements are met if

(1) the two subsequent functions are equal at the junction,
(2) have the same first derivative at the junction,
(3) have the same second derivative at the junction.

There is a lot of matrix algebra involved, but a computer program can do the calculations for you, and provide you with the best fit spline curve.

Even with knots as few as 2, cubic spline regression may provide an adequate fit for the data.

In computer-graphics spline models are popular curves, because of their accuracy and capacity to fit complex data patterns. So far, they are not yet routinely used in clinical research for making predictions from response patterns, but this is a matter of time. Excel provides free cubic spline function software (Cubic spline for excel, www.srs1software.com). The spline model can be checked for its smoothness and fit using lambda-calculus and generalized additive models. Unfortunately, multidimensional smoothing using spline modeling is difficult. Instead you may perform separate procedures for each covariate. Two-dimensional spline modeling is available in the user-friendly SPSS statistical software:

Command:
Graphs. . . .Chart Builder. . . .Basic Elements. . . .Choose Axes. . . .y-x. . . .
Gallery. . . .Scatter/Dot. . . .OK. . . .double click in outcome graph to start Chart Editor. . . .Elements. . . .Interpolate. . . .Properties. . . .mark: Spline. . . .click: Apply. . . . the best fit spline model is in the outcome graph.

8   Loess Modeling

Maybe, the best fit for many types of nonlinear data is offered by still another novel regression method called Loess (locally weighted scatter plot smoothing, http://en. wikipedia.org). This computationally very intensive program calculates the best fit polynomials from subsets of your data set in order to eventually find out the best fit curve for the overall data set, and is related to Monte Carlo modeling. It does not work with knots, but, instead chooses the bets fit polynomial curve for each value, with outlier values given less weight. Loess modeling is available in SPSS:

Command: graphs. . . .chart builder. . . .basic elements. . . .choose axes. . . .y-x. . . . gallery. . . .scatter/dot. . . .ok. . .double click in outcome graph to start chart editor . . . . elements. . . .fit line at total. . . .properties. . . .mark: Loess. . . .click: apply. . . . best fit Loess model is in the outcome graph.

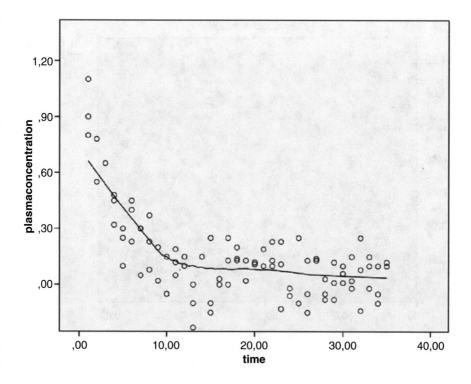

In the output sheets the above best fit Loess model of the data is given.

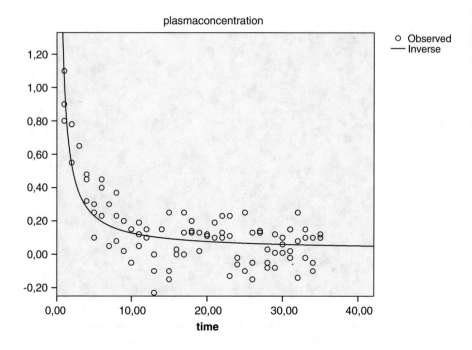

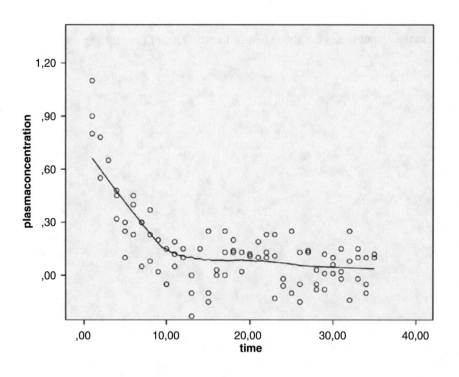

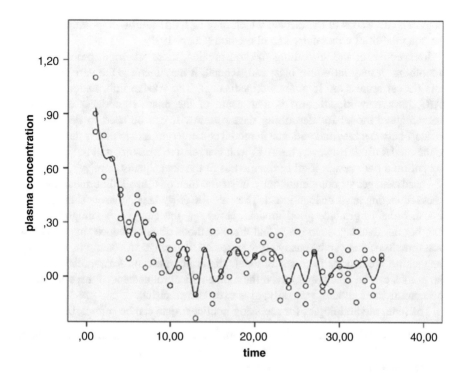

The above 3 graphs compare the best fit Loess model (middle graph) with the best fit cubic spline model (lower graph), and the best fit non-mem exponential model for describing a plasma concentration-time pattern. Both give a better fit for the data than does the traditional exponential modeling with 9 and 29 values in the Loess and spline lines compared to only 5 values in the exponential line (upper graph). However, it is impossible to estimate plasma half life from Loess and spline. We have to admit that, with so much spread in the data like in the given example, the meaning of the calculated plasma half life is, of course, limited.

9 Conclusion

Many tools are available for developing nonlinear models for characterizing data sets and making predictions from them. Sometimes it is difficult, to choose the degree of smoothness of such models: e.g., with polynomial regression, the question is, which order, and with spline modeling the questions are how many knots, which locations, which lambdas.

Another method is kernel frequency distribution modeling, which, unlike histograms, consists of multiple similarly sized Gaussian curves, rather than multiple bins of different length. In order to perform kernel modeling the bandwidth (span) of the

Gaussian curves has to be selected which may be a difficult but important factor of the potential fit of a particular kernel method (Chap. 19).

Irrespective of the smoothing method applied, there are some problems with smoothing: it may introduce bias, and, second, it may increase the variance in the data (Local regression, http://en.wikipedia.org). The Akaike information criterion (AIC, http://en.wikipedia.org) is a measure of the relative goodness of fit of a mathematical model for describing data patterns. It can be used to describe the tradeoff between bias and variance in model construction, and to assess the accuracy of the model used. However, the AIC, as it is a relative measure, will not be helpful to confirm a poor result, if all of the models fit the data equally poorly.

Disadvantages of computationally intensive methods, like spline modeling and Loess modeling must be mentioned. They require fairly large, densely sampled data sets in order to produce good models. However, the analysis is straightforward. Another disadvantage is the fact, that these methods do not produce simple regression functions, that can be easily represented by mathematical equations. However, for making predictions from such models direct interpolations/extrapolations from the graphs can be made, and, given the mathematical refinement of these methods, these predictions should, generally, give excellent precision.

The main methodologies for assessing nonlinear data can be summarized.

1. Logit and probit transformation can sometimes be used to mimic a linear model. Logistic regression, Cox regression, Poisson regression, and Markov modeling are examples of log transformations.
2. Either the x- or y-axis or both of them can be logarithmically transformed. Also Box Cox transformation equation and ACE (alternating conditional expectations) or AVAS (additive and variance stabilization for regression) packages are simple empirical methods often successful for linearly remodeling non linear data.
3. Data that are, obviously, sinusoidal, can, generally, be successfully modeled using polynomial regression or Fourier analysis.
4. For exponential patterns like plasma concentration time relationships exponential modeling with or without Laplace transformations is a possibility.
5. Spline and Loess modeling are modern methods, particularly, suitable for smoothing data patterns, if the data plot leaves you with no idea of the relationship between the y- and x-values. Loess tends to skip outlier data, while spine modeling rather tends to include them. So, if you are planning to investigate the outliers, the spline is your tool.

We have to add that traditional nonlinear modeling produces p-values, and modern methods do not. However, given the poor fit of many traditional models, these p-values do not mean too much. Also, it is reassuring to observe that both Loess and spline provide a better fit to nonlinear data than does traditional modeling.

Appendix

In this appendix the mathematical equations of the non linear models as reviewed are given. They are, particularly, helpful for those trying to understand the assumed relationships between the dependent (y) and independent (x) variables (ln = natural logarithm).

$y = a + b_1 x_1 + b_2 x_2 + \ldots\ldots b_{10} x_{10}$ linear

$y = a + bx + cx^2 + dx^3 + ex^4 \ldots$ polynomial

$y = a + sinus\ x + cosinus\ x + \ldots$ Fourier

$Ln\ odds = a + b_1 x_1 + b_2 x_2 + \ldots\ldots b_{10} x_{10}$ logistic

Instead of ln odds (= logit) also probit ($\approx \pi\sqrt{3}$ x logit) is often used for transforming binomial data.

probit

$Ln\ multinomial\ odds = a + b_1 x_1 + b_2 x_2 + \ldots\ldots b_{10} x_{10}$ multinomial logistic

$Ln\ hazard = a + b_1 x_1 + b_2 x_2 + \ldots\ldots b_{10} x_{10}$ Cox

$Ln\ rate = a + b_1 x_1 + b_2 x_2 + \ldots\ldots b_{10} x_{10}$ Poisson

$\log y = a + b_1 x_1 + b_2 x_2 + \ldots\ldots b_{10} x_{10}$ logarithmic

$y = a + b_1 \log x_1 + b_2 x_2 + \ldots\ldots b_{10} x_{10}$ etc "trial and error"

transformation function of $y = (y^\lambda - 1) / \lambda$ Box-Cox
with λ as power parameter

$y = $ (above transformation function)$^{-1}$ ACE modeling

$y = e^{x_1} x_2^{\sin x_3}$ etc AVAS modeling

$y = a + e^{b_1 x_1} + e^{b_2 x_2}$ multi-exponential modeling

$\theta = $ magnitude of x-value (example)
$\theta_1 < x < \theta_2 \qquad y = a_1 + b_1 x^3$ spline modeling
$\theta_2 < x < \theta_3 \qquad y = a_2 + b_2 x^3$
$\theta_3 < x < \theta_4 \qquad y = a_3 + b_3 x^3$

Reference

To readers requesting more background, theoretical and mathematical information of computations given, several textbooks complementary to the current production and written by the same authors are available: Statistics applied to clinical studies 5th edition, 2012, Machine learning in medicine a complete overview 2nd edition, 2020, SPSS for starters and 2nd levelers 2nd edition, 2015, Clinical data analysis on a pocket calculator 2nd edition, 2016, Understanding clinical data analysis, 2017, all of them edited by Springer Heidelberg Germany.

Chapter 19
Special Forms of Continuous Outcomes Regressions

Kernel, Gamma, Tweedie, and Robust Regressions

Abstract First, kernel frequency distributions consist of multiple identical Gaussian curves rather than histograms consistent of bins with different lengths. Kernel regression measures the relationship between x and y data, where the expectation of y is conditional not on all x-values but on locally weighted averages of subsets of consecutive x-values within a bandwidth. And, together, they tend to produce a simple linear regression model. Any linear regression can be replaced with a kernel regression, but the method is particularly appropriate in case of seemingly nonlinear patterns, that, in the end, are linear after all. With skewed data files a (much) better fit for the data is provided by kernel regression than by linear regression.

Second, gamma and Tweedie regressions are worthwhile analysis models for non-negative data complementary to linear regression, that may elucidate effects unobserved in the traditional linear model. The marginal means procedure enables to observe trends in the data, e.g., decreasing outcome score with increasing predictor scores. In the example given, Tweedie regression provided a somewhat better sensitivity of testing than gamma regression did.

Third, robust regressions are explained. If, with linear regression models, your results are borderline significant and outliers are in the data, then robust regression testing can provide you with better statistics, and, thus, better statistical power of testing, than traditional testing does.

Keywords Kernel regression · Tweedie regression · Gamma regression · Robust regression

Electronic Supplementary Material The online version of this chapter (https://doi.org/10.1007/978-3-030-61394-5_19) contains supplementary material, which is available to authorized users. The videos can be accessed by scanning the related images with the SN More Media App.

1 Introduction, History, and Background

Special forms of continuous outcomes regressions have their own advantages. Kernel regression smoothes the data to create an approximating function that attempts to capture important patterns in the data, leaving out noise and other fine scale structures. Tweedie and gamma regressions are nonnegative regressions already successfully used cancer, genetic research, and critical care research. Robust regressions often based on medians rather than means are particularly suitable for detecting and handling outlier data. The current chapter will review, and assess with examples, these alternative methods for the analysis of regressions with continuous outcomes.

Kernel regression is a nonparametric technique to estimate the conditional expectation of a random variable. The objective is to find a non-linear relation between a pair of random variables X and Y. In any nonparametric regression, the conditional expectation of a variable relative to another variable may be estimated. Nadaraya and Watson, from John Hopkins University, proposed in 1964 to estimate it as a locally weighted average, using a kernel K as a weighting function, Priestly and Chao, from Manchester UK, as well as Gasser and Muller, from Zurich University Switzerland used in 1971 different weighting functions.

1. Kernel regressions have probability distributions in the form of the add-up of multiple sets of identical normal distributions.
2. Gamma regressions have probability distributions of the gamma function type (skewed to the right, invented by Daniel Bernouilli, Dutch mathematician with Swiss ancestors, in 1700), and
3. Tweedie (after M Tweedie, 1984, from Liverpool UK) equally so, but with a spike in the left-end of the probability distributions.
4. Robust regressions (used by Wang and Huang from Taipei University 2011, and soon after by many others) apply maximum likelihood estimators instead of traditional F- and T-tests.

2 Kernel Regressions

With traditional regression methods, the outcome values are assumed to be normally distributed around the regression line/curve. With non-normal outcome data, that remain non-normal in spite of transformations (Likert scales is a notorious example), data distributions may be skewed, and nonparametric regression analysis may provide better data fit than traditional parametric models do. Methods including nonparametric regression are pretty new, and not yet widely applied. They include: kriging, otherwise called Gaussian process regression (Chap. 11), decision trees, and bagged (bootstrap aggregated) regression trees (Chap. 22), kernel regressions (current chapter), and median regression, otherwise called robust regression (current chapter).

An example will be given of data appropriate for kernel regression. Kernel regression is equivalent to radial basis function analysis. Kernel frequency distributions consist of multiple identical Gaussian curves, rather than histograms consistent of bins with different lengths. Kernel regression measures the relationship between x and y data, where the expectation of y is conditional not on all x-values but on locally weighted averages of subsets of consecutive x-values with a predefined bandwidth. The subsets are described by their means and standard errors. And, together, these means tend to produce a simple linear regression model. Obviously, any linear regression can be replaced with a kernel regression, but the method is particularly appropriate in case of seemingly nonlinear patterns, that, in the end, are linear after all. In many cases a (much) better fit for the data is provided by kernel regression than for linear regression.

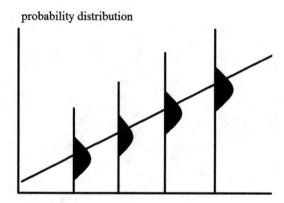

The above graph shows a linear regression model with the x-values sampled without error, and the y-values having uncertainties in the form of identical normal curves (Gaussian curves drawn in black). This linear regression model is parametric, because normal curves are symmetric around their mean.

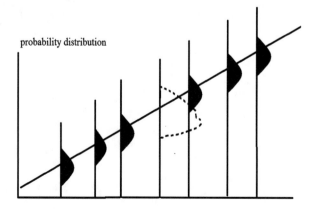

The above graph shows, how kernel regression works. The uncertainty of the y-values is not expressed in the form of Gaussian curves, but rather the add-up sum of multiple Gaussian curves. The area under curve of the dotted curve is the add-up sum of the six Gaussian areas under the curve (in black). It is called, a kernel density distribution. The corresponding x-values are subsets of all x-values of a data file. The subset given is within one bandwidth. Linear modeling of multiple bandwidths results in a linear model, even if the traditional linear model has a pretty poor linear fit.

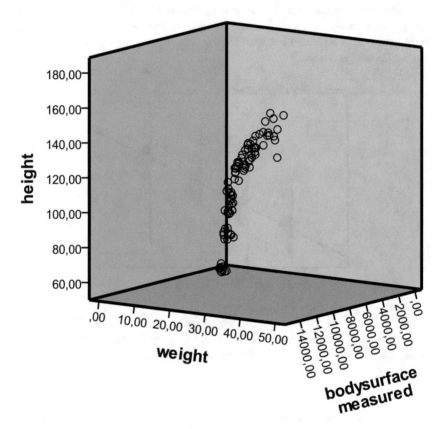

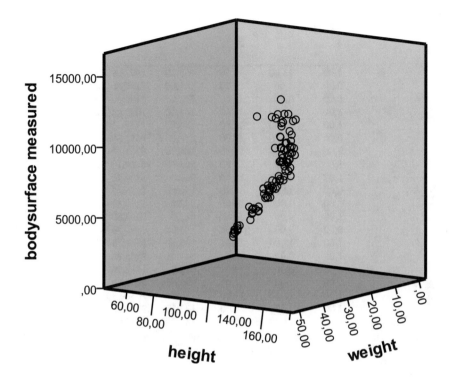

In the above graphs body heights and weights were assessed as possible predictors of the measured body surfaces in 90 persons. The graphs show in the three dimensional scatter plots, that the relationships were, obviously, nonlinear. Body height and weight are generally assumed not to be linear with time but rather s-shape or biphasic. Kernel regression may be particularly advantageous for modeling such data. For analysis in SPSS statistical software we will use, from the module neural networks, radial basis neural (RBF) networks. The body surfaces of the 90 persons were calculated using direct photometric measurements. The data file is entitled "chap19kernel", and is in extras.springer.com. It is previously used by the authors in Machine learning in medicine a complete overview, Chap.63, Springer Heidelberg Germany, 2015.

Variable

1	2	3	4	5
1,00	13,00	30,50	138,50	10072,90
0,00	5,00	15,00	101,00	6189,00
0,00	0,00	2,50	51,50	1906,20
1,00	11,00	30,00	141,00	10290,60
1,00	15,00	40,50	154,00	13221,60
0,00	11,00	27,00	136,00	9654,50
0,00	5,00	15,00	106,00	6768,20
1,00	5,00	15,00	103,00	6194,10
1,00	3,00	13,50	96,00	5830,20
0,00	13,00	36,00	150,00	11759,00
0,00	3,00	12,00	92,00	5299,40
1,00	0,00	2,50	51,00	2094,50
0,00	7,00	19,00	121,00	7490,80
1,00	13,00	28,00	130,50	9521,70
1,00	0,00	3,00	54,00	2446,20
0,00	0,00	3,00	51,00	1632,50
0,00	7,00	21,00	123,00	7958,80
1,00	11,00	31,00	139,00	10580,80
1,00	7,00	24,50	122,50	8756,10
1,00	11,00	26,00	133,00	9573,00
.	.	.	.	.
.	.	.	.	.
.	.	.	.	.

Var 1 gender
Var 2 age
Var 3 weight (kg)
Var 4 height (m)
Var 5 body surface measured (cm^2)

With regression analyses the best fit regression equation is often used for making predictions from your data about future data, as shown in many chapters of this edition. A problem with this procedure is, that the best fit regression equation is used for generating inter- and extrapolated data, and the fit is not always, what it should be, particularly with data, that tend to be nonlinear, like in the above graphs. It is accepted that some kind of goodness of fit test prior to prediction is recommendable, and, that, with multidimensional data, it is even possible to improve the predictive regression equation with the help of training samples, as will be shown underneath.

First, the performance of kernel regression, as compared to traditional log likelihood modeling will be assessed. Akaike's information criterion (AIC) will be applied for the purpose. With regression equations, p-values of regression coefficients tell you something about the goodness of the independent variables. However, it says nothing about the complexity of the regression model applied, and it is easy to find a very complex model, that produces very good p-values. The trick is, thus, to find a regression model, that has the best trade-off between goodness and complexity. This is what the AIC will find out. The smaller the AIC of a model, the better the trade-off. In other words, each regression model suffers from information loss, the better the model, the better information loss is minimized. We will start by opening our data file, entitled "chap19kernel", and stored in extras.springer.com, in our computer with SPSS statistical software installed. Then command.

Command:
click Analyze....click Generalized Linear Models....click Generalized Linear Models
....click Type of Model....click Scale Response....click Linear....click Response....
click Dependent Variable:...enter bodysurface measured....click Predictors....click
Factors: ....enter gender, age, weight, height....click Model....click Main effects....
enter gender, age, weight, height....click Estimation....click Maximal likelihood
estimate....click Statistics....click Log-Likelihood Function....click Full....click OK.

Goodness of Fit[a]

	Value	df	Value/df
Deviance	134551,397	5	26910,279
Scaled Deviance	90,000	5	
Pearson Chi-Square	134551,397	5	26910,279
Scaled Pearson Chi-Square	90,000	5	
Log Likelihood[b]	-456,650		
Akaike's Information Criterion (AIC)	1085,299		
Finite Sample Corrected AIC (AICC)	6073,299		
Bayesian Information Criterion (BIC)	1300,283		
Consistent AIC (CAIC)	1386,283		

Dependent Variable: bodysurface measured
Model: (Intercept), VAR00001, VAR00002, VAR00003, VAR00004

a. Information criteria are in small-is-better form.

b. The full log likelihood function is displayed and used in computing information criteria.

In the output sheets a goodness of fit is given. Particularly, the Akaike's Information Criterion is relevant.

Next, we will assess the goodness of fit of the kernel model. The same commands can be given, except for the three final commands, that should be....click Log-Likelihood Function....click Kernel....click OK.

The underneath goodness of fit table is given.

Goodness of Fit[a]

	Value	df	Value/df
Deviance	134551,397	5	26910,279
Scaled Deviance	90,000	5	
Pearson Chi-Square	134551,397	5	26910,279
Scaled Pearson Chi-Square	90,000	5	
Log Likelihood[b]	-373,945		
Akaike's Information Criterion (AIC)	919,890		
Finite Sample Corrected AIC (AICC)	5907,890		
Bayesian Information Criterion (BIC)	1134,874		
Consistent AIC (CAIC)	1220,874		

Dependent Variable: bodysurface measured
Model: (Intercept), VAR00001, VAR00002, VAR00003, VAR00004

a. Information criteria are in small-is-better form.

b. The kernel of the log likelihood function is displayed and used in computing information criteria.

The AICs of the two models are respectively 1085 and 920, difference 165. This means, that the probability of the kernel regression model to minimize information loss is e $^{(165)/2}$ times less worse than the full loglikelihood model is. Obviously, the kernel model performs endlessly better.

We can also use the kernel regression methodology available in the RBF model of the SPSS module Neural Networks for predictive modeling of our data example. Both a training and outcome predictions are possible. The module uses XML (exTended Markup Language) files to store the neural network. Start by the underneath commands.

Command:
click Transform....click Random Number Generators....click Set Starting Point....click Fixed Value (2000000)....click OK....click Analyze.... Neural Networks....Radial Basis Function....Dependent Variables: enter Body surface measured....Factors: enter gender, age, weight, and height....Partitions: Training 7....Test 3....Holdout 0....click Output: mark Description....Diagram.... Model summary....Predicted by observed chart....Case processing summary ....click Save: mark Save predicted value of category for each dependent variable....automatically generate unique names....click Export....mark Export synaptic weights estimates to XML file....click Browse....File Name: enter "exportradialbasisnn" and save in the appropriate folder of your computer....click OK.

The output warns, that, in the testing sample, some cases have been excluded from analysis, because of values not occurring in the training sample. Minimizing the output sheets shows the data file with predicted values. They are pretty much the same as the measured body surface values. We will use linear regression to estimate the association between the two.

Command:
Analyze....Regresssion....Linear....Dependent: bodysurface ....Independent: RBF_PredictedValue....OK.

The output sheets show that the r-value is 0.931, p < 0.0001. The saved XML file will now be used to compute the body surface in six individual patients.

gender	age	weight	height
1,00	9,00	29,00	138,00
1,00	1,00	8,00	76,00
,00	15,00	42,00	165,00
1,00	15,00	40,00	151,00
1,00	1,00	9,00	80,00
1,00	7,00	22,00	123,00

gender
age (years)
weight (kg)
height (m)

Enter the above data in a new SPSS data file.

Command:
Utilities....click Scoring Wizard....click Browse....click Select....Folder: enter the exportradialbasisnn.xml file....click Select....in Scoring Wizard click Next.... click Use value substitution....click Next....click Finish.

The underneath data file now gives the body surfaces computed by the neural network with the help of the XML file.

gender	age	weight	height	predicted body surface
1,00	9,00	29,00	138,00	9219,71
1,00	1,00	8,00	76,00	5307,81
,00	15,00	42,00	165,00	13520,13
1,00	15,00	40,00	151,00	13300,79
1,00	1,00	9,00	80,00	5170,13
1,00	7,00	22,00	123,00	8460,05

gender
age (years)
weight (kg)
height (m)
predicted body surface (cm^2)

Kernel regression based RBF networks can, thus, be readily trained to provide accurate body surface values of individual patients. Many more examples can be given, but kernel regression seems to works best with big data, not only with radial basis neural networks, but also with other machine learning methodologies like support vector machines, otherwise called the fast cluster methodology.

3   Gamma and Tweedie Regressions

The gamma frequency distribution is suitable for statistical testing of nonnegative data with a continuous outcome variable and fits such data often better than does the normal frequency distribution, particularly when magnitudes of benefits or risks is the outcome, like costs. It is often used in marketing research. It is little used in medical research.

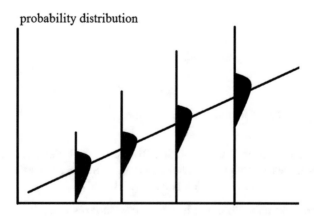

The above graph gives a schematic view of a probability distribution of the gamma type (skewed to the right). Skewed data like quality of life (QOL) scores in sick populations (that are clustered towards low QOL scores) may better fit gamma distributions, than they do normal distributions.

A data example is given. In 110 patients the effects of age class, psychological and social score on health scores were assessed. The first 10 patients are underneath. The entire data file is entitled "chap19gamma", and is in extras.springer.com. It is previously used by the authors in Machine learning in medicine a complete overview, Chap. 30, Springer Heidelberg Germany, 2016.

health score	age class	psychologic score	social score
8	3	5	4
7	1	4	8
4	1	5	13
6	1	4	15
10	1	7	4
6	1	8	8
8	1	9	12
2	1	8	16
6	1	12	4
8	1	13	1

age = age class 1-7
psychologicscore = psychological score 1-20
socialscore = social score 1-20
healthscore = health score 1-20.

Start by opening the data file in your computer installed with SPSS statistical software. We will first perform a linear regression with the 3 predictors as independent variables and health scores as outcome. The data suggest, that both psychological and social scores are significant predictors of health, but age class is not. In order to adjust possible confounding, a multiple linear regression is performed.

Command:
Analyze....Regression....Linear....Dependent: enter healthscore....Independent (s): enter socialscore, psychologicscore, age....click OK.

Coefficients[a]

Model		Unstandardized Coefficients		Standardized Coefficients	t	Sig.
		B	Std. Error	Beta		
1	(Constant)	9.388	.870		10.788	.000
	social score	-.329	.049	-.533	-6.764	.000
	psychological score	.111	.046	.190	2.418	.017
	age class	-.184	.109	-.132	-1.681	.096

a. Dependent Variable: health score

Social score is again very significant. Psychological score also, but after Bonferroni adjustment (rejection p-value = 0,05/4 = 0,0125) it would be no more so, because p= 0,017 is larger than 0,0125. Age class is again not significant. Health score is here a continuous variable of nonnegative values, and, perhaps, better fit of these data might be obtained by a gamma regression. We will use SPSS statistical software again.

For analysis the module Generalized Linear Models is required. It consists of two submodules: Generalized Linear Models and Generalized Estimation Models. The first submodule covers many statistical models like gamma regression and Tweedie regression (the current chapter), Poisson regression (the Chaps. 4 and 5), and the analysis of paired outcomes with predictors (the Chap. 8). The second is for analyzing binary outcomes. We will use the statistical model Gamma Distribution in the submodule Generalized Linear Models.

Command:
Analyze....click Generalized Linear Models....click once again Generalized Linear Models....mark Custom....Distribution: select Gamma....Link function: select Power....Power: type -1....click Response....Dependent Variable: enter healthscore click Predictors....Factors: enter socialscore, psychologicscore, age....Model: enter socialscore, psychologicscore, age....Estimation: Scale Parameter Method: select Pearson chi-square....click EM Means: Displays Means for: enter age, psychologicscore, socialscore....click Save....mark Predict value of linear predictor....Standardize deviance residual....click OK.

Tests of Model Effects

Source	Type III		
	Wald Chi-Square	df	Sig.
(Intercept)	216.725	1	.000
ageclass	8.838	6	.183
psychologicscore	18.542	13	.138
socialscore	61.207	13	.000

Dependent Variable: health score
Model: (Intercept), ageclass, psychologicscore,
socialscore

The above table gives the overall result: it is comparable to that of the multiple linear regression with only social score, as significant independent predictor. In the output additional relevant information is given.

Parameter Estimates

Parameter	B	Std. Error	95% Wald Confidence Interval		Hypothesis Test		
			Lower	Upper	Wald Chi-Square	df	Sig.
(Intercept)	.188	.0796	.032	.344	5.566	1	.018
[ageclass=1]	-.017	.0166	-.050	.015	1.105	1	.293
[ageclass=2]	-.002	.0175	-.036	.032	.010	1	.919
[ageclass=3]	-.015	.0162	-.047	.017	.839	1	.360
[ageclass=4]	.014	.0176	-.020	.049	.658	1	.417
[ageclass=5]	.025	.0190	-.012	.062	1.723	1	.189
[ageclass=6]	.005	.0173	-.029	.039	.087	1	.767
[ageclass=7]	0[a]	.	.	.	.	.	.
[psychologicscore=3]	.057	.0409	-.023	.137	1.930	1	.165
[psychologicscore=4]	.057	.0220	.014	.100	6.754	1	.009
[psychologicscore=5]	.066	.0263	.015	.118	6.352	1	.012
[psychologicscore=7]	.060	.0311	-.001	.121	3.684	1	.055
[psychologicscore=8]	.061	.0213	.019	.102	8.119	1	.004
[psychologicscore=9]	.035	.0301	-.024	.094	1.381	1	.240
[psychologicscore=11]	.057	.0325	-.007	.120	3.059	1	.080
[psychologicscore=12]	.060	.0219	.017	.103	7.492	1	.006
[psychologicscore=13]	.040	.0266	-.012	.092	2.267	1	.132
[psychologicscore=14]	.090	.0986	-.103	.283	.835	1	.361
[psychologicscore=15]	.121	.0639	-.004	.247	3.610	1	.057
[psychologicscore=16]	.041	.0212	-.001	.082	3.698	1	.054
[psychologicscore=17]	.022	.0241	-.025	.069	.841	1	.359
[psychologicscore=18]	0[a]	.	.	.	.	.	.
[socialscore=4]	-.120	.0761	-.269	.029	2.492	1	.114
[socialscore=6]	-.028	.0986	-.221	.165	.079	1	.778
[socialscore=8]	-.100	.0761	-.249	.050	1.712	1	.191
[socialscore=9]	.002	.1076	-.209	.213	.000	1	.988
[socialscore=10]	-.123	.0864	-.293	.046	2.042	1	.153
[socialscore=11]	.015	.0870	-.156	.185	.029	1	.865
[socialscore=12]	-.064	.0772	-.215	.088	.682	1	.409
[socialscore=13]	-.065	.0773	-.216	.087	.703	1	.402
[socialscore=14]	.008	.0875	-.163	.180	.009	1	.925
[socialscore=15]	-.051	.0793	-.207	.104	.420	1	.517
[socialscore=16]	.026	.0796	-.130	.182	.107	1	.744
[socialscore=17]	-.109	.0862	-.277	.060	1.587	1	.208
[socialscore=18]	-.053	.0986	-.246	.141	.285	1	.593
[socialscore=19]	0[a]	.	.	.	.	.	.
(Scale)	.088[b]						

Dependent Variable: health score
Model: (Intercept), ageclass, psychologicscore, socialscore

a. Set to zero because this parameter is redundant.
b. Computed based on the Pearson chi-square.

Various levels of the predictors are given separately. Age class was not a significant predictor. Of the psychological scores, however, no less than 8 scores produced pretty small p-values, even as small as 0,004 to 0,009. Of the social scores now none were significant. In order to better understand what is going on, SPSS also provides marginal means.

Estimates

age class	Mean	Std. Error	95% Wald Confidence Interval Lower	95% Wald Confidence Interval Upper
1	5.62	.531	4.58	6.66
2	5.17	.461	4.27	6.07
3	5.54	.489	4.59	6.50
4	4.77	.402	3.98	5.56
5	4.54	.391	3.78	5.31
6	4.99	.439	4.13	5.85
7	5.12	.453	4.23	6.01

Of the different age classes the mean health scores were, indeed, hardly different.

Estimates

psychological score	Mean	Std. Error	95% Wald Confidence Interval Lower	95% Wald Confidence Interval Upper
3	5.03	.997	3.08	6.99
4	5.02	.404	4.23	5.81
5	4.80	.541	3.74	5.86
7	4.96	.695	3.60	6.32
8	4.94	.359	4.23	5.64
9	5.64	.809	4.05	7.22
11	5.03	.752	3.56	6.51
12	4.95	.435	4.10	5.81
13	5.49	.586	4.34	6.64
14	4.31	1.752	.88	7.74
15	3.80	.898	2.04	5.56
16	5.48	.493	4.51	6.44
17	6.10	.681	4.76	7.43
18	7.05	1.075	4.94	9.15

However, increasing psychological scores seem to be associated with increasing levels of health.

Estimates

social score	Mean	Std. Error	95% Wald Confidence Interval	
			Lower	Upper
4	8.07	.789	6.52	9.62
6	4.63	1.345	1.99	7.26
8	6.93	.606	5.74	8.11
9	4.07	1.266	1.59	6.55
10	8.29	2.838	2.73	13.86
11	3.87	.634	2.62	5.11
12	5.55	.529	4.51	6.59
13	5.58	.558	4.49	6.68
14	3.96	.711	2.57	5.36
15	5.19	.707	3.81	6.58
16	3.70	.371	2.98	4.43
17	7.39	2.256	2.96	11.81
18	5.23	1.616	2.06	8.40
19	4.10	1.280	1.59	6.61

In contrast, increasing social scores are associated with decreasing levels of health, with mean health scores close to 3 in the higher social score patients, and close to 10 in the lower social score patients.

In conclusion, gamma regression is a worthwhile analysis model complementary to linear regression, ands may elucidate effects unobserved in the linear models. The marginal means procedure readily enables to observe trends in the data, e.g., decreasing outcome score with increasing predictor scores.

The underneath graph gives a schematic view of a probability distributions of the gamma type (skewed to the right) with, in addition, a spike at its left end. It is called a Tweedie distribution, and often fits slightly better than gamma distribution skewed quality of life (QOL) data, that are not only clustered towards low QOL scores but also rocket at zero.

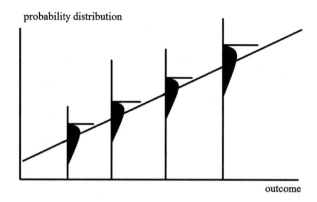

The above data example will be used once more. For analysis the module Generalized Linear Models is again required. It consists of two submodules: Generalized Linear Models and Generalized Estimation Models. We will use, in the submodule Generalized Linear Models, the Tweedie Distribution.

Command:

Analyze....Generalized Linear Model....Generalized Linear Model....mark Tweedie with log link....click Response....Dependent Variable: enter health score....click Predictors....Factors: enter age class, psychological score, social score....click OK.

The underneath table shows the results.

Tests of Model Effects

Source	Type III		
	Wald Chi-Square	df	Sig.
(Intercept)	776,671	1	,000
ageclass	9,265	6	,159
psychologicscore	22,800	13	,044
socialscore	90,655	13	,000

Dependent Variable: health score
Model: (Intercept), ageclass, psychologicscore, socialscore

Parameter Estimates

Parameter	B	Std. Error	95% Wald Confidence Interval		Hypothesis Test		
			Lower	Upper	Wald Chi-Square	df	Sig.
(Intercept)	1,848	,3442	1,173	2,523	28,818	1	,000
[ageclass=1]	,085	,1056	-,122	,292	,647	1	,421
[ageclass=2]	,008	,1083	-,204	,220	,006	1	,940
[ageclass=3]	,113	,1039	-,091	,317	1,185	1	,276
[ageclass=4]	-,072	,1046	-,277	,133	,470	1	,493
[ageclass=5]	-,157	,1087	-,370	,056	2,089	1	,148
[ageclass=6]	-,038	,1043	-,243	,166	,136	1	,712
[ageclass=7]	0ª	.	.	.	.	.	.
[psychologicscore=3]	-,395	,2072	-,801	,011	3,640	1	,056
[psychologicscore=4]	-,423	,1470	-,711	-,135	8,293	1	,004
[psychologicscore=5]	-,503	,1721	-,840	-,166	8,539	1	,003
[psychologicscore=7]	-,426	,2210	-,859	,007	3,713	1	,054
[psychologicscore=8]	-,445	,1420	-,723	-,166	9,807	1	,002
[psychologicscore=9]	-,255	,1942	-,636	,125	1,729	1	,189
[psychologicscore=11]	-,435	,1870	-,802	-,069	5,416	1	,020
[psychologicscore=12]	-,437	,1466	-,725	-,150	8,904	1	,003
[psychologicscore=13]	-,299	,1748	-,641	,044	2,922	1	,087
[psychologicscore=14]	-,522	,4349	-1,374	,330	1,440	1	,230
[psychologicscore=15]	-,726	,2593	-1,234	-,218	7,839	1	,005
[psychologicscore=16]	-,340	,1474	-,629	-,051	5,329	1	,021
[psychologicscore=17]	-,154	,1682	-,484	,175	,842	1	,359
[psychologicscore=18]	0ª	.	.	.	.	.	.
[socialscore=4]	,677	,3062	,077	1,277	4,885	1	,027
[socialscore=6]	,060	,4310	-,785	,905	,020	1	,889
[socialscore=8]	,513	,3062	-,087	1,114	2,811	1	,094
[socialscore=9]	-,017	,4332	-,866	,832	,002	1	,969
[socialscore=10]	,676	,4040	-,115	1,468	2,803	1	,094
[socialscore=11]	-,162	,3587	-,865	,541	,203	1	,652
[socialscore=12]	,289	,3124	-,323	,902	,858	1	,354
[socialscore=13]	,272	,3149	-,345	,889	,745	1	,388
[socialscore=14]	-,036	,3505	-,723	,651	,010	1	,919
[socialscore=15]	,212	,3240	-,423	,848	,430	1	,512
[socialscore=16]	-,089	,3127	-,702	,524	,081	1	,776
[socialscore=17]	,563	,4030	-,227	1,353	1,953	1	,162
[socialscore=18]	,217	,4287	-,623	1,058	,257	1	,612
[socialscore=19]	0ª	.	.	.	.	.	.
(Scale)	,167ᵇ	,0223	,129	,217			

Dependent Variable: health score
Model: (Intercept), ageclass, psychologicscore, socialscore

a. Set to zero because this parameter is redundant.
b. Maximum likelihood estimate.

The Wald test statistics are somewhat better than that of the Gamma regression. The Wald Chi-squares values rose respectively from 8,8 to 9,3, 18,5 to 22,8 and 61,2 to 90,7. The Parameter Estimates table showed that 7 instead 4 p-values for psychological scores were < 0,05.

In conclusion, gamma and Tweedie regressions are worthwhile analysis models complementary to linear regression, that may elucidate effects unobserved in the traditional linear model. The marginal means procedure enables to observe trends in the data, e.g., decreasing outcome score with increasing predictor scores. In the example given, Tweedie regression provided a somewhat better sensitivity of testing than gamma regression did. We should add, that gamma and Tweedie regressions are particularly useful with "necessary condition analysis (NCA)". "Necessary condition analysis" concerns factors, that don't work. Examples are: Aids will not develop without HIV, students will not be admitted to university without a high school diploma. With NCA analysis you may simply turn the variables that do not work into nonnegative variables, and perform a gamma or Tweedie regression.

4 Robust Regressions

Robust regressions. often based on medians rather than means, are particularly suitable for detecting and handling outlier data. Also they tend to better fit non-normal data like current Covid-19 data with excessive age-dependent mortality / morbidity risks. An example of a robust linear regression will be given, with a single outcome and predictor variable, but multiple variables models are also possible. Numbers of stools per month on a new laxative as outcome and the numbers stools on an old laxative (bisacodyl) per month as predictor will be assessed with traditional linear regression analysis using SPSS statistical software. The simple linear regression produced a p-value of 0.049.

The data file is called robustregressession.sav, and is in extras.springer.com.

The data are slightly different from the Chap.1 data, because they were not checked for errors.

newtreat	oldtreat	agecats	patientno
24,00	8,00	2,00	1,00
30,00	13,00	2,00	2,00
25,00	15,00	2,00	3,00
35,00	10,00	3,00	4,00
39,00	9,00	3,00	5,00
30,00	10,00	3,00	6,00
27,00	8,00	1,00	7,00
14,00	5,00	1,00	8,00
39,00	13,00	1,00	9,00
42,00	15,00	1,00	10,00
41,00	11,00	1,00	11,00
38,00	11,00	2,00	12,00
39,00	112,00	2,00	13,00
37,00	10,00	3,00	14,00
47,00	18,00	3,00	15,00
30,00	13,00	2,00	16,00
36,00	12,00	2,00	17,00
12,00	4,00	2,00	18,00
26,00	10,00	2,00	19,00
20,00	8,00	1,00	20,00
43,00	16,00	3,00	21,00
31,00	15,00	2,00	22,00
40,00	114,00	2,00	23,00
31,00	7,00	2,00	24,00
36,00	12,00	3,00	25,00
21,00	6,00	2,00	26,00
44,00	19,00	3,00	27,00
11,00	5,00	2,00	28,00
27,00	8,00	2,00	29,00
24,00	9,00	2,00	30,00
40,00	15,00	1,00	31,00
32,00	7,00	2,00	32,00
10,00	6,00	2,00	33,00
37,00	14,00	3,00	34,00
19,00	7,00	2,00	35,00

newtreat = new treatment
oldtreat = old treatment
agecats = age categories
patientno = patient number

Start by opening the data file in your computer with SPSS installed.

Command:
Analyze....Regression....Linear....Dependent: enter new laxative....Independent: enter old laxative....click OK.

Model Summary

Model	R	R Square	Adjusted R Square	Std. Error of the Estimate
1	,335[a]	,113	,086	9,53445

a. Predictors: (Constant), old treatment

ANOVA[a]

Model		Sum of Squares	df	Mean Square	F	Sig.
1	Regression	380,283	1	380,283	4,183	,049[b]
	Residual	2999,888	33	90,906		
	Total	3380,171	34			

a. Dependent Variable: new treatment

b. Predictors: (Constand), old treatment

Coefficients[a]

Model		Unstandardized Coefficients		Standardized Coefficients	t	Sig.
		B	Std. Error	Beta		
1	(Constant)	28,521	1,951		14,616	,000
	old treatment	,137	,067	,335	2,045	,049

a. Dependent Variable: new treatment

The above tables are in the output sheets. The old treatment is a borderline significant predictor at p = 0.049 of the old treatment. More statistical power is desirable. Instead of traditional linear regression also a GENLIN (Generalized linear regression-generalized linear regression) procedure can be followed in SPSS using maximum likelihood estimators instead of traditional F- and t-tests. The results are likely to produce a bit better precision.

Command:

Generalized Linear Models....Generalized Linear Models....mark: Custom.... Distribution: select Normal....Link function: select identity....Response: Dependent Variable: enter new treatment....Predictors: Factors: enter old treatment....Model: Model: enter oldtreat....Estimation: mark Model-based Estimator....click OK.

The underneath table is in the output.

Parameter Estimates

Parameter	B	Std. Error	95% Wald Confidence Interval		Hypothesis Test		
			Lower	Upper	Wald Chi-Square	df	Sig.
(Intercept)	40,000	4,2650	31,641	48,359	87,958	1	,000
[oldtreat=4,00]	-28,000	6,0317	-39,822	-16,178	21,550	1	,000
[oldtreat=5,00]	-27,500	5,2236	-37,738	-17,262	27,716	1	,000
[oldtreat=6,00]	-24,500	5,2236	-34,738	-14,262	21,999	1	,000
[oldtreat=7,00]	-12,667	4,9248	-22,319	-3,014	6,615	1	,010
[oldtreat=8,00]	-15,500	4,7684	-24,846	-6,154	10,566	1	,001
[oldtreat=9,00]	-8,500	5,2236	-18,738	1,738	2,648	1	,104
[oldtreat=10,00]	-8,000	4,7684	-17,346	1,346	2,815	1	,093
[oldtreat=11,00]	-,500	5,2236	-10,738	9,738	,009	1	,924
[oldtreat=12,00]	-4,000	5,2236	-14,238	6,238	,586	1	,444
[oldtreat=13,00]	-7,000	4,9248	-16,652	2,652	2,020	1	,155
[oldtreat=14,00]	-3,000	6,0317	-14,822	8,822	,247	1	,619
[oldtreat=15,00]	-5,500	4,7684	-14,846	3,846	1,330	1	,249
[oldtreat=16,00]	3,000	6,0317	-8,822	14,822	,247	1	,619
[oldtreat=18,00]	7,000	6,0317	-4,822	18,822	1,347	1	,246
[oldtreat=19,00]	4,000	6,0317	-7,822	15,822	,440	1	,507
[oldtreat=112,00]	-1,000	6,0317	-12,822	10,822	,027	1	,868
[oldtreat=114,00]	0[a]	.	.	.	.	.	.
(Scale)	18,190[b]	4,3484	11,386	29,062			

Dependent Variable: new treatment
Model: (Intercept), oldtreat

a. Set to zero because this parameter is redundant.

b. Maximum likelihood estimate.

A better precision can be obtained by the use of robust standard errors, i.e., Hubert-White estimators for linear models.

Command:
Generalized Linear Models....Generalized Linear Models....mark: Custom.... Distribution: select Normal....Link function: select identity.... Response: Dependent Variable: enter new treatment....Predictors: Factors: enter old treatment....Model: Model: enter newtreat....Estimation: mark Robust Estimator....click OK.

The underneath table is in the output.

Parameter Estimates

Parameter	B	Std. Error	95% Wald Confidence Interval		Hypothesis Test		
			Lower	Upper	Wald Chi-Square	df	Sig.
(Intercept)	40,000	.	.	.	.	1	,000
[oldtreat=4,00]	-28,000	.	.	.	.	1	,000
[oldtreat=5,00]	-27,500	1,0607	-29,579	-25,421	672,222	1	,000
[oldtreat=6,00]	-24,500	3,8891	-32,122	-16,878	39,686	1	,000
[oldtreat=7,00]	-12,667	3,4102	-19,351	-5,983	13,796	1	,000
[oldtreat=8,00]	-15,500	1,4361	-18,315	-12,685	116,485	1	,000
[oldtreat=9,00]	-8,500	5,3033	-18,894	1,894	2,569	1	,109
[oldtreat=10,00]	-8,000	2,1506	-12,215	-3,785	13,838	1	,000
[oldtreat=11,00]	-,500	1,0607	-2,579	1,579	,222	1	,637
[oldtreat=12,00]	-4,000	.	.	.	.	1	,000
[oldtreat=13,00]	-7,000	2,4495	-11,801	-2,199	8,167	1	,004
[oldtreat=14,00]	-3,000	.	.	.	.	1	,000
[oldtreat=15,00]	-5,500	3,4369	-12,236	1,236	2,561	1	,110
[oldtreat=16,00]	3,000	.	.	.	.	1	,000
[oldtreat=18,00]	7,000	.	.	.	.	1	,000
[oldtreat=19,00]	4,000	.	.	.	.	1	,000
[oldtreat=112,00]	-1,000	.	.	.	.	1	,000
[oldtreat=114,00]	0[a]	.	.	.	.	.	.
(Scale)	18,190[b]	4,3484	11,386	29,062			

Dependent Variable: new treatment
Model: (Intercept), oldtreat

a. Set to zero because this parameter is redundant.

b. Maximum likelihood estimate.

Out of the stool scores with old treatment, 6 scores produced p-values of < 0,05 with the Model-based Estimator, while it produced up to 14 p-values < 0,05 with the Robust Estimator. Most of the p-values even as small as < 0,0001.

If your results are borderline significant like in the above case, then loglikelihood regression testing and robust regression testing can provide you with better statistics, and, thus, better statistical power of testing, than traditional testing. More information is given in SPSS for starters and 2nd levelers 2nd edition, the Chap. 29, Springer Heidelberg Germany, 2016, from the same authors.

We should add, that robust regressions in SPSS can also be performed with binary outcomes, like the data used as an example in the Chaps. 2 and 4. The GEE (Generalized linear regression- generalized estimation equations) model is needed. With mixed models (Chap. 8) the GENLINMIXED procedure is possible following the commands: Analyze....Mixed Models....Generalized Linear....

The "complex samples" methodology is a weighting procedure for obtaining unbiased population estimates. It is not covered in the current edition, but more information is given in the edition Machine learning in medicine a complete overview, Springer Heidelberg Germany, 2015, from the same authors. Also "complex samples" can be assessed in SPSS statistical software with robust regressions. You need to use the CSGLM, CSLOGISTIC and CSCOXREG (CS = complex samples) procedures in the appropriate modules.

5   Conclusion

In this chapter various special forms of regression models for continuous outcomes alternative to the traditional ordinary least squares model are given. First, kernel regression is reviewed. It is equivalent to radial basis function analysis. Kernel frequency distributions consist of multiple identical Gaussian curves rather than histograms consistent of bins with different lengths. Kernel regression measures the relationship between x and y data, where the expectation of y is conditional not on all x-values, but on locally weighted averages of subsets of consecutive x-values within a bandwidth. And, together, they tend to produce a simple linear regression model. Any linear regression can be replaced with a kernel regression, but the method is particularly appropriate in case of seemingly nonlinear patterns, that, in the end, are linear after all. In skewed data files a (much) better fit for the data is provided by kernel regression than for linear regression.

Second, gamma and Tweedie regressions are worthwhile analysis models complementary to linear regression, that may elucidate effects unobserved in the traditional linear model. The marginal means procedure enables to observe trends in the data, e.g., decreasing outcome score with increasing predictor scores. In the example given, Tweedie regression provided a somewhat better sensitivity of testing than gamma regression did.

Third, robust regressions are explained. If, with linear regression models, your results are borderline significant like in the above case, then loglikelihood regression testing and robust regression testing may provide you with better statistics, and, thus, better statistical power of testing, than traditional testing may do.

Reference

To readers requesting more background, theoretical and mathematical information of computations given, several textbooks complementary to the current production and written by the same authors are available: Statistics applied to clinical studies 5th edition, 2012, Machine learning in medicine a complete overview, 2015, SPSS for starters and 2nd levelers 2nd edition, 2015, Clinical data analysis on a pocket calculator 2nd edition, 2016, Understanding clinical data analysis, 2017, all of them edited by Springer Heidelberg Germany.

Chapter 20
Regressions for Quantitative Diagnostic Testing

Accounting Uncertainties of Both Dependent and Independent Variable

Abstract Linear regressions are applied for validating quantitative diagnostic tests, but traditional analysis of variance is not good enough for the purpose. Also, traditional linear regression assumes, that, either dependent or independent variable comes with an amount of residual uncertainty. Pretty commonly a novel treatment is tested against a standard treatment, otherwise called the gold standard, which is measured without error. However, in clinical chemistry and laboratory medicine, there may be no gold standard. Not only the x-, but also the y-variables, have been measured with a (same) amount of uncertainty. Traditional linear regression is, then, not appropriate, and regression analyses have to include the uncertainty of both the y- and x-variable. Two methods for the purpose are Deming regression and Passing-Bablok regression. Particularly, the latter of the two has many advantages. It is non-parametric, which means, that non - normal distributions of the data are no problem. It also is robust, because medians rather than means are applied, and, so, outliers are no problem either, and overdispersion is, thus, taken into account.

Keywords Regressions for quantitative diagnostic testing · Accounting uncertainties both dependent and independent ones · Deming regressions · Passing-Bablok regression

1 Introduction, History, and Background

Traditional linear regression assumes, that either dependent or independent variable, comes with an amount of residual uncertainty. Usually, it is reasoned, that the independent x-variable is estimated without error. With quantitative diagnostic testing, it is often reasoned, that a new diagnostic has the error, while the gold standard test is, simply, the complete truth, and they are computed as, respectively, y- and x-variables. In clinical research, particularly, clinical chemistry and laboratory medicine, there may thus be no gold standard. Not only the x-, but also the y-variables have been measured with a (same) amount of uncertainty. Traditional linear regression is, then, not appropriate, and regression analyses have to include the uncertainties of both the y- and x-variable. Two methods for the purpose are Deming

T. J. Cleophas, A. H. Zwinderman, *Regression Analysis in Medical Research*,
https://doi.org/10.1007/978-3-030-61394-5_20

regression and Passing-Bablok regression. The current chapter shows regressions for validating quantitative diagnostic tests including linear regressions, Deming regressions, and Passing-Bablok regressions.

Deming regression is named after W. Edwards Deming, 1900–1993, statistics professor, Sioux City Iowa. It is an errors-in-variables model which tries to find the best fit regression line for a two-dimensional dataset. It differs from the simple linear regression in that it accounts for errors in observations on both the x- and the y- axis. It is a special case of total least squares, which allows for any number of predictors and a more complicated error structure. Deming regression is equivalent to the maximum likelihood estimation of an errors-in-variables model in which the errors for the two variables are assumed to be independent and normally distributed, and the ratio of their variances, denoted δ, is known. In practice, this ratio might be estimated from related data-sources; however the regression procedure takes no account for possible errors in estimating this ratio. The Deming regression is only slightly more difficult to compute compared to the simple linear regression. Most statistical software packages used in clinical chemistry offer Deming regression. The model was originally introduced by Adcock (Ann Math, 1878; 6: 97) who considered the case $\delta = 1$, and then more generally by Kummell (Ann Math 1879; 6: 97) with arbitrary δ. However their ideas remained largely unnoticed for more than 50 years, until they were revived by Koopmans (in Linear regression analysis of economic time series, 1937, edited by De Erven Bohn, Haarlem, Neth), and later propagated even more by Deming (1943). The latter book became so popular in clinical chemistry and related fields, that the method was even dubbed *Deming regression* in those fields.

Passing–Bablok regression is named after Passing, a mathematician at Hoechst AG, Frankfurt Germany, and Bablok, a mathematician at Boehringer, Mannheim Germany. It is a statistical method for non-parametric regression analysis suitable for testing comparison studies, and was published by the authors in J Clin Chem Biochem in 1983; 21: 709. The procedure is symmetrical and is robust in the presence of one or few outliers. The Passing-Bablok procedure fits the parameters a and b of the linear equation y = a + b x using non-parametric methods. The coefficient b is calculated by taking the shifted median of all slopes of the straight lines between any two points, disregarding lines for which the points are identical or b = −1. The median is shifted, based on the number of slopes where b < −1 to create an unbiased estimator. The parameter a is calculated from a = median (y_i -bx_i). Passing and Bablok define a method for calculating a 95% confidence interval (CI) for both a and b in their original paper, though bootstrapping the parameters is the preferred method for in vitro diagnostics (IVD) when using patient samples. The Passing-Bablok procedure is valid only when a linear relationship exists between x and y, which can be assessed by a cusum test. A cusum (cumulative sum control) test is a test used to monitor whether a process is drifting away from its mean. The cusum chart is centered around the mean value of a process. The process is said to be "out of control" if the cumulative sums of the standardized deviations exceed a specified range. It was proposed by Page, a statistician from Cambridge UK in Biometrica 1954). The results from the cusum test are here interpreted as follows. If 0 is in the CI of a, and 1 is in the CI of b, the two methods will be comparable within the investigated concentration range. If 0 is not in the CI of a, there will be a systematic difference, and if 1 is not in the CI of b, then there will be a proportional difference between the two methods. However, the use of Passing-Bablok regression in method comparison studies has been criticized, because it ignores random differences between methods to be compared.

2 Example

With linear regression the x-values are, traditionally, used to predict the y-values. Ordinary least squares (OLS) are used for finding the best fit regression line, and analysis of variance are used to test, whether the y-values are closer to the best fit regression line than could happen by chance (see Chap. 1). These computations are not good enough for validating a quantitative diagnostic test, because there is too much spread in the y-data even with a very significant analysis of variance.

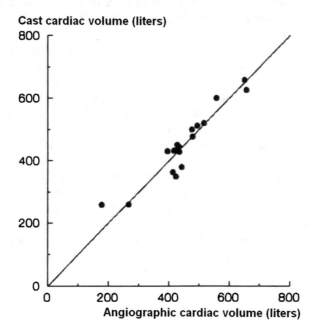

In the above example the data are close to the "b = 1.000 and a = 0.000 line", otherwise called the identity line.

The 95% confidence interval of the b − value is between 0.917 ± 1.96 × 0.083,

is between 0.751 and 1.083,

and it, thus, contains the value 1.000.
The 95%confidence interval of the a − value is between 39.340 ± 1.96 × 38.704,

is between − 38.068 and 116.748,

and it, thus, contains the value 0.000.

The diagnostic test is validated, because the 95% confidence interval of the a- and the b-values contain respectively the values 0.000 and 1.000. And so, this method for validating is almost OK. However, there is still a slight problem, if we reason, that not only the y-values but also the x-values are measured with an amount of uncertainty. Unfortunately, with quantitative diagnostic testing, it is often reasoned, that not only the x- but also the y-variables have been measured with a (same) amount of uncertainty. Traditional linear regression is, then, not appropriate, and regression analyses have to include the uncertainty of both the x- and y-variables. Two methods for the purpose are Deming regression and Passing-Bablok regression. Particularly, The latter of the two has many advantages. It is non-parametric, which means that nonnormal distributions of the data are no problem. It also is robust, because medians rather than means are applied, and so outliers are no problem either, and overdispersion is taken into account.

If we want to account the uncertainty of a control test, which is not a gold standard test, then a better approach will be to test both the new and the control test against the gold standard test. This will unmask which of the two tests performs better. In the situation where there is no certain gold standard test and where it is decided to account uncertainty of the control test to be used, Deming and Passing-Bablok regression are sometimes used instead. They are methods based on linear regression and mathematically more complex than simple linear regression. Deming regression, just like the paired t-test and the Altman-Bland plot, assumes normal distributions of the subsequent x- and y-data. In contrast, Passing-Bablok regression does not. It is a non-parametric method using the Kendall's rank-correlation test to assess the above-described hypotheses that b = 1.000, and a = 0.000. First, one should produce a ranked sequence of all possible slope-values between two x and two y-values (Sij values). We, then, compare the Sij values >1 with those <1, and test, whether there is a significant difference using Kendall's standard error equation, SE (standard error) = √ n(n-1) (2n + 5) /18 with n = number of paired values. If, after continuity correction (add −1 to the difference as calculated), the SE is smaller than half the size of the calculated difference, then the b-value is not significantly different from 1.000. The a-value is calculated from the medians of x and y using the calculated b, its SE from the upper and lower limit of the confidence intervals of the b-value. The method is laborious, particularly, with large samples, but available through S-plus, Analyse-it, EP Evaluator, and MedCalc and other software programs.

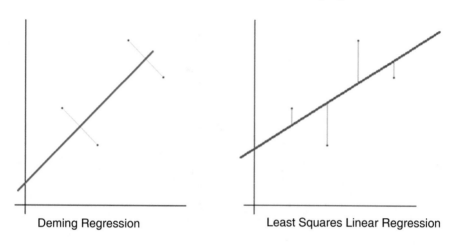

Deming Regression   Least Squares Linear Regression

Traditional linear regression finds the best fit regression line by minimizing the sum of the least squared distances, otherwise called residuals, from the points to the best fit line. The above right graph give a simple example. Deming regression works slightly different. It finds the residuals from the best fit Deming-regression line as the perpendicular distances from the points to their fitted values on the line. The above left graph gives a simple example.

Deming and Passing-Bablok regressions are mathematically more complex than linear regression. Fortunately, the internet provides automatic calculators both for Deming and for Passing-Bablok regressions. We should add, that Deming regression is parametric, and, in addition, it includes uncertainties of both the x-and y-variables.

3 Deming Regression

The data from the first graph in Sect. 2 are given.

494	512
395	430
516	520
434	428
476	500
557	600
413	364
442	380
650	658
433	445
417	432
656	626
267	260
478	477
178	259
423	350
427	451

Many regression calculators are available either online or at scientific pocket calculators. They will produce the regression equation y = 0,9709 x + 15,2316.

If you wish information on the spread, then the regression equation is not enough. We will use Med Calc Statistical Software for the purpose. It can be downloaded very easily at www.medcalc.org. It can be used for free for a period of 6 weeks. Once downloaded, commands can be readily given. MedCalc requires coefficients of variation (CVs). CV = 100 x (mean / standard deviation) %.

Command
click file....enter data in the data screen in the row under **A** and **B**....under A fill out "cast"....under B fill out "angio"....click Statistics....click Method comparison & evaluation....click Deming regression....method y: enter cast....CV: enter 26%.... method x: enter angio....CV: enter 25%....click OK.

The underneath table and graphs is in the output. It is shown, that the 95% confidence intervals of the intercept contains the value 0, and the 95% confidence intervals of the slope contains the value 1. We can conclude, that angiography is a valid method for predicting cardiac volumes.

Method X	angio
Method Y	cast

Method	Mean	Coefficient of variation (%)
X	452,4706	25,00
Y	450,3529	26,00

Sample size	17
Variance ratio	0,9333

Regression Equation

y = -14,7220 + 1,0279 x

Parameter	Coefficient	Std. Error	95% CI
Intercept	-14,7220	66,9929	-156,7405 to 127,2966
Slope	1,0279	0,1338	0,7442 to 1,3115

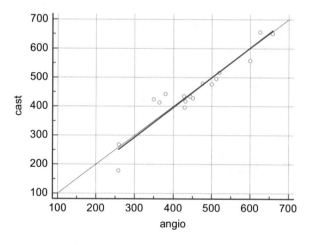

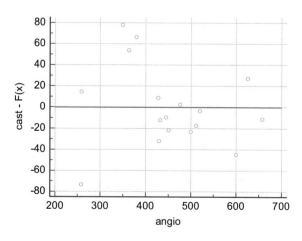

4   Passing-Bablok Regression

The above example is used once more. We will apply Med Calc Statistical Software again.

Command

click file....enter data in the data screen in the row under **A** and **B**....under A fill out "cast"....under B fill out angio....click Statistics....click Method comparison & evaluation....click Passing & Bablok regression....click OK.

The underneath table and graphs is in the output. Again, the 95% confidence intervals of the intercepts contain the value 0, and 95% confidence intervals of the slope contains the value 1. Also Passing-Bablok regression demonstrates, that angiography is a valid predictor of cardiac volumes.

Sample size	17

	Variable X	Variable Y
Lowest value	259,0000	178,0000
Highest value	658,0000	656,0000
Arithmetic mean	452,4706	450,3529
Median	445,0000	434,0000
Standard deviation	113,1151	116,3126
Standard error of the mean	27,4344	28,2099

Regression Equation

$y = 22{,}826087 + 0{,}939130\ x$	
Systematic differences	
Intercept A	22,8261
95% CI	-99,0805 to 127,4545
Proportional differences	
Slope B	0,9391
95% CI	0,7159 to 1,1946
Random differences	
Residual Standard Deviation (RSD)	29,2596
± 1.96 RSD Interval	-57,3488 to 57,3488
Linear model validity	
Cusum test for linearity	No significant deviation from linearity (P=0,95)

Spearman rank correlation coefficient

Correlation coefficient	0,900
Significance level	P<0,0001
95% CI	0,738 to 0,964

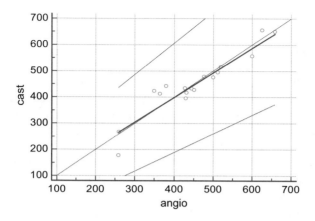

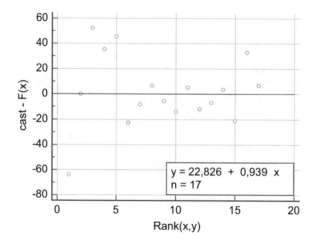

The Passing-Bablok regression equation is different from the Deming regression equation. The Passing-Bablok regression uses rankcorrelation testing, and is statistically significant just like the traditional OLS regression analysis is.

5 Conclusion

Traditional linear regression assumes, that, either dependent or independent variable, comes with an amount of residual uncertainty. In clinical research, particularly, clinical chemistry and laboratory medicine, there is no gold standard. Not only the x- but also the y-variables have been measured with a (same) amount of uncertainty. Traditional linear regression is, then, not appropriate, and regression analyses have to include the uncertainty of the y-variable. Two methods for the purpose are Deming regression and Passing-Bablok regression. Particularly, The latter of the two has many advantages. It is non-parametric, which means that nonnormal distributions of the data are no problem. It also is robust, because medians rather than means are applied, and so outliers are no problem either, and overdispersion is taken into account. Instead of the "Spearman-rho" rank correlation testing, the Kendall-Tau rank correlation test can be used. It does provide slightly better statistics (Clinical data analysis on a pocket calculator 2nd edition, Chap. 9, Springer Heidelberg Germany, 2016, from the same authors). It is not like Spearman based on the rho-squared-value = "fraction of variance explained by the regression"methodology. But it performs better with nonlinear data. Usually Passing-Bablok software uses unfortunately Spearman tests, despite the notorious nonlinearity of laboratory data files.

References

To readers requesting more background, theoretical and mathematical information of computations given, several textbooks complementary to the current production and written by the same authors are available: Statistics applied to clinical studies 5th edition, 2012, Machine learning in medicine a complete overview, 2015, SPSS for starters and 2nd levelers 2nd edition, 2015, Clinical data analysis on a pocket calculator 2nd edition, 2016, Understanding clinical data analysis, 2017, all of them edited by Springer Heidelberg Germany

We should add, that the Med Calc Statistical Software used in this chapter is provided by Microsoft Partner located at Ostend Belgium. It is a member of the International Association of Statistical Computing founded in London 1885 by 81 statisticians, current address ISI Permanent Office The Hague Netherlands

Chapter 21
Regressions, a Panacee or at Least a Widespread Help for Clinical Data Analyses

Exploring Changes in Experimental Settings, Sensitivity Assessments, Categorical Outcomes and Predictors, Goodness of Qualitative Diagnostic Tests, Heteroscedasticity Issues

Abstract In current clinical trials, like pharmaceutical trials, data driven decisions are considered the most important decisions, and regression analysis is the most frequently used methodology for the purpose. Why so? First, regression analysis is not hypothetical, but data driven. Second, it assesses, which factors matter, and which do not. Third, it assesses and quantifies error terms, and does not necessarily assess causality. However, regression analyses easily include mistakes. For example, regression analyses are, virtually always, at least partly explorative, and, therefore, carry the risk of fishing data, and bad data entering the analysis. Nonetheless, regressions are a widespread help for data analyses. This chapter gives examples of: how regressions help make sense of the effects of small changes in experimental settings, how they can assess the sensitivity of your predictors, how they can produce meaningful analyses even with multiple exposure and outcome categories, how they are used for assessing the goodness of your novel diagnostic tests, how they can help you find out about data subsets with unusually large variance spread. Methods for adjusting unusually large variance spread are given as well, and they include.

Keywords Exploring changes in experimental settings · Sensitivity assessments · Categorical outcomes and predictors · Goodness of qualitative diagnostic tests · Heteroscedasticity issues · Maximum likelihood estimations · Autocorrelations · Weighted least squares · Two stage least squares · Robust standard errors · Generalized least squares

Electronic Supplementary Material The online version of this chapter (https://doi.org/10.1007/978-3-030-61394-5_21) contains supplementary material, which is available to authorized users. The videos can be accessed by scanning the related images with the SN More Media App.

1   Introduction, History, and Background

In current clinical trials, like pharmaceutical trials, data driven decisions at work are considered the most important types of decisions, and regression analysis is the most frequently used methodology for the purpose (Alemayehu et al., Data driven approach to quality care, Perspect Clin Res 2013; 4: 221–6). However, today, just like 50 years ago, "garbage in garbage out" is still true (Gallo, Harvard Business Review, November 2015). Regression analysis may still be the most important type of data analysis. Why so?

- First, it is not hypothetical but data driven.
- Second, it assesses, which factors matter, and which ones do not.
- Third, it assesses and quantifies error terms, and does not necessarily assess causality, although, with structural equation modelings (described in the Chap. 12), causality is close by.

However, regression analyses easily include mistakes as well. Regression analyses are, virtually always, at least partly explorative, and, therefore, carry the risk of fishing data, and bad data entering your analysis. These data may be exciting up to a level, that makes you forget about the best functions of your brain, meaning functions like common sense reasoning, intuition, gut feeling, functions that should be on top of your data and mathematical models. Nonetheless, regressions are a widespread help for data analyses. This chapter will give examples of:

- how regressions help you make sense of the effects of small changes in experimental settings,
- how they can assess the sensitivity of your predictors,
- how they can produce meaningful analyses even with multiple exposure and outcome categories,
- how they are used for assessing the goodness of your novel diagnostic tests,
- how they can help you find out about data subsets with unusually large spread.

Particularly, the last point has been given much attention lately. Ordinary least squares (OLS) has traditionally been the basis for regression analyses (Chaps. 1 and 2) under the assumption of homoscedasticity (a single standard deviation for all of the outcome data). A problem with current regression data is, that they are increasingly complex, and contain data subsets with unusually large spread. Variances of the outcome data are not homogeneous anymore. And it is time, that we considered methods for taking these heterogeneities of variances into account. Modern regression analyses can now be performed with alternative computational methods, adjusting for variance heterogeneities, such as robust regressions, auto-regressions, weighted least squares, multiple stage least squares, and generalized least squares. All of this is at the dawn of an entirely new and more appropriate world of data analysis.

Traditional clinical research is, rather than hypothesis-generating (A), preferred to be hypothesis-confirming (B). The latter addresses the field of controlled clinical trials, the former that of observational studies, while combination of the two, as observed in the science lines at investigation centers, is called hypothesis driven clinical research. Part A has applied regression analyses all the time, while part B has mainly involved prospective randomized double blind clinical trials, until recently. As will be addressed in the next Chap., the European Medicine Administration EMA), with headquarters in Amsterdam Netherlands, has implicated the limited use of regression analyses into the statistical analysis of controlled clinical trials. Another issue is that of pursuing hypothesis generation and confirmation in a single research project with one study subsequent to the other. Such projects are recognized to rapidly include data-driven means of progress of science. The adjective *data-driven* here means, that progress in a scientific activity is compelled by data, rather than by intuition or by personal experience. Data-driven may here refers to all of the underneath points.

- Data-driven programming (computer programming in which program statements describe data to be matched and the processing required).
- Data-driven testing (computer software testing done using a table of conditions directly as test inputs and verifiable outputs).
- Data-driven learning (a learning approach driven by research-like access to data).
- Data-driven science (an interdisciplinary field of scientific methods to extract knowledge from data).

How much the term has penetrated into some field of clinical research, is illustrated in the underneath example from the field of genomic research. Genomics has profoundly changed biology by grand-scaling data acquisition, which has provided researchers with the opportunity to interrogate biological issues in novel and creative ways. No longer constrained by low-throughput assays, researchers have developed hypothesis-generating approaches in order to understand the molecular basis of nature, both normal and pathological. The paradigm of hypothesis-generating research does not replace or undermine hypothesis-testing modes of research; instead, it complements them and has facilitated discoveries that may not have been possible with hypothesis-testing research. The hypothesis-generating mode of research has been primarily practiced in basic science, but has recently been extended to clinical-translational work as well. Just as in basic science, this approach to research can facilitate insights into human health and disease mechanisms, and provide the crucially needed data set of the full spectrum of, for example, genotype–phenotype correlations. Finally, the paradigm of hypothesis-generating research is conceptually similar to the underpinning of predictive genomic medicine, which has the potential to shift medicine from a primarily population- or cohort-based activity to one that, instead, uses individual susceptibility, and prognostic and pharmacogenetic profiles for maximizing the efficacy, and minimizing the iatrogenic effects of medical interventions.

A more general example of the data-driven approach to clinical research is in the above recent-article of Alemayehu et al. (A data-driven approach to quality risk management, Ind Soc Clin Res 2013; 4: 2013). The authors demonstrated, that an effective clinical trial strategy to ensure trial safety, as well as trial quality and efficacy assessment, involved an integrated approach, including a prospective identification of risk factors, and a mitigation of risks through proper study design and trial execution, and through the assessment of quality metrics in real-time. Such an integrated quality management plan is also enhanced by the use of data-driven techniques for indentifying the risk factors, that are most relevant to the predictive issues of a trial. Using a meta-analysis of actual controlled trials, the above authors demonstrated, that the use of traditional statistical significance tests like Wilcoxon rank-sum tests and logistic regressions were helpful to represent associations between risk factors, and the significant presence of trial quality issues. Such issues included the presence of a placebo, as well as over 25 planned study procedures, like, for example, package labeling, and complex dosage schedules.

   We should emphasize the difference between the terms data-driven and data-informed. The former term sensu strictu only includes the brainless activity of data cumulation, while the latter is probably much better, because it includes activities of the human brain, and includes checks based on understanding and/or intuition. The current Chap. will give additional examples of regression methodologies helpful to the latter issue.

2   How Regressions Help You Make Sense of the Effects of Small Changes in Experimental Settings

As an example, the air quality of operation rooms will be used. It is important for infection prevention. Particularly, the factors (1) humidity (30–60%), (2) filter capacity (70–90%), and (3) air volume change (20–30% per hour) are supposed to be important determinants. Can evolutionary operation (Evop) methodology be used for process improvement. Eight operation room air condition settings were investigated, and the results are underneath.

Operation Setting	humidity (30% = 1, 60% = 4)	filter capacity (70% = 1, 90% = 3)	air volume change (20% = 1, 30% = 3)	infections number of
1	1	1	1	99
2	2	1	1	90
3	1	2	1	75
4	2	2	1	73
5	1	1	2	99
6	2	1	2	99
7	1	2	2	61
8	2	2	2	52

We will use linear regression in SPSS with the number of infections as outcome and the three factors as predictors to identify the significant predictors. First, the data file available as "chap21evops" is in extras.springer.com. It has been previously used by the authors in the Machine learning in medicine a complete overview, Chap.69, Springer Heidelberg Germany, 2015. Start by opening the data file in your computer installed with SPSS statistical software.

Command

Analyze....Regression....Linear....Dependent: enter "Var00004".... Independent(s): enter "Var00001–00003)"....click OK.

The underneath table in the output shows that all of the determinants are statistically significant at $p < 0.10$. A higher humidity, filtering level, and air volume change will better prevent infections.

Coefficients[a]

Model		Unstandardized Coefficients		Standardized Coefficients	t	Sig.
		B	Std. Error	Beta		
1	(Constant)	103,250	18,243		5,660	,005
	hunidity1	-12,250	3,649	-,408	-3,357	,028
	filter capacity1	-21,250	3,649	-,707	-5,824	,004
	airvolume change1	15,750	3,649	,524	4,317	,012

a. Dependent Variable: infections1

Subsequently, higher settings can be tested until no further improvements of infection prevention is obtained. This method helps you make sense of the effects of small changes in experimental settings, and ultimately helps you find the optimal setting, i.e. the setting, that will produce the best result at lowest costs.

3   How Regressions Can Assess the Sensitivity of your Predictors

A sensitive study with multiple predictors will produce virtually the same result, if one or more predictors are removed. We say, the study is robust against leaving out one or two predictors out of say 12 predictors. As an example, we will use the data from the example from the Chap. 13 once more. The underneath table show the first 13 patient data of the 250 patient data file including 16 variables, G1 to G 27 were the predictors, highly expressed gene scores predicting clinical outcomes, O1 to O4 were outcome variables, efficacy scores of cytostatic treatments. The entire data file is in extras.springer.com, and is entitled "chap13manova". It was previously used in the Chap. 13 of this edition.

G1	G2	G3	G4	G16	G17	G18	G19	G24	G25	G26	G27	O1	O2	O3	O4
8	8	9	5	7	10	5	6	9	9	6	6	6	7	6	7
9	9	10	9	8	8	7	8	8	9	8	8	8	7	8	7
9	8	8	8	8	9	7	8	9	8	9	9	9	8	8	8
8	9	8	9	6	7	6	4	6	6	5	5	7	7	7	6
10	10	8	10	9	10	10	8	8	9	9	9	8	8	8	7
7	8	8	8	8	7	6	5	7	8	8	7	7	6	6	7
5	5	5	5	5	6	4	5	5	6	6	5	6	5	6	4
9	9	9	9	8	8	8	8	9	8	3	8	8	8	8	8
9	8	9	8	9	8	7	7	7	7	5	8	8	7	6	6
10	10	10	10	10	10	10	10	10	8	8	10	10	10	9	10
2	2	8	5	7	8	8	8	9	3	9	8	7	7	7	6
7	8	8	7	8	6	6	7	8	8	8	7	8	7	8	8
8	9	9	8	10	8	8	7	8	8	9	9	7	7	8	8

Traditional linear regression in SPSS statistical software, installed in your computer, with the summary O1 to O4 as outcome and the gene scores as predictors was performed.

Command

Analyze....Regression....Linear....Dependent: enter Summary outcome.... Independent: enter G1 to G 27....click OK.

The output sheets show the results table given below.

Coefficients^a

Model		Unstandardized Coefficients		Standardized Coefficients	t	Sig.
		B	Std. Error	Beta		
1	(Constant)	3.293	1.475		2.232	.027
	geneone	-.122	.189	-.030	-.646	.519
	genetwo	.287	.225	.078	1.276	.203
	genethree	.370	.228	.097	1.625	.105
	genefour	.063	.196	.014	.321	.748
	genesixteen	.764	.172	.241	4.450	.000
	geneseventeen	.835	.198	.221	4.220	.000
	geneeighteen	.088	.151	.027	.580	.563
	genenineteen	.576	.154	.188	3.751	.000
	genetwentyfour	.403	.146	.154	2.760	.006
	genetwentyfive	.028	.141	.008	.198	.843
	genetwentysix	.320	.142	.125	2.250	.025
	genetwentyseven	-.275	.133	-.092	-2.067	.040

a. Dependent Variable: summaryoutcome

The table shows, that 6 of the 12 predictor variables were significant predictors of the summary outcome with p-values 0.04 or less. We will now investigate, whether this analysis is sensitive against leaving out one or two predictors (independent variables). This would mean, that, after removing these one or two from the analysis, the remainder can still be used as a valid diagnostic test for the complete set of predictors. First, genetwentyseven is left out.

Command
Analyze....Regression....Linear....Dependent enter summaryoutcome.... Independent (s) enter all of the predictors except genetwentyseven....click OK.

Coefficients[a]

Model		Unstandardized Coefficients		Standardized Coefficients	t	Sig.
		B	Std. Error	Beta		
1	(Constant)	2,914	1,474		1,977	,049
	geneone	-,118	,190	-,029	-,618	,537
	genetwo	,257	,226	,070	1,136	,257
	genethree	,443	,227	,116	1,954	,052
	genefour	-,011	,194	-,003	-,058	,954
	genesixteen	,810	,171	,256	4,730	,000
	geneseventeen	,788	,198	,209	3,980	,000
	geneeighteen	,045	,151	,014	,298	,766
	genenineteen	,575	,155	,188	3,720	,000
	genetwentyfour	,330	,143	,126	2,315	,021
	genetwentyfive	-,025	,139	-,008	-,180	,857
	genetwentysix	,284	,142	,111	2,001	,047

a. Dependent Variable: summaryoutcome

The above table is in the output. Genetwentyseven is no more present. We will now perform a simple one by one linear regression of the standardized regression coefficients of the above two tables. First enter in a new data view the standardized coefficients of the above two tables as respectively the variable 1 and 2.

Command

Analyze....Regression....Linear....Dependent enter variable 1....Independent (s) enter variable 2...click OK.

Coefficients[a]

Model		Unstandardized Coefficients		Standardized Coefficients	t	Sig.
		B	Std. Error	Beta		
1	(Constant)	,012	,006		1,916	,088
	VAR00002	,947	,046	,990	20,604	,000

a. Dependent Variable: VAR00001

The 95% confidence intervals of B should not cross zero, that of B (VAR00002) should not cross one (unit).

$$B \text{ (Constant)} = ,012 \pm 2 \times ,006$$

$$= \text{between}, 000 \text{ and}, 024$$

$$B \text{ (VAR00002)} = ,947 \pm 2 \times ,046$$

$$= \text{between}, 855 \text{ and } 1,039$$

This means, that B (Constant) does contain 0, and B (VAR00002) contains 1, and, that the regression model as applied is sensitive (robust) against leaving out the predictor variable genetwentyseven.

All of the predictor variables can be assessed similarly, and all of them are similarly sensitive. However, when we leave out two of the most important predictors, namely the ones with the largest standardized betas (the genesixteen and geneseventeen), the remainder of the model may not be sensitive anymore. We will use the data file from the example once more.

Command

Analyze....Regression....Linear....Dependent enter summaryoutcome.... Independent (s) enter all of the predictor variables except genesixteen and geneseventeen.... click OK.

Coefficients[a]

Model		Unstandardized Coefficients		Standardized Coefficients	t	Sig.
		B	Std. Error	Beta		
1	(Constant)	6,173	1,594		3,873	,000
	geneone	-,163	,210	-,041	-,777	,438
	genetwo	,427	,248	,116	1,725	,086
	genethree	,567	,253	,149	2,239	,026
	genefour	,062	,218	,014	,283	,778
	geneeighteen	,458	,155	,143	2,957	,003
	genenineteen	,830	,164	,271	5,064	,000
	genetwentyfour	,608	,161	,232	3,789	,000
	genetwentyfive	,032	,158	,010	,203	,839
	genetwentysix	,461	,157	,180	2,931	,004
	genetwentyseven	-,292	,147	-,097	-1,981	,049

a. Dependent Variable: summaryoutcome

The above table is in the output. Genesixteen and seventeen are no more present. We will now perform a simple one by one linear regression of the standardized regression coefficients from the above table versus those of the initial full table. Enter in a new data view the standardized coefficients of the above two tables as respectively the variable 1 and 2.

Then Command
Analyze....Regression....Linear....Dependent enter variable1....Independent (s) enter variable 2....click OK.

Coefficients[a]

Model		Unstandardized Coefficients		Standardized Coefficients	t	Sig.
		B	Std. Error	Beta		
1	(Constant)	-,011	,010		-1,075	,314
	VAR00006	,694	,067	,964	10,325	,000

a. Dependent Variable: VAR00001

The 95% confidence intervals of B (Constant) should not cross zero, that of B (VAR00002) should not cross one (unit).

$$B \text{ (Constant)} = -,011 \pm 2 \times ,010$$

$$= \text{between} -,031 \text{ and}, 009$$

$$B \text{ (VAR00002)} = ,694 \pm 2 \times ,067$$

$$= \text{between}, 560 \text{ and}, 828$$

This mean, that B(Constant) contains 0, but B(VAR00002) does not contain 1, and, that the regression model applied is, thus, not entirely sensitive (robust) against leaving out the predictor variables genesixteen, and geneseventeen.

Sensitivity can be assessed many ways. The current example shows, that multiple predictor models can be pretty robust against slight alterations of the list of independent variables, but in a multiple variables model the most important predictors are almost always lacking robustness, and are, thus, somewhat fragile.

4   How Regressions Can be Used for Data with Multiple Categorical Outcome and Predictor Variables

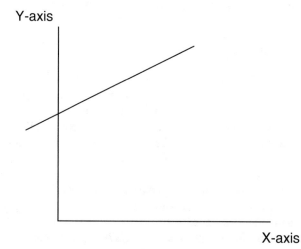

Simple linear regression uses the equation y = a + bx,
Use the equation for make predictions from x values about y values.
If we fill out x-value = 0 = > then the equation turns into y = a

$x = 1 \Rightarrow y = a + b$
$x = 2 \Rightarrow y = a + 2b$

For each x value the equation gives best predictable y-value, all y values constitute a regression line = the best fit for the data (with the shortest distance to the y-values). In this model only one a-value, otherwise called the intercept, exists.

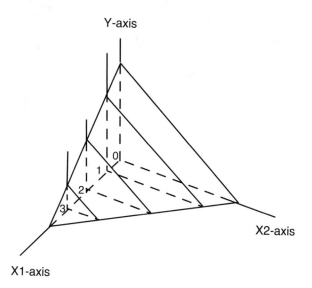

Multiple regression with 3 variables uses the equation $y = a + b_1 x_1 + b_2 x_2$.
We can use a 3-axes-model with y-axis, x_1-axis and x_2-axis:

If we fill out $x_1 = 0$, then the equation turns into $y = a + b_2 x_2$
$x_1 = 1$ $y = a + b_1 + b_2 x_2$
$x_1 = 2$ $y = a + 2b_1 + b_2 x_2$
$x_1 = 3$ $y = a + 3b_1 + b_2 x_2$

Each x_1 –value has its own regression line, all lines constitute a regression plane: this is the best fit plane (with the shortest distance to values in space). All of the regression lines have identical regression coefficients, the b_2 - values, but the intercepts, given by a, $a + b_1$, $a + 2b_1$, $a + 3b_1$, change continually. This is called multiple linear regression, but can also be called *fixed effect* intercept modeling.

In practice, the regression plane has a lot of uncertainty, caused by the standard deviations of x_1 and x_2, and, in addition, some unexplained uncertainty, otherwise called residual uncertainty. With null hypothesis testing, we, usually, test, whether the amount of uncertainty due to the variables x_1 and x_2 is larger than that due to residual uncertainty. This is called a fixed effect intercept testing. Sometimes, a better sensitivity of testing is obtained by a *random effect* intercept modeling or random intercept modeling. Here the standard deviation of x_1 is assumed to be unexpected, not something we could predict, and it is, therefore, added to the residual uncertainty of the model. With truly unexpected effects, often, indeed, a better data fit is obtained, and therefore better p-values of testing. An example will be given. The three rows of graphs giving underneath show histograms of patients by ageclasses. The effect of ageclass on fall out of bed by hospital department is studied.

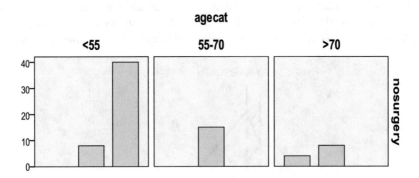

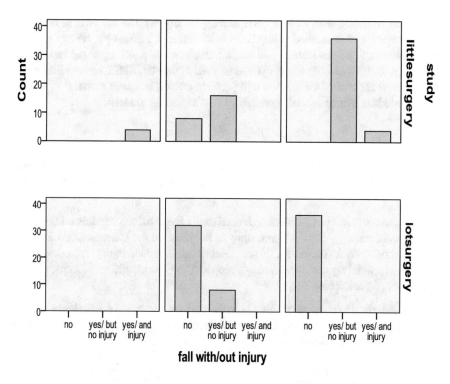

fall with/out injury

The above graphs of bars of counts of falloutofbed shows a lot of heterogeneity. Some cells were empty or nearly empty. The type of patients, from no surgery to a lot of surgery, might be responsible, because surgery in older patients may be accompanied by increased risks of fall out of bed. Therefore, both a fixed and a random intercept analysis for categorical outcome and predictor variables will be performed. With multiple categories or with categories both in the outcome and as predictors, random intercept models may provide better sensitivity of testing. The latter models assume, that, for each predictor category or combination of categories x_1, x_2,..., slightly different a-values can be computed with a better fit for the outcome category y, than a single a-value.

$$y = a + b_1 x_1 + b_2 x_2 + \ldots.$$

We should add that, instead of the above linear equation, even better results are often obtained with log-transformed outcome variables (log = natural logarithm).

$$\log y = a + b_1 x_1 + b_2 x_2 + \ldots.$$

Are in a study of exposure and outcome categories the exposure categories significant predictors of the outcome categories, and does a random intercept provide

better test-statistics, than does a fixed intercept analysis? The data file is in extras.
springer.com, and is entitled "chap21randomintercepts". It was previously used by
the authors in SPSS for starters and second levelers, Chap. 45, Springer Heidelberg
Germany, 2016. The data file is opened in your computer. SPSS version 20 and up
can be used for analysis. First, we will perform a fixed intercept model.

The module Mixed Models consists of two statistical models:

Linear,
Generalized Linear.

For analysis the statistical model Generalized Linear Mixed Models is required.
First, we will perform a fixed effects model analysis, then a random effects model.

Command
click Analyze....Mixed Models....Generalized Linear Mixed Models....click Data
Structure....click "patient_id" and drag to Subjects on the Canvas....click Fields
and Effects....click Target....Target: select "fall with/out injury"....click Fixed
Effects ....click "agecat", and "study" and drag to Effect Builder:....mark Include
intercept ....click Run.

The underneath results show, that both the various regression coefficients, and the
overall correlation coefficients between the predictors and the outcome are, generally, statistically significant.

Source	F	df1	df2	Sig.
Corrected Model ▼	9,496	4	10	,002
agecat	6,914	2	10	,013
study	9,924	2	10	,004

Probability distribution:Multinomial
Link function:Cumulative logit

In the mixed model output sheets an interactive graph is observed with predictors
shown as lines with thicknesses corresponding to their predictive power and the
outcome in the form of a histogram. There is no best fit Gaussian curve in the above
histogram, because a nongaussian multinomial distribution has been chosen. Overall, both the predictors "agecat and study" are statistically significant, respectively at
p-values of 0.013 and 0.004.

The underneath graph shows the effects of the single categorical variables. Again
many variables were significant.

Model Term		Coefficient ▶	Sig.
Threshold for falloutofbed=	0	2,133	,027
	1	7,203	,000
agecat=0		5,243	,005
agecat=1		-0,016	,986
agecat=2		0,000ᵃ	
study=0		3,627	,008
study=1		4,260	,002
study=2		0,000ᵃ	

Probability distribution:Multinomial
Link function:Cumulative logit

[a]This coefficient is set to zero because it is redundant.

Subsequently, a random intercept analysis will be performed.

Command

Analyze....Mixed Models....Generalized Linear Mixed Models....click Data Structure....click "patient_id" and drag to Subjects on the Canvas....click Fields and Effects....click Target....Target: select "fall with/out injury"....click Fixed Effects ....click "agecat"and "study" and drag to Effect Builder:....mark Include intercept.... click Random Effects....click Add Block...mark Include intercept ....Subject combination: select patient_id....click OK....click Model Options.... click Save Fields...mark PredictedValue....mark PredictedProbability....click Save ....click Run.

The underneath results show the test-statistics of the random intercept model. The random intercept model shows better statistics:

p = 0.000 and 0.013	overall for age,
p = 0.000 and 0.004	Overall for study,
p = 0.000 and 0.005	Regression coefficients for age class 0 versus 2,
p = 0.814 and 0.998	For age class 1 versus 2,
p = 0.000 and 0.008	For study 0 versus 2, and
p = 0.000 and 0.0002	For study 1 versus 2.

Source	F	df1	df2	Sig.
Corrected Model ▼	31,583	4	213	,000
agecat	21,803	2	213	,000
study	30,181	2	213	,000

Probability distribution:Multinomial
Link function:Cumulative logit

If, as adjustment for multiple testing, a Bonferroni p-value of 0.05 x 2/[k(k-1)], instead of 0.050 was calculated, then the rejecting p-value with 6 p-values would turn out to be 0.0033. The overall test for predicting age and the study, and the regression coefficient for age class 0 versus 2, and for study 0 versus 2 would no longer be statistically significant. In contrast, with the random effect model the adjustment would not affect any of these statistical significance tests.

Model Term		Coefficient ▶	Sig.
Threshold for falloutofbed=	0	2,079	,000
	1	5,447	,000
agecat=0		3,872	,000
agecat=1		0,090	,814
agecat=2		0,000[a]	
study=0		3,208	,000
study=1		3,559	,000
study=2		0,000[a]	

Probability distribution:Multinomial
Link function:Cumulative logit

[a]This coefficient is set to zero because it is redundant.

We can conclude that, regressions can produce meaningful analyses, even with multiple exposure and outcome categories.

5 How Regressions Are Used for Assessing the Goodness of Novel Qualitative Diagnostic Tests

Regression analysis is routinely used with quantitative diagnostic tests (Chap. 20). The goodness of novel quantitative diagnostic tests against a standard test can be assessed with linear regression with the diagnostic result as predictor (independent) variable and the severities of disease as outcome (dependent) variable: if the R-square value of the novel test is significantly closer to 1.00 than the standard test, then, it is concluded that the novel test performs significantly better than the standard test does. However, unfortunately, in clinical research many diagnostic tests have *qualitative* rather than quantitative outcome variables, e.g., a clinical event / disease or not, and linear regression is not applicable for judging the goodness of such tests. Logistic regression with the odds of disease as outcome (dependent) variable and the test-scores as covariate (independent variable) could be used as a method to model such data. The following reasoning is applied. If the threshold for a positive test is taken high, then the proportion of false positives will be small. The steeper the logistic regression line, the faster this will happen. In contrast, if the threshold is taken low, then the proportion of false negatives will be small. The steeper the logistic regression line, the faster also this will happen. This would mean, that the steeper the logistic regression line, the fewer false positives and false negatives, and, thus, the better the diagnostic test is. A pleasant aspect of this approach is, that absolute instead of relative risks are measured. As example a non-invasive test for the diagnosis of peripheral vascular disease and a modified version of the test was used in respectively 640 and 587 patients.

The underneath figure shows the histograms of vascular lab scores patients with peripheral vascular disease and healthy controls. The figure shows, that the test is not perfect at all with considerable overlap between the patients with and without vascular disease.

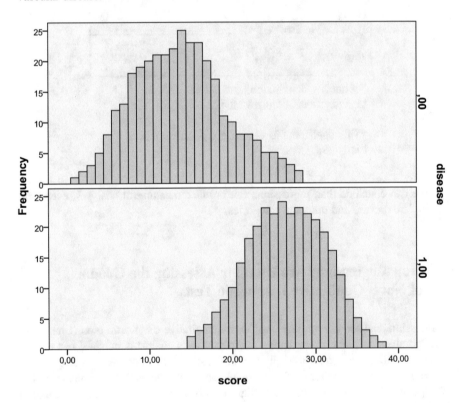

The underneath figure shows the result of a modified, and, possibly, improved test performed in 587 patient group with similar patient characteristics. Again the test is not perfect, but the patterns of the curves have slightly changed. We can not observe from the figures which of the two tests is the best one.

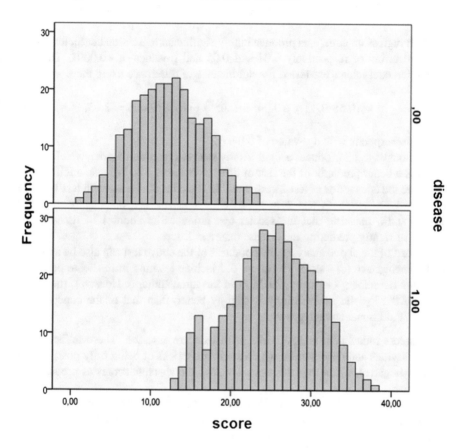

We will use a model similar to the linear regression model, used for assessing the goodness of quantitative diagnostic tests. However, because the outcome (dependent) variable is a binary rather than continuous variable, logistic regression instead of linear regression has to be applied. Again SPSS is used.

Command
Analyze....Regression.....click Binary Logistic.... Dependent variable: disease.... Covariate: score....click OK.

The best fit regression equation for test 1 is given underneath:

log odds of having the disease $= -9.20 + 0.45$ times the score

The best fit regression equation for test 2 is below:

log odds of having the disease $= -9.31 + 0.58$ times the score.

Both regression equations produce highly significant regression coefficients with standard errors of respectively 0.04 and 0.05 and p-values of <0.0001. The two regression coefficients are tested for significance of difference using the z – test:

$$z = (0.58\text{-}0.45)/\surd(0.04^2 + 0.05^2) = 0.13/0.064 = 2.03,$$

which corresponds with a p-value of 0.04.

Obviously, test 2 produces a significantly steeper regression model, which means that it is a better predictor of the risk of disease than test 1 is. We can, additionally, calculate the odds ratios of test 2 versus test 1. The odds of disease with test 1 equals $e^{0.45} = 1.57$, and with test 2 it equals $e^{0.58} = 1.79$. The odds ratio $= 1.79 / 1.57 = 1.14$, meaning that the second test produces an about 1.14 times better chance of rightly predicting the disease than test 1 does.

Instead of the above analysis, the goodness of the above test can also be assessed with concordance (c) - statistics (Chap. 6, Machine learning in medicine part two, Springer Heidelberg Germany, 2013, from the same authors). However, the sensitivity of the logistic methodology is slightly better, than that of the concordance models. Additional advantages are

1. Absolute rather than relative risks of disease are assessed. The c-statistic uses sensitivities and specificities, which are relative risks of being truly positive and truly negative, while logistic regression uses the absolute scores as predictor of disease.
2. An additional limitation of c-statistic is the following. The increase of area under the curve, to judge a new (and better) diagnostic test is very small, if the standard test already produced a large area under the curve.

6   How Regressions Can Help you Find out about Data Subsets with Unusually Large Spread

Heteroscedasticity is a term used for describing the phenomenon of different variances within a subset of data samples. It comes from the Greek: hetero means different, skedasis means dispersion. The phenomenon causes invalidity of statistical tests, if subsets of data samples are incorrectly assumed to have a uniform spread (dispersion). In the Chap. 18 and many more chapters mathematical models were applied for modeling nonlinear regressions. After mathematical remodeling of the exposure and / or outcome values, a best fit regression line or curve is usually obtained with the help of some form of the ordinary least squares principle, i.e., the squared distances from all of the outcome values to the regression line or curve are minimal, meaning that no better result could have been obtained. However, in some regressions files something very special is going on. Subsets of outcome data may have a variance different from that of other subsets.

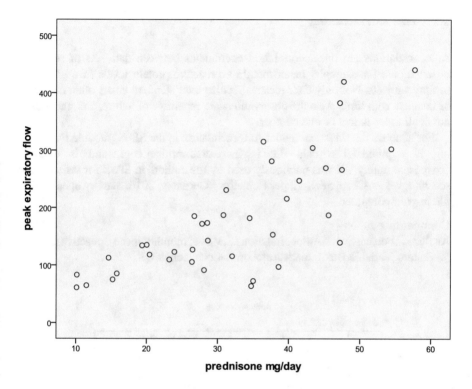

For example, in the above graph the spread of the outcome data in the low dose prednison treatment group is much smaller than it is in the higher dose prednison treatment groups. Another example of subsets outcome data with a large and those with a small variance is often observed with time series for the study of seasonalities. In these kinds of studies ordinary least squares analysis is inadequate, and alternative assessments are required. In this chapter we will address some of the analytic solutions currently available for the assessment of heteroscedastic data samples.

6.1 Maximum Likelihood Estimation

Maximum likelihood estimations are virtually always OK with heteroscedasticities in regression data, because it uses standardized regression coefficients, that have a standard error of 1. It is routinely used with structural equation modeling, which computes the standardized instead of the unstandardized regression coefficients (the path statistics). In Chap. 12 examples are given.

6.2   Autocorrelations

Autocorrelations are linear correlation coefficients between data sets of seasonal observations, for example, mean monthly c reactive protein levels of a group of healthy subjects. Not only their means, but also their standard errors often decrease or increase with time. Also the observations are repetitive of nature, and, therefore, not at all independent of one another.

For analysis the statistical model Autocorrelation in the SPSS' module Forecasting is required. The data file is in extras.springer.com, and is entitled "chap6seasonality". It was previously used by the authors in SPSS for starters and second levelers, Chap. 58, Springer Heidelberg Germany, 2016. Start by opening the file in your computer.

Command
Analyze....Forecasting....Autocorrelations....Move: monthly mean c-reactive protein levels into Variable Box....mark Autocorrelations....click OK.

Autocorrelations

Series:mean CRP 1month

Lag	Autocorrelation	Std. Error[a]	Box-Ljung Statistic		
			Value	df	Sig.[b]
1	.618	.192	10.364	1	.001
2	.347	.188	13.774	2	.001
3	-.072	.183	13.928	3	.003
4	-.364	.179	18.054	4	.001
5	-.664	.174	32.531	5	.000
6	-.603	.170	45.154	6	.000
7	-.502	.165	54.407	7	.000
8	-.239	.160	56.626	8	.000
9	-.017	.155	56.638	9	.000
10	.262	.150	59.704	10	.000
11	.419	.144	68.141	11	.000
12	.452	.139	78.750	12	.000
13	.300	.133	83.857	13	.000
14	.194	.127	86.217	14	.000
15	.005	.120	86.218	15	.000
16	-.179	.113	88.716	16	.000

a. The underlying process assumed is independence (white noise).
b. Based on the asymptotic chi-square approximation.

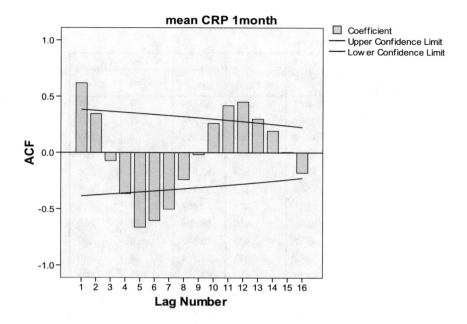

The above table gives the statistics. The graph shows that in the month 11 (auto-correlation coefficient 0,419, standard error 0,144, t-value 0,419/0,144 = 2910, p < 0,01) autocorrelation is statistically significant, supporting the seasonality of the observed 16 month mean c-reactive protein pattern. However, independence of the monthly observations versus one another is assumed. However, in practice seasonal data are, generally, very much dependent on one another. Therefore, accounting this is welcome. For that purpose SPSS does not include an entire generalized least squares procedure, but, instead, it includes a HAC (heteroscedasticity and autocorrelation consistent estimator). Bartlett estimator is here used for the purpose.

Autocorrelations

Series:mean CRP 1month

Lag	Autocorrelation	Std. Error[a]	Box-Ljung Statistic		
			Value	df	Sig.[b]
1	.618	.204	10.364	1	.001
2	.347	.271	13.774	2	.001
3	-.072	.289	13.928	3	.003
4	-.364	.290	18.054	4	.001
5	-.664	.308	32.531	5	.000
6	-.603	.363	45.154	6	.000
7	-.502	.403	54.407	7	.000
8	-.239	.428	56.626	8	.000
9	-.017	.433	56.638	9	.000
10	.262	.433	59.704	10	.000
11	.419	.440	68.141	11	.000
12	.452	.456	78.750	12	.000
13	.300	.475	83.857	13	.000
14	.194	.482	86.217	14	.000
15	.005	.486	86.218	15	.000
16	-.179	.486	88.716	16	.000

a. The underlying process assumed is MA with the order equal to the lag number minus one. The Bartlett approximation is used.
b. Based on the asymptotic chi-square approximation.

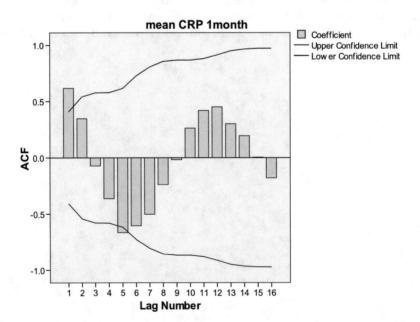

The above tables and graph shows the result of the adjusted procedure. The 95% confidence intervals are wider, and the presence of significant autocorrelation can no further be confirmed.

6.3   Weighted Least Squares

The spread of the outcome data may be smaller with low dosage treatments, than it is with high dosage treatment. The effect of prednisone on peak expiratory flow was assumed to be more variable with increasing dosages. Can it, therefore, be measured with more precision, if linear regression is replaced with weighted least squares procedure.

Var 1	Var 2	Var 3	Var 4
1	29	1,40	174
2	15	2,00	113
3	38	0,00	281
4	26	1,00	127
5	47	1,00	267
6	28	0,20	172
7	20	2,00	118
8	47	0,40	383
9	39	0,40	97
10	43	1,60	304
11	16	0,40	85
12	35	1,80	182
13	47	2,00	140
14	35	2,00	64
15	38	0,20	153
16	40	0,40	216

Var 1 Patient no
Var 2 prednisone (mg/24h)
Var 3 peak flow (ml/min)
Var 4 beta agonist (mg/24h)

Only the first 16 patients are given, the entire data file is entitled "chap21weightedleastsquares" and is in extras.springer.com. The data have been previously used by the authors in Machine learning in medicine a complete overview, Chap.25, Springer Heidelberg Germany, 2015. SPSS 19.0 and up can be used for data analysis. We will first make a graph of prednisone dosages and peak expiratory flows. Start with opening the data file in your computer.

Command

click Graphs....Legacy Dialogs....Scatter/Dot....click Simple Scatter....click Define....Y Axis enter peakflow....X Axis enter prednisone....click OK.

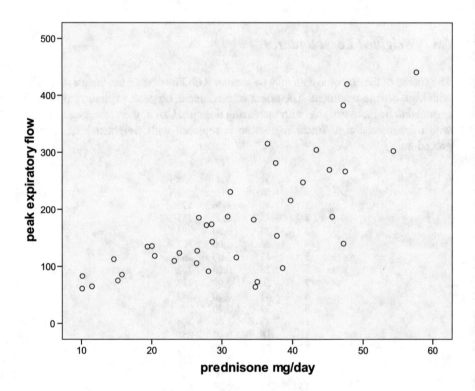

The output sheet shows, that the spread of the y-values is small with low dosages and gradually increases. We will, therefore, perform both a traditional, and a weighted least squares analysis of these data.

Command

Analyze....Regression....Linear....Dependent: enter peakflow....Independent: enter prednisone, betaagonist....click OK.

Model Summary[b]

Model	R	R Square	Adjusted R Square	Std. Error of the Estimate
1	,763[a]	,582	,571	65,304

a. Predictors: (Constant), beta agonist mg/24h, prednisone mg/day

b. Dependent Variable: peak expiratory flow

Coefficients[a]

Model		Unstandardized Coefficients B	Std. Error	Standardized Coefficients Beta	t	Sig.
1	(Constant)	-22,534	22,235		-1,013	,314
	prednisone mg/day	6,174	,604	,763	10,217	,000
	beta agonist mg/24h	6,744	11,299	,045	,597	,552

a. Dependent Variable: peak expiratory flow

In the output sheets an R value of 0.763 is observed, and the linear effects of prednisone dosages are a statistically significant predictor of the peak expiratory flow, but, surprisingly, the beta agonists dosages are not. We will, subsequently, perform a WLS analysis.

Command

Analyze....Regression....Weight Estimation.... select: Dependent: enter peakflow .... Independent(s): enter prednisone, betaagonist....select prednisone also as Weight variable....Power range: enter 0 through 5 by 0.5....click Options....select Save best weights as new variable....click Continue....click OK.

In the output sheets it is observed, that the software has calculated likelihoods for different powers, and the best likelihood value is chosen for further analysis. When returning to the data file again a novel variable is added, the WGT_1 variable (the weights for the WLS analysis). The next step is, to perform again a linear regression, but now with the weight variable included.

Command

Analyze....Regression....Linear.... select: Dependent: enter peakflow.... Independent(s): enter prednisone, betaagonist....select the weights for the wls analysis (the GGT_1) variable as WLS Weight....click Save....select Unstandardized in Predicted Values....deselect Standardized in Residuals....click Continue.... click OK.

Model Summary[b,c]

Model	R	R Square	Adjusted R Square	Std. Error of the Estimate
1	,846[a]	,716	,709	,125

a. Predictors: (Constant), beta agonist mg/24h, prednisone mg/day

b. Dependent Variable: peak expiratory flow

c. Weighted Least Squares Regression - Weighted by Weight for peakflow from WLS, MOD_6 PREDNISONE** -3,500

Coefficients[a,b]

Model		Unstandardized Coefficients B	Std. Error	Standardized Coefficients Beta	t	Sig.
1	(Constant)	5,029	7,544		,667	,507
	prednisone mg/day	5,064	,369	,880	13,740	,000
	beta agonist mg/24h	10,838	3,414	,203	3,174	,002

a. Dependent Variable: peak expiratory flow

b. Weighted Least Squares Regression - Weighted by Weight for peakflow from WLS, MOD_6 PREDNISONE** -3,500

The output table now shows an R value of 0.846. It has risen from 0.763, and provides thus more statistical power. The above lower table shows the effects of the two medicine dosages on the peak expiratory flows. The t-values of the medicine predictors have increased from approximately 10 and 0.5 to 14 and 3.2. The p-values correspondingly fell from 0.000 and 0.552 to respectively 0.000 and 0.002. Larger prednisone dosages and larger beta agonist dosages significantly and independently increased peak expiratory flows. After adjustment for heteroscedasticity, the beta agonist became a significant independent determinant of peak flow.

6.4 Two Stage Least Squares

Two stage least squares can be applied, if heterogeneous variances due to interactions are expected. Heteroscedasticity is often due to error in the outcome from an independent variable, if another independent variable is interacting. As an example, the effects of counseling and non-compliance (pills not used) on the efficacy of a novel laxative drug is studied in 35 patients. The first 10 patients of the data file is given below.

Var 1	var 2	var 3 (var = variable)
Outcome	Pt Instrumental variable	Problematic predictor
Frequency counseling	Pills not used	Efficacy estimator of new laxative (stools/month)
1. 24	8	25
2. 30	13	30
3. 25	15	25
4. 35	14	31
5. 39	9	36
6. 30	10	33
7. 27	8	22
8. 14	5	18
9. 39	13	14
10. 42	15	30

The data file is in extras.springer.com and is entitled "chap21twostageleastsquares". It was previously used by the authors in Machine learning in medicine a complete overview, Chap. 34, Springer Heidelberg Germany, 2015. Start by opening the file in your computer.

Command

Analyze....Regression....Linear....Dependent enter ther eff (stools/month).... Independent(s) counseling and non-compliance....click OK.

Coefficients[a]

Model		Unstandardized Coefficients		Standardized Coefficients	t	Sig.
		B	Std. Error	Beta		
1	(Constant)	2,270	4,823		,471	,641
	counseling	1,876	,290	,721	6,469	,000
	non-compliance	,285	,167	,190	1,705	,098

a. Dependent Variable: ther eff

The above table shows, that non-compliance is a borderline significant predictor of stools.

Next a one by one linear regession shows, that non-compliance is also a significant predictor of counseling.

Coefficients[a]

Model		Unstandardized Coefficients		Standardized Coefficients	t	Sig.
		B	Std. Error	Beta		
1	(Constant)	4,228	2,800		1,510	,141
	non-compliance	,220	,093	,382	2,373	,024

a. Dependent Variable: counseling

This would mean, that non-compliance may work two ways: it predicts thera-peutic efficacy *directly* and *indirectly* through counseling. However, the indirect way is not taken into account in the usual one step linear regression. Two stage least square (2LS) method is possible, and is available in SPSS. It works as follows. First, a simple regression analysis, with counseling as outcome and non-compliance as predictor, is performed. Then the outcome values of the regression equation are used as predictor of therapeutic efficacy.

Command

Analyze....Regression....2 Stage Least Squares....Dependent: therapeutic efficacy....Explanatory: non-compliance....Instrumental:counseling ....OK.

Model Description

		Type of Variable
Equation 1	VAR00001	dependent
	VAR00003	predictor
	VAR00002	instrumental

MOD_1

Coefficients

		Unstandardized Coefficients		Beta	t	Sig.
		B	Std. Error			
Equation 1	(Constant)	-61,095	37,210		-1,642	,110
	VAR00003	3,113	1,256	2,078	2,478	,019

The above tables show the results of the 2LS method. As expected the final p-value is smaller than the simple linear regression p-value of the effect of non-compliance on therapeutic efficacy.

6.5 Robust Standard Errors

Robust standard errors are another methodology for handling data subsets with unusually large spread. For example, in a parallel-group study of 52 patients the presence of torsade de pointes was measured during two treatment modalities.

treatment modality	presence torsade de pointes
,00	1,00
,00	1,00
,00	1,00
,00	1,00
,00	1,00
,00	1,00
,00	1,00
,00	1,00
,00	1,00
,00	1,00

The first 10 patients are above. The entire data file is in extras.springer.com, and is entitled "chap21rates". SPSS statistical software will be used for analysis. Start by opening the file in your computer. First, we will perform a traditional binary logistic regression with torsade de pointes as outcome, and treatment modality as predictor.

Command

Analyze....Regression....Binary Logistic....Dependent: torsade....Covariates: treatment.... OK.

Variables in the Equation

		B	S.E.	Wald	df	Sig.	Exp(B)
Step 1[a]	VAR00001	1,224	,626	3,819	1	,051	3,400
	Constant	-,125	,354	,125	1	,724	,882

a. Variable(s) entered on step 1: VAR00001.

The above table shows that the treatment modality does not significantly predict the presence of torsades de pointes. The numbers of torsades in one group is not significantly different from the other group. A rate analysis is performed subsequently.

Command

Generalized Linear Models ….mark Custom….Distribution: Poisson ….Link Function: Log….Response: Dependent Variable: torsade…. Predictors: Main Effect: treatment…..Estimation: mark Robust Tests….OK.

Parameter Estimates

Parameter	B	Std. Error	95% Wald Confidence Interval		Hypothesis Test		
			Lower	Upper	Wald Chi-Square	df	Sig.
(Intercept)	-,288	,1291	-,541	-,035	4,966	1	,026
[VAR00001=,00]	-,470	,2282	-,917	-,023	4,241	1	,039
[VAR00001=1,00]	0ª	.	.	.	.	.	.
(Scale)	1ᵇ						

Dependent Variable: torsade
Model: (Intercept), VAR00001

a. Set to zero because this parameter is redundant.

b. Fixed at the displayed value.

The predictor treatment modality is now statistically significant at p = 0.039. And so, using the Poisson distribution in Generalized Linear Models, we found, that treatment one performed significantly better in predicting numbers of torsades de pointe than did treatment zero at p = 0.039. We will check with a 3-dimensional graph of the data, if this result is in agreement with the data as observed.

Command

Graphs….Legacy Dialog….3-D Bar: X-Axis mark: Groups of Cases, Z-Axis mark: Groups of Cases…Define 3-D Bar: X Category Axis: treatment, Z Category Axis: torsade….OK.

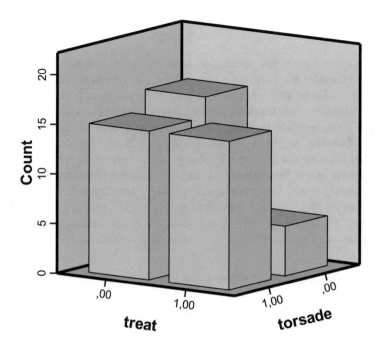

The above graph shows, that in the 0-treatment (placebo) group the number of patients with torsades de pointe is virtually equal to that of the patients without. However, in the 1-treatment group it is smaller. The treatment seems to be efficacious. Rate analysis using a robust Poisson regression is different from logistic regression, because it uses a log transformed dependent variable. For the analysis of robust Poisson regression is very sensitive, and, thus, better than standard logistic regression.

6.6   Generalized Least Squares

Generalized least squares is without doubt the most appropriate method, if the assumptions, particularly those of homoscedasticity can not be met, but it is not in SPSS, although it is in many software programs like R, SAS, Matlab etc. The problem with these software programs is, that they, unlike SPSS, require a lot of syntax knowledge, and health professionals do generally have a lousy syntax knowledge. Fortunately, generalized least squares for the purpose of adjusting heterogeneity of variances can be replaced with alternative methods, like all of the ones mentioned in the current section of this chapter.

7   Conclusion

In current clinical trials, like pharmaceutical trials, data driven decisions at work are considered the most important types of decisions, and regression analysis is the most frequently used methodology for the purpose. Why so? First, it is not hypothetical but data driven. Second, it assesses, which factors matter, and which ones do not.

Third, it assesses and quantifies error terms, and does not necessarily assess causality. However, regression analyses easily include mistakes as well. For example, regression analyses are, virtually always, at least partly explorative and, therefore, carry the risk of fishing data, and bad data entering your analysis. These data may be exciting up to a level, that makes you forget about the best functions of your brain, meaning functions like common sense reasoning, intuition and gut feeling, functions that should be on top of your data and mathematical models.

Nonetheless, regressions are a widespread help for data analyses. This chapter gives examples of:

how regressions help you make sense of the effects of small changes in experimental settings,
how they can assess the sensitivity of your predictors,
how they can be used for meta-analyses of diagnostic tests,
how they are used for assessing the goodness of your novel diagnostic tests,
how they can help you find out about data subsets with unusually large variance spread.

Particularly, the above last point has been given much attention lately. Methods for adjusting unusually large variance spread are.

maximum likelihood estimations,
autocorrelations,
weighted least squares,
two stage least squares,
robust standard errors,
generalized least squares.

Reference

To readers requesting more background, theoretical and mathematical information of computations given, several textbooks complementary to the current production and written by the same authors are available: Statistics applied to clinical studies 5th edition, 2012, Machine learning in medicine a complete overview, 2015, SPSS for starters and 2nd levelers 2nd edition, 2015, Clinical data analysis on a pocket calculator 2nd edition, 2016, Understanding clinical data analysis, 2017, all of them edited by Springer Heidelberg Germany

Chapter 22
Regression Trees

Classification and Regression Tree (CART) Models

Abstract With decision trees, data samples of patients with and without the presence of a disease can be assessed for subgroup properties. Usually binary variables are used for assessment, but binary cut-off values of continuous variables can also be used. The current chapter reviews the use of linear regression for finding the optimal cut-offs of subgroups, i.e., the cut-offs with the linear regression equation, that produces the largest test statistic. It is a lot of work, and it is, sometimes, called exhaustive searching, but for a computer it is not hard to do. We should add, that the computer is even capable of finding best cut-offs with partitioning into more than two subgroups, if required. But we stuck to two subgroups, in order to avoid too much complexity in the model.

Keywords Regression trees · Classification and regression trees (CART trees)

1 Introduction, History, and Background

One of the most important and original applications of binary partitioning was, to develop data-based decision cut-off levels, that can assist physicians in diagnosing patients potentially suffering heart attacks (Wasson et al. N Engl J Med 1985; 313: 793). Traditionally, the physicians made decisions, based on their clinical experience. Also, laboratories developed diagnostic tests with normal values based on rather intuitive grounds. Classifications based on representative historical data has the advantage of added empirical information from large numbers of patients. This will be, particularly, important if symptoms, signs and diagnostic procedures give rise to a substantial number of false positive and false negative results, as often observed in clinical practice. The main purpose of the data-based methods is, to reduce the latter number. The book by Breiman, from Berkeley University (ed by Chapman & Hall, NY, 1984) on classification and regression trees is a milestone on

Electronic Supplementary Material The online version of this chapter (https://doi.org/10.1007/978-3-030-61394-5_22) contains supplementary material, which is available to authorized users. The videos can be accessed by scanning the related images with the SN More Media App.

binary partitioning, and shows that it is closely related to cut-off decision trees, otherwise, called CART (classification and regression) trees. Breiman, 1928–2005, was a distinguished professor in statistics at Berkeley CA. He is particularly known for his continued efforts to bridge the gap between statistics and machine learning. With decision trees, data samples of patients with and without the presence of a disease can be assessed for subgroup properties. Usually, binary variables are used for assessment, but binary cut-off values of continuous variables can also be used. More information is given in the Chap. 53, Statistics applied to clinical studies, Springer Heidelberg Germany, 2012, from the same authors. This chapter will review the use of linear regression for finding the optimal cut-offs of subgroups, i.e., the cut-offs with the linear regression equation, that produces the largest test statistic. It is a lot of work, and it is sometimes called exhaustive searching, but for a computer it is not hard to do. We should add, that the computer will even be capable of finding best cut-offs with partitioning into more than two subgroups, if required.

For the purpose Breiman used the entropy method. This method has an interesting history. It received its name, because it makes use of an equation that was formerly applied in science to estimate the amount of energy loss in thermodynamics, but, otherwise, has no connection with its application to science. A pleasant thing about this method is that, unlike the traditional ROC (receiver operating) method, it can be easily adjusted for magnitude of the samples. This is also the reason that the result of the ROC method often slightly differs from that of the entropy method.

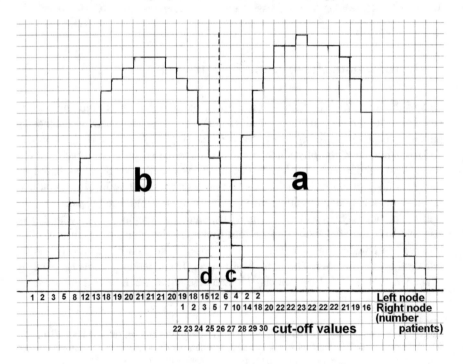

A sample of 1004 patients at risk of infarct is used as an example with a yes no outcome (yes or no infarct in the past). In entropy-method-terminology such sample is called the parent node, which can, subsequently, be repeatedly split, partitioned if you will, into binary internal nodes. Mostly, internal nodes contain false positive or negative patients, and are, thus, somewhat impure. The magnitude of their impurity is assessed by the log likelihood method. Impurity equals the maximum log likelihood of the y-axis-variable by assuming that the x-axis-variable follows a Gaussian (i.e. binomial) distribution and is expressed in units, sometimes called bits (a shortcut for "binary digits"). All of this sounds rather complex, but it works smoothly.

The x − axis variable for the right node $= x_r = a/(a+d)$,
for the left node $= x_l = b/(b+c)$.

If the impurity equals 1.0 bits, then it will be maximal, if it equals 0.0, then it will be minimal.

Impurity node either right or left $= -x \ln x - (1-x) \ln (1-x)$,
where ln means natural logarithm.

The impurities of the right and left node are calculated separately. Then, a weighted overall impurity of each cut-off level situation is calculated according to:

Weighted impurity cut-off = 1/2 *
[(a + d)/(a + b + c + d)*impurity-right-node] +
[(b + c)/(a + b + c + d)*impurity-left-node],
with the symbol of multiplication as given by *.

Underneath, an overview is given of the calculated impurities at the different cut-off levels. The cut-off percentage of 27 gives the smallest weighted impurity, and is, thus, a better predictor for vascular disease than the value of 20% as previously agreed on intuitive grounds.

Cut-off	impurity right node	impurity left node	impurity weighted
22 %	0.5137	0.0000	0.3180
23 %	0.4392	0.0559	0.3063
24 %	0.4053	0.0982	0.2766
25 %	0.3468	0.1352	0.2711
26 %	0.1988	0.1688	0.1897
27 %	**0.1352**	**0.2268**	**0.1830**
28 %	0.0559	0.3025	0.1850
29 %	0.0559	0.3850	0.2375
30 %	0.0000	0.4690	0.2748

Also, it should be a better predictor of vascular disease than the 26% value as established by the ROC method, because, unlike the ROC method, the entropy method takes into account, and adjusts the differences in sample sizes of the nodes.

2   Data Example with a Continuous Outcome, More on the Principles of Regression Trees

A 953 patient data file is used of various predictors of ldl (low-density-lipoprotein)-cholesterol reduction including weight reduction, gender, sport, treatment level, diet. The SPSS data file is in extras.springer.com and is entitled "chap22regressiontree. sav". It is previously used by the authors in Machine learning in medicine a complete overview, Chap. 53, Springer Heidelberg Germany, 2015. The file is opened in your computer with SPSS statistical software installed. The first 13 patients are given underneath.

Variables (Var)					
1	2	3	4	5	6
3,41	0	1	3,00	3	0
1,86	-1	1	2,00	3	1
,85	-2	1	1,00	4	1
1,63	-1	1	2,00	3	1
6,84	4	0	4,00	2	0
1,00	-2	0	1,00	3	0
1,14	-2	1	1,00	3	1
2,97	0	1	3,00	4	0
1,05	-2	1	1,00	4	1
,63	-2	0	1,00	3	0
1,18	-2	0	1,00	2	0
,96	-2	1	1,00	2	0
8,28	5	0	4,00	2	1

Var 1 ldl_reduction
Var 2 weight_reduction (kg)
Var 3 gender (0-1)
Var 4 sport (1-4)
Var 5 treatment_level (1-4)
Var 6 diet (0-1)

We will assess the effect of weight reduction on ldl cholesterol reduction, and try and find the best cut-off for assessment. First, we will transform the continuous weight reduction variable into various binary variables with different levels of cut-offs and use linear regression with ldl reduction as outcome for identifying the cut-off with the largest test statistic assessed with analysis of variance.

Command
in Data View click weight_reduction....click Transform....click Compute Variable.... in Numeric Expression: enter weight_reduction<= 0....in Target Variable enter weight_reduction2....click OK....in Variable View under Name enter weightreduction2.... click Analyze....Regression....Linear.... Independent Variables: enter weight_reduction....Dependent Variable: enter ldl_reduction....click OK.

The same commands have to be executed for different weight reduction cut-offs, for example:

weight_reduction<= 0
weight_reduction<= 1
weight_reduction<= 2
weight_reduction<= ...

In the output sheets of the regressions the underneath F-statistics are given.

with weight_reduction<= 0 the F-statistic equals   1992,721
with weight_reduction<= 1   2963,858
with weight_reduction<= 2   2267,989.

Obviously, weight reduction $< = 1$ produces the best statistic, and is, thus, the cut-off level, which provides the most significant difference between ldl reductions, that can be obtained from differences in weight reductions. And, so, it is the best level for making predictions about the effect of weight reduction on ldl reduction.

3 Automated Entire Tree Regression from the LDL Cholesterol Example

We will now perform an automated entire regression tree analysis from the ldl cholesterol example.

Command
Analyze....Classify...Tree.... Dependent Variable: enter ldl_reduction.... Independent Variables: enter weight reduction, gender, sport, treatment level, diet .... Growing Methods: select CRT ....click Criteria: enter Parent Node 300, Child Node 100....click Output: Tree mark Tree in table format....click OK.

ldl reduction

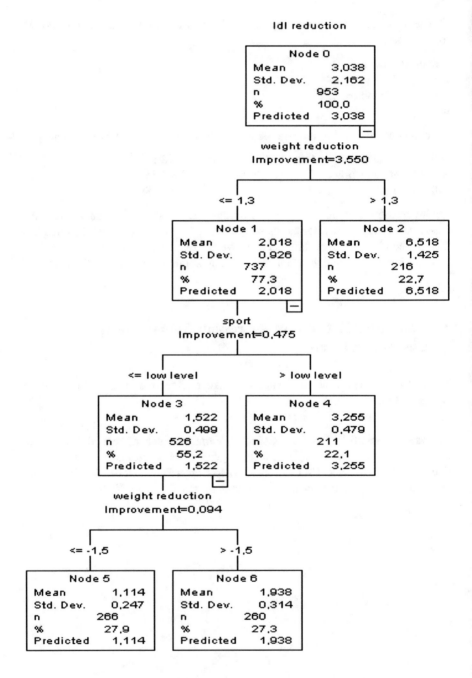

Tree Table

Node	Mean	Std. Deviation	N	Percent	Predicted Mean	Parent Node	Primary Independent Variable		
							Variable	Improvement	Split Values
0	3,0379	2,16238	953	100,0%	3,0379				
1	2,0178	,92633	737	77,3%	2,0178	0	weight reduction	3,550	<= 1,3
2	6,5184	1,42469	216	22,7%	6,5184	0	weight reduction	3,550	> 1,3
3	1,5216	,49947	526	55,2%	1,5216	1	sport	,475	<= low level
4	3,2549	,47895	211	22,1%	3,2549	1	sport	,475	> low level
5	1,1143	,24685	266	27,9%	1,1143	3	weight reduction	,094	<= -1,5
6	1,9383	,31410	260	27,3%	1,9383	3	weight reduction	,094	> -1,5

Growing Method: CRT
Dependent Variable: ldl reduction

The output sheets show the above regression tree, and tree table with computed summaries. Only weight reduction and sport significantly contributed to the model, with the overall mean and standard deviation dependent variable ldl cholesterol in the parent (root) node. Weight reduction with a cut-off level of 1.3 units is the best predictor of ldl reduction. In the little weight reduction group, sport is the best predictor. In the low sport level subgroup, again weight reduction is a predictor, but here is a large difference between weight gain (<-1.5 units) and weight loss (> -1.5 units). Also in the output are the gain summary for nodes and the risk estimate.

Gain Summary for Nodes

Node	N	Percent	Mean
2	216	22,7%	6,5184
4	211	22,1%	3,2549
6	260	27,3%	1,9383
5	266	27,9%	1,1143

Growing Method: CRT
Dependent Variable: ldl reduction

Risk

Estimate	Std. Error
,552	,042

Growing Method: CRT
Dependent Variable: ldl
reduction

The "gains for nodes table" provides the data from the terminal nodes of the model. The mean ldl cholesterol reductions are given. The node 2 has a mean reduction of 6.518 mmol/l, while the reduction in the remainder was only 2.018 mmol/l. The node 4 for has a mean reduction of 3.255 versus 1.522 mmol/l in the worst sport level category. The node 6 has a mean ldl cholesterol reduction of 1.938 versus 1.114 mmol/l in the worst weight reduction category. The risk table shows the risk of misclassification. It is very large in our example with an overall estimate of 0,552, meaning 55% misclassifications.

4   Conclusion

An important application of binary partitioning was, to develop data-based decision cut-off levels, that can assist physicians in making diagnoses. Traditionally, the physicians made decisions based on their clinical experience. Also, laboratories developed diagnostic tests with normal values based on rather intuitive grounds.

Classifications based on representative historical data has the advantage of added empirical information from large numbers of patients. This is, particularly, important if symptoms, signs and diagnostic procedures give rise to a substantial number of false positive and false negative results, as often observed in clinical practice. With decision trees, data samples of patients with and without the presence of a disease can be assessed for subgroup properties. Usually binary variables are used for assessment, but binary cut-off values of continuous variables can also be used. The current chapter reviewed the use of linear regression for finding the optimal cut-offs of subgroups, i.e., the cut-offs with the linear regression equation, that produces the largest test statistic. It is a lot of work, and it is sometimes called exhaustive searching, but, for a computer, it is not hard to do. We should add, that the computer is even capable of finding best cut-offs with partitioning into more than two subgroups if required. But we stuck to two subgroups, in order to avoid too much complexity in the model.

Reference

More background, theoretical and mathematical information of decision trees as well as the steps for utilizing syntax files is available in Machine learning in medicine part three, Chap. 14, entitled "Decision trees", pp 153–168, Springer Heidelberg, Germany 2013. Better accuracy from decision trees is sometimes obtained by the use of a training sample (Chap. 8 of the same edition).

Chapter 23
Regressions with Latent Variables

Factor Analysis, Partial Least Squares, Discriminant Analysis

Abstract Latent variables are unmeasured variables, inferred from measured variables. The main purpose for using them is data variables reduction. Current research increasingly involves multiple variables, and traditional statistical models tend to get powerless with too many variables included. Multiple analyses is, generally, no solution, because of increased type I errors due to multiple testing. In contrast, a few latent variables replacing multiple manifest variables can be applied. However, a disadvantage is, that latent variables are rather subjective, because they are dependent on subjective decisions to cluster some measured variables and remove others. The current chapter reviews, how to construct high quality latent variables, and how they can be successfully implemented in many modern methodologies for data analysis. Three of them will be reviewed.

First, factor analysis: it is an unsupervised learning methodology, i.e., it has no dependent variable.

Second, partial least squares: it is a supervised learning methodology, where outcomes are separately included.

Third, discriminant analysis: it is very similar to the above two, but goes one step further. It includes a grouping predictor variable, e.g., treatment modality.

All of the three methodologies are multivariate methods. They are explained in the current chapter with data examples.

Keywords Regressions with latent variables · Factor analysis · Partial least squares · Discriminant analysis

Electronic Supplementary Material The online version of this chapter (https://doi.org/10.1007/978-3-030-61394-5_23) contains supplementary material, which is available to authorized users. The videos can be accessed by scanning the related images with the SN More Media App.

1   Introduction, History, and Background

Latent variables are unmeasured variables inferred from directly measured variables. The main purpose for using them is variables reduction, otherwise called data dimension reduction. Current research increasingly involves multiple variables data sets, and traditional statistical models tend to get powerless with too many variables included. Multiple analyses is, generally, no solution, because of increased type I errors due to multiple testing. In contrast, a few latent variables instead of multiple manifest variables does have advantages: (1) less risk of power loss due to too many variables in a model, (2) no risk of false positive results like with multiple testing. However, a disadvantage is, that latent variables are rather subjective, because they are dependent on subjective decisions to cluster some measured variables and remove others.

Some introductory remarks regarding latent variables have already been given in the Chap. 13. The current chapter will review, how to construct high quality latent variables, and how they can be successfully implemented in many forms of modern data analyses, including factor analysis, partial least squares, and discriminant analysis.

Multivariate analysis of variance is based on matrix algebra, particularly, cross matrices, invented by two English mathematicians in 1850, James Sylvester and Arthur Cayley. It may have a long history, but Theodore Wilber Anderson' s 1958 textbook, *An Introduction to Multivariate Statistical Analysis*, has been published several places and times, and educated generations of theorists and statisticians. Anderson, a statistician from Minnesota, emphasized hypothesis testing via likelihood ratio tests and power functions. In the first years multivariate analyses were considered as a field inadmissible to non-mathematicians (Anderson 1958). The 1964 book entitled "Statistical Methodologies, of Harry Seal, edited by Methuen & Co London UK, for example, recommended the geometric approach rather algebraic approach (Seal 1964). Multivariate analysis once solely stood in the statistical theory realms due to the size, complexity of underlying data sets and high computational consumptions. With the tremendous growth of computational power, it now plays an increasingly important role in data analysis and has wide application in OMICS fields, where Omics is a neologism related to the suffix -oma, frequently applied in medicine to signify nouns, that emphasize tumors or swellings. Omics now refers to current names of scientific fields in biology ending on omics, including terms like genomics, transcriptomics, proteonomics, metabolomics, lipidomics, glucomics, foodomics. With the recent advent of big data in clinical studies, multivariate statistical analysis (which was already the domain of other scientific applications including psychology, genetics, chemistry, image analysis, nutrition, economics and social science) is now increasingly of interest in medical applications. Standard application of statistical methods in medicine traditionally covers multiple regression, leaving out the plethora of methods that fall under the general title of 'multivariate methods'. These methods have in common the attempt to model mathematically or statistically a set of variables measured on the same observations using matrix algebra and statistical and computational models and algorithms.

Current multivariate statistical methods are also applied with Descriptive (mathematical/geometric) methods like:

1) Principal component analysis;
2) Correspondence analysis;
3) Multidimensional scaling;
4) Cluster analysis;

with Methods based on a statistical model (i.e., with a model having a probability distribution) like:

5) Factor analysis;
6) Discriminant analysis;
7) Partial least squares;
8) Reduced rank regression;
9) Simultaneous equations and instrumental variable models with multiple instruments;
10) Mediation analysis with high-dimensional mediators. -Machine learning methods like (9), Classification and regression trees (recursive partitioning (Chap. 22));
11) Neural networks;
12) Support vector machines.

The current chapter will, particularly, address multivariate analysis in latent variables models, but this edition also applied multivariate regressions elsewhere, like in the above Chaps. 8, 10, and 13.

2 Factor Analysis

Many factors in life are complex and difficult to measure directly. Charles Spearman, a London UK psychometrician in the 40s, searched for a method to measure intelligence (Barthelemew, Br J Math Stat Psychol 1995; 48: 211). Intelligence has many aspects, that can be measured and modeled together. The simplest model is the use of add-up scores. However, add-up scores do not account for the relative importance of the separate aspects, their interactions and differences in units. All of this is accounted for by a technique called factor analysis: two or three unmeasured factors are identified to explain a much larger number of measured variables. Although factor analysis is a major research tool in behavioral sciences, social sciences, marketing, operational research, and other applied sciences, it is rarely applied in clinical research. When searching the internet we found, except for a few genetic studies (Meng J, www.cmsworldwide.com, 2011, Hochreiter et al. Bioinformatics 2006; 22: 943) no clinical studies applying factor analysis. This is a pity given the presence of large numbers of variables in this field, particularly, in diagnostic research.

In this section we will assess, whether the performance of a diagnostic battery for making clinical predictions can be improved by using factor analysis. We will also assess, whether factor analysis enables to make predictions about individuals. We hope, that this chapter will stimulate clinical investigators to start using this method. For factor analysis internal consistency between the original variables contributing to a factor is required. There should be a strong correlation between the answers given to questions within one factor: all of the questions should, approximately, predict one and the same thing. The level of correlation is expressed as Cronbach' s alpha: 0 means poor, 1 perfect relationship. The test-retest reliability of the original variables should be assessed with one variable missing: all of the data files with one missing variable should produce at least for 80% the same result, as that of the non-missing data file (alphas > 80%).

Cronbach's alpha

$$\text{alpha} = \frac{k}{(k-1)} \cdot \left(1 - \sum \frac{s_i^2}{s_T^2}\right)$$

K = number of original variables
s_i^2 = variance of ith original variable
s_T^2 = variance of total score of the factor obtained by summing up all of the original variables

Also, there should not be a too strong correlation between different original variable values in a conventional linear regression. Correlation coefficient (R) > 0.80 means the presence of multicollinearity and, thus, of a flawed multiple regression analysis. R is the Pearson's correlation coefficient, and has the underneath mathematical equation with x and y, as the variables of the x- and y-axes of a linear regression.

$$R = \frac{\sum(x - \bar{x})(y - \bar{y})}{\sqrt{\sum(x - \bar{x})^2 \sum(y - \bar{y})^2}}$$

R is a measure for the strength of association between two variables. The stronger the association, the better one variable predicts the other. It varies between -1 and $+1$, zero means no correlation at all, -1 means 100% negative correlation, $+1$ 100% positive correlation.

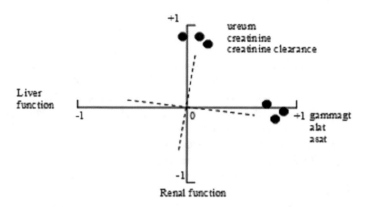

The factor analysis theory will be explained using the data of the above graph. ALAT (alanine aminotransferase), ASAT (aspartate aminotransferase) and gammaGT (gamma glutamyl tranferase) are a cluster of variables telling us something about a patient's liver function, while ureum, creatinine (creat) and creatininine clearance (c-clear) tell us something about the same patient's renal function. In order to make morbidity/mortality predictions from such variables, often, multiple regression is used. However, with multicollinearity, the variables cannot be used simultaneously in a regression model, and an alternative method has to be used. With factor analysis, all of the variables are replaced with a limited number of novel variables, that have the largest possible correlation coefficients with all of the original variables. As a multivariate technique, it is somewhat similar to Manova (multivariate analysis of variance), with the novel variables, otherwise called the factors, as outcomes, and the original variables, as predictors. However, it is less affected by multicollinearity, because the y- and x-axes are used to present the novel factors in an orthogonal way, and it can be shown, that, with an orthogonal relationship between two variables, the magnitude of their covariance will be zero, and will not have to be taken into account. The magnitude of the latent factor values for individual patients are calculated as shown below:

$$\text{Factor}_{\text{liver function}} = \quad 0.87 \times \text{ASAT} + 0.90 \times \text{ALAT} + 0.85 \times \text{GammaGT} +$$
$$0.10 \times \text{creatinine} + 0.15 \times \text{creatininine clearance}$$
$$-0.05 \times \text{reum}$$

$$\text{Factor}_{\text{renal function}} = \quad -0.10 \times \text{ASAT} - 0.05 \times \text{ALAT} + 0.05 \times \text{GammaGT} +$$
$$0.91 \times \text{creatinine} + 0.88 \times \text{creatininine clearance}$$
$$+0.90 \times \text{ureum}$$

The term factor loadings is given to the linear correlation coefficients between the original variables and the estimated novel variable, the latent factor, adjusted for all of the manifest variables, and adjusted for eventual differences in units.

It can be demonstrated in a "2 factor" factor analysis, that, by slightly rotating both x and y-axes, the model can be fitted even better. When the y- and x-axes are rotated simultaneously, the two novel factors are assumed to be 100% independent of one another, and this rotation method is called varimax rotation. Independence needs not be true, and, if not true, the y-axis and x-axis can, alternatively, be rotated separately, in order to find the best fit model for the data given. Eigenvectors is a term often used with factor analysis. The R-values of the manifest variables versus latent factors are the eigenvalues of the original variables, their place in the above graph the eigenvectors. A scree plot is used to compare the relative importance of the latent factors with that of the manifest variables using eigenvector values.

Complex mathematical models are often laborious, so that even modern computers have difficulty to process them. Software packages currently make use of a technique called iterations: five or more calculations are estimated, and the one with the best fit will be chosen.

The term components is often used to indicate the factors in a factor analysis, e.g., in rotated component matrix and in principal component analysis. The term latent factors is used to indicate the factors in a factor analysis. They are called latent, because they are not directly measured, but, rather, derived from the original variables. An y- and x-axis are used to represent them in a two factor model. If a third factor existed in your data model, it could be represented by a third axis, a z-axis creating a 3-d graph. Also additional factors can be added to the model, but they cannot be presented in a 2- or 3-d drawing anymore, but, just like with multiple regression modeling, the software programs have no problem with multidimensional computations similar to the above 2-d calculations.

Variables										
1	2	3	4	5	6	7	8	9	10	11
sv	gam	asat	alat	bili	ureum	creat	creat-cl	esr	c-react	leuc
,00	20,00	23,00	34,00	2,00	3,40	89,00	-111,00	2,00	2,00	5,00
,00	14,00	21,00	33,00	3,00	2,00	67,00	-112,00	7,00	3,00	6,00
,00	30,00	35,00	32,00	4,00	5,60	58,00	-116,00	8,00	4,00	4,00
,00	35,00	34,00	40,00	4,00	6,00	76,00	-110,00	6,00	5,00	7,00
,00	23,00	33,00	22,00	4,00	6,10	95,00	-120,00	9,00	6,00	6,00
,00	26,00	31,00	24,00	3,00	5,40	78,00	-132,00	8,00	4,00	8,00
,00	15,00	29,00	26,00	2,00	5,30	47,00	-120,00	12,00	5,00	5,00
,00	13,00	26,00	24,00	1,00	6,30	65,00	-132,00	13,00	6,00	6,00
,00	26,00	27,00	27,00	4,00	6,00	97,00	-112,00	14,00	6,00	7,00
,00	34,00	25,00	13,00	3,00	4,00	67,00	-125,00	15,00	7,00	6,00

sv = survival / septic death
gam = gamma-gt
bili = bilirubine
creat = creatinine
creat-cl = creatinine clearance
esr = erythrocyte sedimentation rate
c-react = c-reactive protein
leuc = leucocyte count

As a data example, a simulated data set of 200 patients at risk of septic death is used.

The first 10 patients are in the table above. The individual patient data are given in the SPSS data file "chap23factoranalysis", and is in extras.springer.com. It is previously used by the authors in Machine learning in medicine, Chap. 14, Springer Heidelberg Germany, 2013. We will first test the test-retest reliability of the original variables. The test-retest reliability of the original variables should be assessed with Cronbach' s alphas using the correlation coefficients after deletion of one variable: all of the data files should produce by at least 80% the same result, as that of the non-deleted data file (alphas > 80%). Open the data file in your computer installed with SPSS statistical software.

Command:
Analyze....Scale....Reliability Analysis....transfer original variables to Variables box....click Statistics....mark Scale if item deleted....mark Correlations.... Continue ....OK.

The underneath table is in the SPSS output sheets. The test-retest reliability of the manifest variables as assessed with Cronbach's alphas using the correlation coefficients with one variable missing are given. All of the missing data files should produce at least by 80% the same result as those of the non-missing data files (alphas > 80%).

Item-Total Statistics

	Scale Mean if Item Deleted	Scale Variance if Item Deleted	Corrected Item-Total Correlation	Squared Multiple Correlation	Cronbach's Alpha if Item Deleted
gammagt	650,6425	907820,874	,892	,827	,805
asat	656,7425	946298,638	,866	,772	,807
alat	660,5975	995863,808	,867	,826	,803
bili	789,0475	1406028,628	,938	,907	,828
ureum	835,8850	1582995,449	,861	,886	,855
creatinine	658,3275	1244658,421	,810	,833	,814
creatinine clearance	929,7875	1542615,450	,721	,688	,849
esr	812,2175	1549217,863	,747	,873	,850
c-reactive protein	827,8975	1590791,528	,365	,648	,857
leucos	839,0925	1610568,976	,709	,872	,859

The table shows, that, indeed, none of the manifest variables after deletion reduced the test-retest reliability. The data are reliable. We will now perform the factor analysis with three factors in the model, one for liver, one for renal and one for inflammatory function.

Command:
Analyze....Dimension Reduction....Factor....enter variables into Variables box.... click Extraction....Method: click Principle Components....mark Correlation Matrix, Unrotated factor solution....Fixed number of factors: enter 3....Maximal Iterations for Convergence: enter 25....Continue....click Rotation.... Method: click Varimax.... mark Rotated solution....mark Loading Plots....Maximal Iterations: enter 25.... Continue....click Scores.... mark Display factor score coefficient matrix ....OK.

The underneath table is in the output.

Rotated Component Matrix[a]

	Component		
	1	2	3
gammagt	,885	,297	,177
alat	,878	,297	,167
bili	,837	,423	,215
asat	,827	,339	,206
creatinine clearance	,373	,819	,204
creatinine	,555	,694	,262
ureum	,590	,677	,285
c-reactive protein	,105	,102	,964
leucos	,325	,572	,699
esr	,411	,546	,658

Extraction Method: Principal Component Analysis.
Rotation Method: Varimax with Kaiser
Normalization.

a. Rotation converged in 5 iterations.

Gammagt = gamma glutamyl tranferase (N<50 u/l); alat = alanine aminotransferase (N<41 U/l); bili = total bilirubine (N<5 mmumol/l); asat = aspartate aminotransferase (N<37 U/l); leucos = leucocyte count (N<10.10^9/l; esr = erythrocyte sedimentation rate (N<20 mm).

The table shows the best fit coefficients of the manifest variables constituting 3 components (otherwise called latent factors here), that can be interpreted as overall liver function (1), renal function (2), and inflammatory function (3) as calculated by the factor analysis program. The component 1 has a very strong correlation with gammaGT, ALAT, and ASAT, the component 2 with creatinine clearance, creatinine, and ureum, the component 3 with c-reactive protein, leucos (leucocyte count), and esr (erythrocyte sedimentation rate). The 3 components can, thus, be interpreted as overall liver function (1), renal function (2), and inflammatory function (3).

When minimizing the outcome file, and returning to the data file, we now observe, that, for each patient, the software program has produced the individual values of the factors 1 (liver function), 2 (renal function), and 3 (inflammatory function). Using the manifest variables as predictors of death risk in a multiple binary logistic regression none of them was statistically significant (not shown). However, using the latent factors as predictors the effect on death risk was very significant. The table below shows the results.

Variables in the Equation

		B	S.E.	Wald	df	Sig.
Step 1ª	FAC1_1	3,032	,627	23,393	1	,000
	FAC2_1	1,900	,560	11,503	1	,001
	FAC3_1	8,933	1,713	27,196	1	,000
	Constant	2,597	,623	17,391	1	,000

a. Variable(s) entered on step 1: FAC1_1, FAC2_1, FAC3_1.

We may conclude, that factor analysis in clinical trials serves two purposes. It enables to produce an improved performance of conventional diagnostic tests. It enables to use the conventional diagnostic tests for health risk profiling of individual patients. Factor analysis is rarely used in clinical research. This is a pity, since factor analysis has been demonstrated to be easy and inexpensive to do, and, because it increases the statistical power of testing by reducing a multiple variables model into a sparse variables model, thereby reducing the risk of false positive effects. Another advantage of factor analysis is, that it can, sometimes, discover hidden interrelationships between variables.

We should add some limitations.

1. The original variables in a factor should be strong-positive correlated, and this issue should be tested prior to the analysis, e.g. by using Cronbach' s alphas of the complete data versus the data after one by one deletion of the original variables.
2. Background knowledge or a theory about the mechanisms causing the strong correlations is required; high correlation without an apparent reason or without an adequate underlying theory, is, clinically, generally not very relevant, and could easily be due to type I errors.

3. Correlations should not be taken equal for causalities. In order to prove causality, the introduction of a factor, and the occurrence of a subsequent event is considered better proof for that purpose.

The example used in this chapter must be confirmed with real data. Often the correlations between the clusters of original variables are smaller than in the example given. It is, then, difficult to choose, which variables to maintain in the model, and which to remove, also, which variables to add to what factor. Like with other aspects of statistics, choices to be made are rather subjective. Statistics requires a lot of biological knowledge, and a bit of maths used to answer biological questions. Particularly factor analysis is a good example of this principle.

3   Partial Least Squares (PLS)

In this section the example from the Chap. 13 reviewing multivariate analysis of variance (Manova) will be used once more. Manova is the standard method for the analysis of multivariate data. Just like Anova it is based on sums of squares (Chap. 7), and, in addition, it computes SSCP matrices (sums of squares and cross products matrices. A problem with Manova is, that it rapidly loses statistical power with increasing numbers of variables, and that computer commands may not be executed due to numerical problems with higher order calculations among components. Also, clinically, we are often more interested in the combined effects of the clusters of variables than in the separate effects of the different variables. As a simple solution composite variables can be used as add-up sums of separate variables, but add-up sums do not account the relative importance of the separate variables, their interactions, and differences in units. Instead a high quality data dimension reduction can be applied with the help of latent variables.

As an example, twelve highly expressed genes were used to predict four drug efficacy outcomes.

G1	G2	G3	G4	G16	G17	G18	G19	G24	G25	G26	G27	O1	O2	O3	O4
8	8	9	5	7	10	5	6	9	9	6	6	6	7	6	7
9	9	10	9	8	8	7	8	8	9	8	8	8	7	8	7
9	8	8	8	8	9	7	8	9	8	9	9	9	8	8	8
8	9	8	9	6	7	6	4	6	6	5	5	7	7	7	6
10	10	8	10	9	10	10	8	8	9	9	9	8	8	8	7
7	8	8	8	8	7	6	5	7	8	8	7	7	6	6	7
5	5	5	5	5	6	4	5	5	6	6	5	6	5	6	4
9	9	9	9	8	8	8	8	9	8	3	8	8	8	8	8
9	8	9	8	9	8	7	7	7	7	5	8	8	7	6	6
10	10	10	10	10	10	10	10	10	8	8	10	10	10	9	10
2	2	8	5	7	8	8	8	9	3	9	8	7	7	7	6
7	8	8	7	8	6	6	7	8	8	8	7	8	7	8	8
8	9	9	8	10	8	8	7	8	8	9	9	7	7	8	8

The variables G1-27 are highly expressed genes estimated from their arrays' normalized ratios. The variables O1-4 are drug efficacy outcome scores (the variables 20–23 from the initial data file) The data from the first 13 patients are shown only (see extras.springer.com for the entire data file entitled "chap13manova"). First the reliability of the model was assessed by assessing the test-retest reliability of the manifest predictor variables using the correlation coefficients after deletion of one variable: all of the data files should produce at least by 80% the same result as that of the non-deleted data file (alphas > 80%). SPSS 19.0 is used. Start by opening the data file in your computer with SPSS mounted.

Command:

Analyze....Scale....Reliability Analysis....transfer original variables to Variables box....click Statistics....mark Scale if item deleted....mark Correlations .... Continue....OK.

Item-Total Statistics

	Scale Mean if Item Deleted	Scale Variance if Item Deleted	Corrected Item-Total Correlation	Squared Multiple Correlation	Cronbach's Alpha if Item Deleted
geneone	80,8680	276,195	,540	,485	,902
genetwo	80,8680	263,882	,700	,695	,895
genethree	80,7600	264,569	,720	,679	,895
genefour	80,7960	282,002	,495	,404	,904
genesixteen	81,6200	258,004	,679	,611	,896
geneseventeen	80,9800	266,196	,680	,585	,896
geneeighteen	81,5560	263,260	,606	,487	,899
genenineteen	82,2040	255,079	,696	,546	,895
genetwentyfour	81,5280	243,126	,735	,632	,893
genetwentyfive	81,2680	269,305	,538	,359	,902
genetwentysix	81,8720	242,859	,719	,629	,894
genetwentyseven	81,0720	264,501	,540	,419	,903

None of the original variables after deletion reduced the test-retest reliability. The data are reliable. We will, now, perform a principal components analysis with three components, otherwise called latent variables.

Command:

Analyze....Dimension Reduction....Factor....enter variables into Variables box.... click Extraction....Method: click Principle Components....mark Correlation Matrix, Unrotated factor solution....Fixed number of factors: enter 3....Maximal Iterations plot Convergence: enter 25....Continue....click Rotation.... Method: click Varimax ....mark Rotated solution....mark Loading Plots....

Maximal Iterations: enter 25.... Continue....click Scores.... mark Display factor
score coefficient matrix ....click OK.

Rotated Component Matrix[a]

	Component		
	1	2	3
geneone	,211	,810	,143
genetwo	,548	,683	,072
genethree	,624	,614	,064
genefour	,033	,757	,367
genesixteen	,857	,161	,090
geneseventeen	,650	,216	,338
geneeighteen	,526	,297	,318
genenineteen	,750	,266	,170
genetwentyfour	,657	,100	,539
genetwentyfive	,219	,231	,696
genetwentysix	,687	,077	,489
genetwentyseven	,188	,159	,825

Extraction Method: Principal Component
Analysis.
 Rotation Method: Varimax with Kaiser
Normalization.

a. Rotation converged in 8 iterations.

   The best fit coefficients of the manifest variables constituting 3 new factors
(unmeasured, otherwise called latent, factors) are given. The latent factor 1 has a
very strong correlation with the genes 16–19, the latent factor 2 with the genes 1–4,
and the latent factor 3 with the genes 24–27. When returning to the data file, we now
observe, that, for each patient, the software program has produced the individual
values of these novel predictors. In order to fit these novel predictors with the
outcome variables, i.e., the drug efficacy scores (variables O1-4), multivariate
analysis of variance (Manova) should be appropriate. However, the large number
of columns in the design matrix caused integer overflow, and the command was not
executed. Instead we performed a univariate multiple linear regression with the
add-up scores of the outcome variables (using the Transform and Compute Variable
command) as novel outcome variable.

Command:

Transform....Compute Variable....transfer outcomeone to Numeric Expression box ....click +....outcometwo idem....click+....outcomethree idem....click +.... outcomefour idem....Target Variable: enter "summaryoutcome".... click OK.

In the data file the summaryoutcome values are displayed as a novel variable. Subsequently, a multiple linear regression with the latent factors as independent variables and the summary outcome as dependent variable will be performed.

Command:

Analyze....Regression....Dependent: enter summaryoutcome....Independent: enter Fac 1, Fac 2, and Fac 3....click OK.

Coefficients^a

Model		Unstandardized Coefficients		Standardized Coefficients	t	Sig.
		B	Std. Error	Beta		
1	(Constant)	27,332	,231		118,379	,000
	REGR factor score 1 for analysis 1	5,289	,231	,775	22,863	,000
	REGR factor score 2 for analysis 1	1,749	,231	,256	7,562	,000
	REGR factor score 3 for analysis 1	1,529	,231	,224	6,611	,000

a. Dependent Variable: summaryoutcome

All of the 3 latent predictors were, obviously, very significant predictors of the summary outcome variable. This is fine, but no information of the separate outcome variables were in the analysis. Partial least squares (Pls) is parsimonious to linear regressions with latent variables, because it, separately, includes outcome variables in the model. Because pls is not available in the basic and regression modules of SPSS, the software program R Partial Least Squares, a free statistics and forecasting software available on the internet as a free online software calculator was used (www.wessa.net/rwasp). The data file is imported directly from the SPSS file with cut/past commands.

Command:

List the selected clusters of variables: latent variable 2 (here G16-19), latent variable 1 (here G24-27), latent variable 4 (here G1-4), and latent outcome variable 3 (here O 1-4).

A square boolean matrix is constructed with "0 or 1" values, if fitted correlation coefficients to be included in the model were "no or yes", according to the underneath table.

	Latent variable	1	2	3	4
Latent variable	1	0	0	0	0
	2	0	0	0	0
	3	1	1	0	0
	4	0	0	1	0

Click "compute". After 15 seconds of computing the program produces the results. First, the data were validated using the GoF (goodness of fit) criteria. GoF = √(mean of r-square values of comparisons in model * r-square overall model), where * is the sign of multiplication. A GoF value varies from 0 to 1, and values larger than 0.8 indicate, that the data are adequately reliable for modeling.

GoF value

Overall	0.9459
Outer model (including manifest variables)	0.9986
Inner model (including latent variables)	0.9466.

The data are, thus, adequately reliable. The calculated best fit r-values (correlation coefficients) are estimated from the model, and their standard errors would be available from second derivatives. However, the problem with the second derivatives is, that they require very large data files in order to be accurate. Instead, distribution free standard errors are calculated using bootstrap resampling.

Latent Variables	Original r-value	Bootstrap r-value	Standard error	t-value	p-value
1 versus 3	0.57654	0.57729	0.08466	6.8189	0.0000
2 versus 3	0.67322	0.67490	0.04152	16.2548	0.0000
4 versus 3	0.18322	0.18896	0.05373	3.5168	0.0010

All of the three correlation coefficients (r-values) are very significant predictors of the latent outcome variable. When using a traditional multiple linear regression with the summary outcome as dependent variable and the three latent variables as predictors, the effects remained statistically significant, however, at lower levels of significance.

Command:
Analyze....Regression....Linear....Dependent: enter summaryoutcome.... Independent: enter the three summary factors 1-3....click OK.

Coefficients[a]

Model		Unstandardized Coefficients		Standardized Coefficients	t	Sig.
		B	Std. Error	Beta		
1	(Constant)	1,177	1,407		,837	,404
	summaryfac1	,136	,059	,113	2,316	,021
	summaryfac2	,620	,054	,618	11,413	,000
	summaryfac3	,150	,044	,170	3,389	,001

a. Dependent Variable: summaryoutcome

The partial least squares method produced smaller t-values, than did factor analysis (t = 3.5–16.3 versus 6.6–22.9) , but it is less biased, because it is a multivariate analysis adjusting relationships between the outcome variables. Both methods provided better t-values than did the above traditional regression analysis of summary variables (t = 2.3–11.4).

Pls can handle many more variables than the standard methods, and account the relative importance of the separate variables, their interactions and differences in units. Partial least squares method is parsimonious to traditional linear regressions with latent variables, because it can separately include outcome variables in the model.

We should add, that canonical regression is a worthwhile alternative to Pls. It computes overall linear correlation coefficients between clusters of predictor and those of outcome variables, and, with the help of repeated testing after removal of separate variables, models with fewer variables, and much better power are obtained. Canonical regression might be scientifically more rigorous, than Pls is. It is reviewed in the Chap.13.

4   Discriminant Analysis

Novel medical treatments are assessed in controlled clinical trials with health recovery as outcome. Usually, health recovery is measured as a simple outcome measure, like the normalization of body temperature, or erythrocyte sedimentation rate. However, health recovery is not a simple entity, but a rather complex one with a multifactorial nature. And current clinical trials increasingly include multiple outcome variables, like a battery of physical, laboratory and imaging tests. As the evaluation of novel treatments lies at the heart of health research for the benefit of mankind, statistical models, that can handle multiple outcome variables are required. Traditionally, Manova (multivariate analysis of variance) is used for the analysis of multiple outcome variables. However, a problem is that its performance is largely dependent on the levels of correlation between the outcome variables (Cole et al, Psychol Bull 1994; 115: 465). If negative, it will perform best, if positive it will do so worst. Unfortunately, in practice mostly strong positive correlations are present.

Discriminant analysis, although described as a method for binary data as early as 1935 by Sir Ronald Fisher (The design of experiments, McMillan London 9th edition 1971) professor of statistics Cambridge UK, was proposed in its current version by William Klecka (Discriminant analysis, Google books, 1973), professor of behavioral sciences, University of Cincinnati USA 1973 (Fisher 1935; Klecka 1973). It eliminates the effects of the correlations between multiple outcome variables from the analysis by a technique called orthogonal linear modeling. It is, currently, widely used in the field of behavioral sciences like social sciences, marketing, operational research and applied sciences, but is, virtually, unused in medicine, despite the omnipresence of multiple outcome variables in this branch of science. When searching Medline, we only found 2 genetic studies (Ogah et al, Egypt Poult Sci 2011; 31: 429, Guo et al, Biostatistics 2005; 1: 1), 4 diagnostic studies (Wernecke, Clustering with parameters from blood tests, Wiley encyclopedia of clinical trials 2007a, Wernecke, Diagnosis of neuroborreliosis burgdorferi, Wiley encyclopedia of clinical trials 2007b, Feigner et al., J Appl Res 2001; 21: 1, Adams J Clin Neuropsychol 1979; 1: 259), and 2 clinical treatment trials (Glasson et al., Investr Ophthalmol Vis Sci 2003; 44: 5116), Fens et al., Am J Respir Crit Care Med 2009; 180: 1076).

Like with factor analysis (see Sect. 2), orthogonal latent variables are inferred from the manifest variables in the data set given, and they are here called functions. The current section, using a simulated case-study, assesses the performance of discriminant analysis versus traditional analysis in a clinical trial, comparing the effect of different treatment modalities on a battery of laboratory outcome variables. This is an introduction to discriminant analysis, and was written as a hand-hold presentation accessible to clinicians, and as a must-read publication for those new to the methods. It is the authors' experience, as master class professors, that students are eager to master adequate command of statistical software. For their benefit all of the steps of the novel method from logging in to the final result using SPSS statistical software will be given. As an example, three treatment regimens for the treatment of sepsis with multi-organ failure were compared. The data file is entitled "chap23discriminantanalysis", and is in extras.springer.com. It is previously used by the authors in Machine in medicine, Chap. 17, Springer Heidelberg Germany, 2013. The laboratory values after treatment were used as measure for treatment success. Multivariate analysis of variance (Manova) with the treatment modality as predictor and the laboratory values as multiple outcome variables was not significant with a Pillai's test with p = 0.082. Roy's largest root test was statistically significant, but did not meet its assumption of adequate power and accurate F approximation, and, so, the Manova concluded, that there was not a significant difference between the three treatments. When performing multiple Anovas (analyses if variance) with a single lab value as outcome and the three treatment modalities as predictors, several (6 out of nine were statistically significant with p-values between 0.049 and 0.010. However, Anova does not account the relationship between the dependent variables. Also, we were more interested in the effect of treatment on the combined outcome result, than the effects on the separate outcome variables. Therefore, discriminant analysis was performed using SPSS statistical software. Open the data file in your computer.

Command:
Analyze....Classify....Discriminant Analysis....enter Grouping Variable: treatment modality....Define Range 1 to 3....enter Independents: variables 1-10.... Statistics: mark Unstandardized....mark Separate-groups covariance....Continue ....Classification: mark All groups equal....mark Summary table....mark Within-groups....mark Combined groups....Continue....Save: mark Discriminant scores....Continue....click OK.

The underneath table gives the b-values (regression coefficients) of two functions.

	Function	
	1	2
gammagt	,000	-,007
asat	,001	,002
alat	,002	,003
bili	-,006	,008
ureum	,055	,019
creatinine	,004	-,003
creatinine clearance	-,023	,005
esr	,001	,005
c-reactive protein	-,024	,026
leucos	-,015	-,057
(Constant)	-3,062	,405

Unstandardized coefficients

The two above functions are orthogonal discriminant functions (latent variables) of 10 outcome variables.

Function 1 = $-3.062 + 0.000$ gammagt $+ 0.001$ asat $+ 0.002$ alat $- 0.006$ bili$+$
              0.055 ureum $+ 0.004$ creatinine $- 0.023$ creatinine clearance$+$
              0.001 esr $- 0.015$ c-reactive protein $- 0.015$ leucos.

Function 2 can be described similarly using the results of the function 2 of the above table. the two functions are the best fit multiple linear regression models to summarize the outcome data.

Test of Function(s)	Wilks' Lambda	Chi-square	df	Sig.
1	,432	31,518	20	,049
2	,865	5,432	9	,795

Of the two functions, which are, otherwise, called variates or latent variables, also test statistics are given. The table shows, that, according to function 1, there is a significant difference between the three treatment modalities at $p = 0.049$, according to function 2 the differences are not statistically significant at $p = 0.795$.

treatmentmodality	Function	
	1	2
1,00	-,541	,392
2,00	-,641	-,508
3,00	1,604	-,028

Also the mean function scores for each treatment is given. The underneath graph gives treatment-group plots, called the treatment-group centroids (dark squares). They are the mean function scores for each group. The circles are the individual scores. The scores of function 1 is estimated along the x –axis, function 2 along the y-axis. The differences in mean scores along the x-axis is significantly different according to Wilk's test (above table). However, with 3 groups, we are unable to tell, whether treatment 3 is different from 1, from 2, or from both 1 and 2. The difference in mean scores along the y-axis is small and according to Wilk's test insignificant.

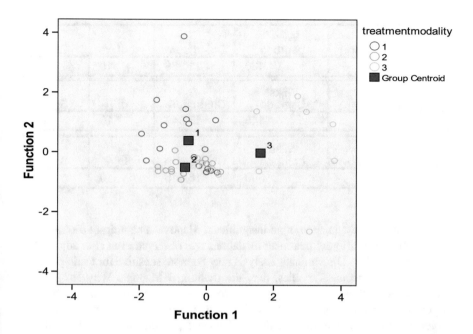

In order to test where the differences between the treatment groups are significant, 3 subgroup analyses are required.

Command:

Analyze....Classify....Discriminant Analysis....enter Grouping Variable: treatment modality....Define Range 2 to 3....enter Independents: variables 1-10.... Statistics: mark Unstandardized....mark Separate-groups covariance....Continue ....

Classification: mark All groups equal....mark Summary table....mark Within-groups....mark Combined groups....Continue....Save: mark Discriminant scores....Continue....OK.

The same procedure must be performed for comparing treatments 2 versus 1, and 1 versus 3. Subgroup discriminant analyses of respectively treatment 2 versus 3, treatment 2 versus 1, and treatment 1 versus 3 are in the table below. Bonferroni adjustment of the rejection p-value with three tests equals

p − value = 0.05 ∗ 2/(3(3 − 1)) = 0.017 (∗ = symbol of multiplication).

Discriminant analysis, thus, shows that, even after correction for multiple testing, treatment 2 performs significantly better than treatment 3. The mean difference between treatment 1 and treatment 3 is virtually similar in magnitude, but probably due to the small sample size (treatment 2, n = 19; treatment 1, n = 14) it did not obtain statistical significance.

Test of Function(s)	Wilks' Lambda	Chi-square	df	Sig.
1	,436	21,586	10	,017

Test of Function(s)	Wilks' Lambda	Chi-square	df	Sig.
1	,446	15,340	10	,120

Test of Function(s)	Wilks' Lambda	Chi-square	df	Sig.
1	,500	16,645	10	,083

In this case-study, in spite of an insignificant Manova, a significant discriminant analysis between different treatment modalities was observed even after adjustment for multiple testing. Discriminant analysis may be more sensitive for testing studies with multiple outcome variables, than the traditional Manova. Why so? This is probably, because the functions produced by the orthogonal linear modeling may better fit multivariate data. Also a problem with Manova is, that its performance is largely dependent on the correlations between the dependent variables. If negative it performs best, if positive it performs worst. In discriminant analysis the effect of the correlations between the dependent variables is eliminated.

Discriminant analysis is very similar to factor analysis (principal components analysis). Both apply orthogonal modeling in order to eliminate the effect of correlations between variables, but the former does so using the dependent variables, while the latter using the independent variables. Also, discriminant analysis goes one step further. It includes a grouping predictor variable in the statistical model, e.g. treatment modality. The scientific question "is the treatment modality a significant predictor of a clinical improvement" is, subsequently, assessed by the question "is the outcome clinical improvement a significant predictor of the odds of having had a particular treatment. This reasoning may seem incorrect, using an outcome for making predictions, but, mathematically, it is no problem. It is just a matter of linear cause-effect relationships, but just the other way around, and it works very conveniently with "messy" outcome variables, like in the example given.

Both methods are currently often listed as machine learning methods, because, unlike traditional statistical tests, they are able to handle large numbers of variables. They are modern computer intensive methods, sometimes also listed as artificial intelligence methods, that allow computers to develop algorithms useful for the benefit of mankind, and based on empirical data (the training data).

Because of the presence of a grouping predictor variable, discriminant analysis is sometimes called supervised machine learning, whereas factor analysis is called unsupervised learning.

These methods are also able to account many limitations of the traditional methods, like accounting the relative importance of the multiple variables, their interactions and differences in units. And, in addition, as observed in the present report, they may be more sensitive than traditional statistical methods.

Although developed and currently widely used in the fields of behavioral sciences, social sciences, marketing, operational research and applied sciences, these methods are little used in clinical research so far.

We have to add, that discriminant analysis is, like Manova, based on linear regression, and most of its prior assumptions are similar, i.e., Gaussian data, homogeneity of variances, random data. However, a strong relationships between variables, sometimes called collinearity, is prohibitive for linear regression, but for discriminant analysis, and many other machine learning methods it is not, either of them, because they are not based on correlations, but distances between individuals (Cluster analysis, 2012, http://en.wikipedia.org, Sequence clustering, 2012, http://en.wikipedia.org) or because the effects of correlations is minimized by orthogonal modeling of variables (Barthelemew, Br J Math Stat Psychol 1995; 48: 211).

The main points of this section.

1. Current clinical trials increasingly include multiple outcome variables.
2. Manova (multivariate analysis of variance) is suitable for analysis, but suffers from power loss, as the correlation between the outcome variables is commonly positive.
3. Just like factor analysis (used for predictor variables), discriminant analysis (used for outcome variables) identifies 2 or more latent variables to explain all of the observed variables. Because both methods linearly models multiple variables along an x and y-axis, intervariable relationships do not further have to be taken into account.
4. Discriminant analysis goes one step further than factor analysis, and includes a grouping predictor variable, namely treatment modality, and answers the scientific question "is the outcome significantly related to the odds of having had a particular treatment".
5. Discriminant analysis is more sensitive for testing studies with multiple outcome variables than the traditional Manova.
6. With multiple treatment modalities subgroup analyses with or without Bonferroni adjustment are possible to find out, where the differences between the treatment groups are.
7. Discriminant analysis is available in SPSS statistical software and a wonderful and welcome methodology for the analysis of clinical trials with multiple outcome variables.

5 Conclusion

Latent variables are unmeasured variables, inferred from directly measured variables. The main purpose for using them is data variables reduction, otherwise called data dimension reduction. Current research increasingly involves multiple variables, and traditional statistical models tend to get powerless with too many variables included. Multiple analyses is, generally, no solution, because of increased type I errors due to multiple testing. In contrast, a few latent variables, instead of multiple manifest variables, does have advantages:

(1) less risk of power loss due to too many variables in a model,
(2) no risk of false positive results like with multiple testing.

However, a disadvantage is, that latent variables are rather subjective, because they are dependent on subjective decisions to cluster some measured variables, and remove others.

The current chapter reviewed, how to construct high quality latent variables, and how they can be successfully implemented in many forms of modern data analyses, including factor analysis, partial least squares, and discriminant analysis.

References

Adams KM (1979) Linear discriminant analysis in clinical neuropsychology. J Clin Neuropsychol 1:259–272

Anderson TW (1958) An introduction to multivariate statistical analysis. Wiley, New York

Barthelemew DJ (1995) Spearman and the origin and development of factor analysis. Br J Math Stat Psychol 48:211–220

Cole DA, Maxwell SE, Arvey R, Salas E (1994) How the power of MANOVA can both increase and decrease as a function of the intercorrelations among the dependent variables. Psychol Bull 115:465–474

Feigner JP, Sverdlov L (2001) The use of discriminant analysis to separate a study population by treatment subgroups in a clinical trial with a new pentapeptide antidepressant. J Appl Res 21:1–6

Fens N, Zwinderman AH, Van der Schee MP, De Nijs SB, Dijkers E, Roldaan AC, Cheung D, Bel EH, Sterk PJ (2009) Exhaled breath profiling enables discrimination of chronic obstructive pulmonary disease and asthma. Am J Respir Crit Care Med 180:1076–1082

Fisher RA. The design of experiments. McMillan, London, first edition 1935, ninth edition 1971

Glasson MJ, Stapleton F, Keay L, Sweeney D, Willcox MD (2003) Differences in clinical parameters and tear film of tolerant and intolerant contact lens wearers. Invest Ophthalmol Vis Sci 44:5116–5124

Guo Y, Hastle T, Tibshirani R (2005) Regularized discriminant analysis and its application in microarrays. Biostatistics 1:1–18

Hochreiter S, Clevert DA, Obermayer K (2006) A new summarization method for affymetrix probe level data. Bioinformatics 22:943–949

Klecka WR (1973) Discriminant analysis. Google books, Google, 08-01-2012

Meng J (2011) Uncover cooperative gene regulations by microRNAs and transcription factors in glioblastoma using a nonnegative hybrid factor. www.cmsworldwide.com?ICASS2011

Ogah DM, Momoh OM, Dim NI (2011) Application of canonical discriminant analysis for assessment of genetic variation in Muscovy duck ecotypes in Nigeria. Egypt Poult Sci 31:429–436

Seal H (1964) Statistical methodologies. Methuen & Co, London

Wernecke KD (2007a) Clustering with parameters from blood tests. Wiley Encyclopedia of clinical trials, 2007. Wiley, London

Wernecke KD (2007b) Diagnosis of neuroborreliosis burgdorferi. Wiley encyclopedia of clinical trials, 2007. Wiley, London

Chapter 24
Partial Correlations

Removing Interaction Effects from Linear Data

Abstract The outcome of clinical research is, generally, affected by many more factors than a single one, and multiple regression assumes, that these factors act independently of one another, but why should they not affect one another. This chapter is to assess how partial correlation can be used to remove interaction effects from linear data.

Without the partial correlation approach the conclusion from studies might have been: no definitive conclusion about the effects of factors is possible, because of a significant interaction between such factors. The partial correlation analysis allows to conclude that multiple interacting factors have a significant linear relationship with a single outcome variable.

Keywords Partial correlations · Removing interactions from linear data · Partial regression analysis

1 Introduction, History, and Background

Partial correlations were first described by Ronald Fisher (1890–1962), the inventor of the F-test (see also the Chap. 7). Nowadays, we have come a long way to recognize that cardiovascular factors are affected by many more factors, than a single one, and multiple regression assumes, that all of these factors act independently of one another, but why should that be true (Cleophas, Sense and nonsense of regression modeling for increasing precision of clinical trials, Clin Pharmacol Ther 2003; 74: 295). If all of these factors affect the outcome, why should they not affect one another. The point is, that, if you don't have an adequate method to study this issue, you will never know.

As an example, both calorie intake and exercise are predictors of weight loss. However, exercise makes you hungry and patients on weight training may be

Electronic Supplementary Material The online version of this chapter (https://doi.org/10.1007/978-3-030-61394-5_24) contains supplementary material, which is available to authorized users. The videos can be accessed by scanning the related images with the SN More Media App.

inclined to reduce their calorie intake. So, there may be an interaction between calorie intake and exercise on weight loss. Suppose an interaction variable (x = calorie intake * exercise, with * symbol of multiplication) is included in the analysis, and a significant interaction is established. With a significant interaction the overall analysis of these data is rather meaningless. The best method to find the true effect of calorie intake on weight loss or of exercise on weight loss, would be, to repeat the study with one of the two variables held constant, and the other allowed to change. However, this would be laborious and costly.

Professor Udny Yule, professor of statistics Cambridge University UK in 1897 (On the theory of correlation, J R Stat Soc 1897; 60: 812), invented the next best method for that purpose, and called it partial regression analysis: the effect of interaction is removed from the data by artificially holding one or more independent variables constant during the analysis (Yule 1897). The great merit of partial correlation analysis is, that it cannot only establish the amount of interaction, but, that it can also remove its effects, and, thus, establish, what would have happened, if there had been no interaction.

The method of partial regression never received much attention, because it is computationally intensive, and, without the help of a computer, hard to perform, particularly, if you wish to hold *multiple* variables simultaneously constant. Instead, simpler methods are, generally, performed. Even in 1986, Willett and Stampfer (Am J Epidemiol 1986; 124: 17) assessed the effects of calorie intake and multiple nutrients intake on health by a two stage procedure (Willett and Stampfer 1986). First, they adjusted calorie intake for nutrients intake, and, then, they used the adjusted values for assessing the effect on health. This procedure is not adequate, because nutrients intake does not only affect calorie intake but also health, and partial correlation analysis would have been the appropriate approach.

Partial correlation, despite its potential to remove the effect of interactions from the data, is little used so far: searching the Internet, we found mainly studies in the field of behavioral sciences like social sciences, marketing, operational research and applied sciences (Waliczek, A primer on partial correlation coefficients, Governmental Editions Washington DC, 1996, ED393882). In clinical research it is virtually unused: when searching Medline, we only found a few time series assessments (Box et al. 2008; Brockwell and Davis 2009), two psychiatric trials (Kazdin et al. 1983; Mattick and Clarke 1998), one genetic study (Paez et al. 2004), and one cardiovascular risk factor study (Larsson et al. 1984). But this lack of use is probably a matter of time, now that it is available in SPSS (since 2005, SPSS 13) and many other software packages.

2 Data Example

The outcome of cardiovascular research is, generally, affected by many more factors than a single one, and multiple regression assumes that these factors act independently of one another, but why should they not affect one another. This chapter is to assess whether partial correlation can be used to remove interaction effects from linear data. Both calorie intake and exercise are significant independent predictors of weight loss. However, exercise makes you hungry and patients on weight training are inclined to reduce (or increase) their calorie intake. Can partial correlations methods adjust the interaction between the two predictors.

Var 1	Var 2	Var 3	Var 4	Var 5
1,00	0,00	1000,00	0,00	45,00
29,00	0,00	1000,00	0,00	53,00
2,00	0,00	3000,00	0,00	64,00
1,00	0,00	3000,00	0,00	64,00
28,00	6,00	3000,00	18000,00	34,00
27,00	6,00	3000,00	18000,00	25,00
30,00	6,00	3000,00	18000,00	34,00
27,00	6,00	1000,00	6000,00	45,00
29,00	0,00	2000,00	0,00	52,00
31,00	3,00	2000,00	6000,00	59,00
30,00	3,00	1000,00	3000,00	58,00
29,00	3,00	1000,00	3000,00	47,00
27,00	0,00	1000,00	0,00	45,00
28,00	0,00	1000,00	0,00	66,00
27,00	0,00	1000,00	0,00	67,00

Var 1 weight loss (kg)
Var 2 exercise (times per week)
Var 3 calorie intake (cal)
Var 4 interaction
Var 5 age (years)

Only the first fifteen patients are given in the above table, the entire file is entitled "chap24partialcorrelations" and is in extras.springer.com. It is previously used by the authors in Machine learning in medicine a complete overview, Chap. 26, Springer Heidelberg Germany, 2015. We will, first, perform a linear regression of these data. SPSS 19.0 is used for the purpose. Open the data file in your computer with SPSS installed.

Command:
Analyze....Regression....Linear....Dependent variable: enter weightloss.... Independent variables: enter exercise and calorieintake....click OK.

Coefficients[a]

Model		Unstandardized Coefficients		Standardized Coefficients	t	Sig.
		B	Std. Error	Beta		
1	(Constant)	29,089	2,241		12,978	,000
	exercise	2,548	,439	,617	5,802	,000
	calorieintake	-,006	,001	-,544	-5,116	,000

a. Dependent Variable: weightloss

The output sheets show, that both calorie intake, and exercise are significant independent predictors of weight loss. However, interaction between exercise and calorie intake is not accounted. In order to check, an interaction variable (x_3 = calorie intake * exercise, with * symbol of multiplication) is added to the model.

Command:
Transform data....Compute Variable....in Target Variable enter the term "interaction"....to Numeric Expression: transfer from Type & Label "exercise" ....click * ....transfer from Type & Label calorieintake....click OK.

The interaction variable is added by SPSS to the data file and is entitled "interaction". After the addition of the interaction variable to the regression model as third independent variable, the analysis is repeated.

Coefficients[a]

Model		Unstandardized Coefficients		Standardized Coefficients	t	Sig.
		B	Std. Error	Beta		
1	(Constant)	34,279	2,651		12,930	,000
	interaction	,001	,000	,868	3,183	,002
	exercise	-,238	,966	-,058	-,246	,807
	calorieintake	-,009	,002	-,813	-6,240	,000

a. Dependent Variable: weightloss

The output sheet now shows, that exercise is no longer significant, and interaction on the outcome is significant at p = 0.002. There is, obviously, interaction in the study, and the overall analysis of the data is, thus, no longer relevant. The best method to find the true effect of exercise would be, to repeat the study with calorie intake held constant. Instead of this laborious exercise, a partial correlation analysis with calorie intake held artificially constant can be adequately performed, and would provide virtually the same result. Partial correlation analysis is performed using the SPSS module Correlations.

Command:

Analyze....Correlate....Partial....Variables: enter weight loss and calorie intake
....Controlling for: enter exercise....OK.

Correlations

Control Variables			weightloss	calorieintake
exercise	weightloss	Correlation	1,000	-,548
		Significance (2-tailed)	.	,000
		df	0	61
	calorieintake	Correlation	-,548	1,000
		Significance (2-tailed)	,000	.
		df	61	0

Correlations

Control Variables			weightloss	exercise
calorieintake	weightloss	Correlation	1,000	,596
		Significance (2-tailed)	.	,000
		df	0	61
	exercise	Correlation	,596	1,000
		Significance (2-tailed)	,000	.
		df	61	0

The upper table shows, that, with exercise held constant, calorie intake is a significant negative predictor of weight loss with a correlation coefficient of −0.548 and a p-value of 0.0001. Also partial correlation with exercise as independent and calorie intake as controlling factor can be performed.

Command:

Analyze....Correlate....Partial....Variables: enter weight loss and exercise....
Controlling for: enter calorie intake....OK.

The lower table shows that, with calorie intake held constant, exercise is a significant positive predictor of weight loss with a correlation coefficient of 0.596 and a p-value of 0.0001.

Why do we no longer have to account interaction with partial correlations. This is simply because, if you hold a predictor fixed, this fixed predictor can no longer change and interact in a multiple regression model.

Also higher order partial correlation analyses are possible. E.g., age may affect all of the three variables already in the model. The effect of exercise on weight loss with calorie intake and age fixed can be assessed.

Command:

Analyze....Correlate....Partial....Variables: enter weight loss and exercise.... Controlling for: enter calorie intake and age....OK.

Correlations

Control Variables			weightloss	exercise
age & calorieintake	weightloss	Correlation	1,000	,541
		Significance (2-tailed)	.	,000
		df	0	60
	exercise	Correlation	,541	1,000
		Significance (2-tailed)	,000	.
		df	60	0

In the above output sheet, it can be observed, that the correlation coefficient is still very significant. With age and calorie intake held constant, exercise is a very significant predictor of weight loss.

3 Conclusion

Without the partial correlation approach the conclusion from this study would have been: no definitive conclusion about the effects of exercise and calorie intake is possible, because of a significant interaction between exercise and calorie intake. The partial correlation analysis allows to conclude, that both exercise and calorie intake have a very significant linear relationship with weight loss effect.

References

Box GEP, Jenkins GM, Reinsel GC (2008) Time series analysis, forecasting and control, 4th edn. Wiley, New York

Brockwell P, Davis R (2009) Time series: theory and methods, 2nd edn. Springer, Heidelberg

Cleophas TJ (2003) Sense and nonsense of regression modeling for increasing precision of clinical trials. Clin Pharmacol Ther 74:295–297

Kazdin AE, French NH, Unis AS, Esveldt-Dawson K, Sherick RB (1983) Hopelessness, depression, and suicidal intent among psychiatrically disturbed inpatient children. J Cons Clin Psychol 51:504–510

Larsson B, Svardsudd K, Welin L, Wilhelmsen L, Bjorntorp P, Tibblin G (1984) Abdominal adipose tissue distribution, obesity, and risk of cardiovascular disease and death: 13 year follow up of participants in the study of men born in 1913. BMJ 288:1401–1409

Mattick RP, Clarke JC (1998) Development and validation of measures of social phobia scrutiny fear and social interaction anxiety. Beh Res Ther 36:455–470

Paez JG, Jänne PA, Lee JC et al (2004) EGFR mutations in lung cancer: correlations with clinical response to gefitimib therapy. Science 304:1497–1482

Waliczek TM (1996) A primer on partial correlation coefficients. Governmental Editions, Washington, DC, p ED393882

Willett W, Stampfer MJ (1986) Total energy intake: implications for epidemiological analyses. Am J Epidemiol 124:17–22

Yule GU (1897) On the theory of correlation. J Roy Statist Soc 60:812–854

Chapter 25
Functional Data Analysis (FDA) Basis

Gene Expression Levels for Predicting Drug Efficacy Outcomes

Abstract Functional data analyis (FDA) analyzes data, that provide information of curves, and surfaces. Often, it is used to analyze, and predict times series data. Traditional analyses of time series are mostly based on multivariate analyses of variance, that completely ignore the effect of data smoothing, and suffer from power loss due to a positive correlation between the outcome variables. FDA uses smoothed curves for obtaining novel functions, that can be applied for making predictions in general practice. And, so, FDA may look like traditional multivariate analysis, but it is more powerful, because results from smoothed data are the start of something new. FDA, if used as a data generating activity, is, of course, pretty meaningless, and sound prior hypotheses are an essential background. In this chapter an example is given of an FDA procedure, successfully including: (1) data dimension reduction with principal components analysis, (2) optimal scaling of the reduced data with spline smoothing, (3) ridge, lasso, and elastic net regularization for increased precision, in a single analysis, and leading to a final analysis result with a better precision, than that of the traditional regressions of the same data.

Keywords Functional data analysis · Gene expression levels · Drug efficacy outcomes · Principal components · Optimal scaling with spline smoothing · Regularized regression · Ridge · Lasso · Elastic net shrinkages

1 Introduction, History, and Background

Functional data analysis (FDA) started in 1995. It is a branch of statistics, that analyzes data, providing information of curves, surfaces, or anything else varying over a continuum. It integrates various methodologies, like data smoothings, principal components analyses, optimal scaling into a single data analysis. Often, it is being used to better analyze, model, and predict times series data. In a systematic review of Ullah and Finch in BMC Medical Research Methodology (2013; 313: 43), entitled "Applications of functional data analysis" the key aspects have been summarized:

data smoothing techniques,
data reduction,
data clustering,
functional linear modeling,
forecasting methods.

Traditional analyses of time series are mostly based on multivariate analyses of variance, that completely ignore the effect of data smoothing, and suffer from power loss due to a positive correlation between the outcome variables. In addition, currently we live in an era of data driven models, and hypothesis driven research with proper focus on the risks of interactions is not usual anymore. Ramsay, from McGill University Montreal Canada, one of the founders of FDA, stated in his edition entitled "Functional data analysis" (2005, Springer Heidelberg Germany):

1. that smoothing enables to obtain functional representations from a finite set of data,
2. it is human nature to think of functionalities through data models, and
3. functionalities is the objective of many, if not all, analyses.

FDA uses smoothed curves for obtaining novel functions, that can, subsequently, be applied for making predictions in general practice. And, so, FDA may look like traditional multivariate analysis, but it is more powerful, because results from smoothed data are the start of something new. FDA, if used as a data generating activity is pretty meaningless, and sound prior hypotheses are an essential background. Also, and even particularly so, in this field of statistics. In this chapter examples are given of an FDA procedure successfully including:

1. data dimension reduction,
2. optimal scaling with spline smoothing,
3. increasing precision with ridge, lasso, and elastic net regularization,

in a single analysis, and leading to a final analysis result with a better precision than that of the traditional regressions of the same data. In the current chapter principal components analysis, optimal scaling with spline smoothing, regularization with ridge regression are applied in a combined analysis of a study of gene expression levels and drug efficacy outcomes.

2 Principles of Principal Components Analysis and Optimal Scaling, a Brief Review

Principal components methodology uses factor analysis theory (see also the Chap. 23). Two or three unmeasured factors are identified to explain a much larger number of measured variables. As an example, in a patient data set the relationships of 6 original variables with two novel variables, the liver and renal function are measured. It can be observed that ureum, creatinine and creatinine clearance have a very strong positive correlation with renal function and the other three with liver function. It can be demonstrated, that, by slightly rotating both x and y-axes, the model can be fitted even better. The correlation coefficients between the original variables and the estimated novel variables are called factor loadings. The estimated novel variables are called latent variables. Principal components analysis is some-what similar to multivariate analysis, with the latent variables as outcomes, and the original variables as predictors. However, unlike multivariate analysis, it is not influenced by multicollinearity, because of the orthogonal relationship between the original and novel (latent) variables, as shown in the underneath graph.

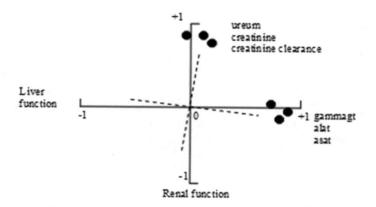

The principles of optimal scales are also briefly reviewed (see also Chap. 16). In clinical trials the research question is, often, measured with multiple variables. For example, the expressions of a number of genes can be used to predict the efficacy of cytostatic treatment, repeated measurements can be used in randomized longitudinal trials, and multi-item personal scores can be used for the evaluation of antidepres-sants. Many more examples can be given. Multiple linear regression analysis is, often, used for analyzing the effect of predictors (x-axis variables) on outcome variables (y-axis variables). A problem with linear regression is, that consecutive levels of the predictor variables are assumed to be equal, while in practice this is virtually never true.

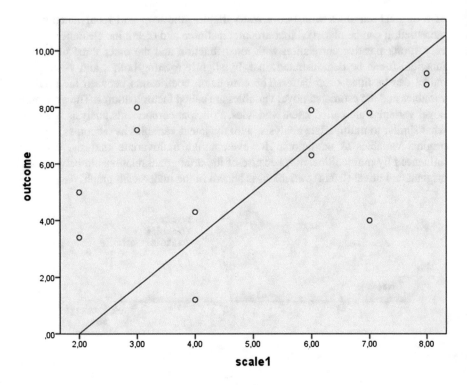

The above graph gives an example of a continuous predictor variable (x-variable) on a scale 0–10. Patients with the predictor values 0, 1, 5, 9 and 10 are missing. Instead of a scale of integers between 0 and 10, other scales are possible, e.g. a scale of two or four parts (Table 3.1). Any scale used is, of course, arbitrary, and can be replaced with another one. In the example the following scales are used.

Scale 1: 0, 1, 2, 3, 4, 5, 6, 7, 8, 9, 10.
Scale 2: 1, 2 (1 = 0–5 from scale 1; 2 = 5–10).
Scale 3: 1, 2, 3, 4 (1 = 0–2.5 from scale 1; 2 = 2.5–5; 3 = 5–7.5; 4 = 7.5–10).

The underneath table of linear regressions from the above graph-data shows, that each scale produced a different pattern of results with one result better than the other.

Coefficients[a]

Model		Unstandardized Coefficients		Standardized Coefficients	t	Sig.
		B	Std. Error	Beta		
1	(Constant)	3,351	1,647		2,034	,069
	scale1	,548	,302	,497	1,813	,100

	Unstandardized Coefficients		Standardized Coefficients		
Model	B	Std. Error	Beta	t	Sig.
1 (Constant)	2,367	2,032		1,165	,271
scale2	,497	,257	,521	1,932	,082

	Unstandardized Coefficients		Standardized Coefficients		
Model	B	Std. Error	Beta	t	Sig.
1 (Constant)	2,217	1,647		1,346	,208
scale3	,620	,246	,623	2,520	,030

a. Dependent Variable: outcome

With the scales 2 and 3 a gradual improvement of the t-values and p-values (Sig.) is observed. Optimal scaling is a method designed to maximize the relationship between a predictor, and an outcome variable.

3 Functional Data Analysis, Data Example

We will use a 250 patient data file for perforrning an entire FDA procedure. The data file has already been used in the Chap. 13 of this edition. The 250 patients' data-file was supposed to include 27 variables consistent of both patients' microarray gene expression levels and their drug efficacy scores. Gene expression levels were used for predicting drug efficacy outcomes. All of the variables were standardized by scoring them on an 11 points linear scale (0–10). The following clusters of genes were highly correlated with one another: the variables 1–5, the variables 16–19, and the variables 24–27. The variables 20–23 were supposed to represent drug efficacy scores and were clustered as the outcome variables. The data file is in extras.springer. com, and is entitled "chap13manova". Before opening the file in SPSS statistical software has to be installed in your computer.

3.1 Principal Components Analysis

We will start with a principal components analysis. SPSS' Data Dimension Reduction module was used. First, the reliability of the model was assessed by testing the test-retest reliability of the original variables. The test-retest reliability of the original variables should be assessed with Cronbach's alphas using the correlation coefficients after deletion of one variable: all of the data files should produce at least by 80% the same result, as that of the non-deleted data file (alphas >80%).

Command

Analyze....Scale....Reliability Analysis....transfer original variables to Variables box....click Statistics....mark Scale if item deleted....mark Correlations.... Continue ....click OK.

In the output sheets several tables are given. The data can be validated, if there is a strong correlation between the scores within the clusters (strong collinearity). The test-retest reliability of the variables as assessed with Cronbach's alphas will be given. All of the reliability assessments should produce at least for 80% the same result (Cronbach's alphas >80%).

Item-Total Statistics

	Scale Mean if Item Deleted	Scale Variance if Item Deleted	Corrected Item-Total Correlation	Squared Multiple Correlation	Cronbach's Alpha if Item Deleted
VAR00001	79,6200	277,273	,547	,486	,903
VAR00002	79,6200	263,980	,724	,700	,896
VAR00003	79,5120	264,749	,743	,671	,895
VAR00004	79,5480	284,361	,477	,385	,906
VAR00005	81,0720	264,501	,566	,386	,903
VAR00016	80,3720	257,166	,714	,623	,895
VAR00017	79,7320	268,494	,665	,582	,898
VAR00018	80,3080	265,869	,588	,477	,902
VAR00019	80,9560	255,038	,719	,555	,895
VAR00024	80,2800	245,696	,719	,611	,895
VAR00025	80,0200	272,702	,507	,340	,905
VAR00026	80,6240	244,581	,714	,627	,896

The above table shows, that, indeed, none of the original variables after deletion, reduced the test-retest reliability. The data are reliable. We will now perform the principal components analysis with three components, otherwise called latent variables.

Command

Analyze....Dimension Reduction....Factor....enter variables into Variables box.... click Extraction....Method: click Principle Components....mark Correlation Matrix, Unrotated factor solution....Fixed number of factors: enter 3....Maximal Iterations plot Convergence: enter 25....Continue....click Rotation.... Method: click Varimax ....mark Rotated solution....mark Loading Plots....

Maximal Iterations: enter 25.... Continue....click Scores.... mark Display factor score coefficient matrix ....click OK.

Again several tables are given in the output sheets. The results of principal components analysis with 3 components is in the output. Component 3 has a strong positive correlation with the MVs (manifest variables) 1–4, component 2 with MVs 24–27, and component 1 with MVs 16–19.

Rotated Component Matrix[a]

	Component		
	1	2	3
VAR00001	,249	,136	,797
VAR00002	,582	,128	,652
VAR00003	,616	,163	,586
VAR00004	,003	,364	,770
VAR00005	,711	,063	,211
VAR00016	,819	,242	,127
VAR00017	,500	,516	,217
VAR00018	,379	,482	,306
VAR00019	,719	,289	,235
VAR00024	,585	,617	,079
VAR00025	,160	,675	,228
VAR00026	,634	,563	,050
VAR00027	,084	,823	,172

Extraction Method: Principal Component Analysis.
Rotation Method: Varimax with Kaiser Normalization.

a. Rotation converged in 8 iterations.

The above table shows the best fit coefficients of the original variables constituting the 3 components, The component 1 has a very strong correlation with the variables 16–19, the component 2 with the variables 24–27, and the component 3 with the variables 1–4. These 3 components can, thus, be interpreted as the latent predictor variables. When minimizing the outcome file, and returning to the data file, we now observe, that, for each patient, the software program has produced the individual values of these three novel predictors.

In order to fit these novel predictors with the outcome variables, the drug efficacy outcome scores (variables 20–23), multivariate analysis of variance (Manova) should be appropriate, given the continuous nature of the 4 outcome variables. However, due to the large number of columns, the design matrix caused integer overflow, and the large number of columns caused too many levels within some components, as well as numerical problems with higher order interactions among components, and the command was not executed. Instead, we performed an univariate multiple linear regression with the add-up scores of the outcome variables, as novel outcome variable, the summaryoutcome values. The underneath tables give the results. All of the 3 latent predictors were very significant independent predictors of the add-up outcome variable.

Model Summary

Model	R	R Square	Adjusted R Square	Std. Error of the Estimate
1	.847[a]	.717	.714	3.65061

a. Predictors: (Constant), REGR factor score 3 for analysis 1, REGR factor score 2 for analysis 1, REGR factor score 1 for analysis 1

ANOVA[b]

Model		Sum of Squares	df	Mean Square	F	Sig.
1	Regression	8311.009	3	2770.336	207.874	.000[a]
	Residual	3278.435	246	13.327		
	Total	11589.444	249			

a. Predictors: (Constant), REGR factor score 3 for analysis 1, REGR factor score 2 for analysis 1, REGR factor score 1 for analysis 1
b. Dependent Variable: summaryoutcome

Coefficients[a]

Model		Unstandardized Coefficients		Standardized Coefficients	t	Sig.
		B	Std. Error	Beta		
1	(Constant)	27,332	,231		118,379	,000
	REGR factor score 1 for analysis 1	5,289	,231	,775	22,863	,000
	REGR factor score 2 for analysis 1	1,749	,231	,256	7,562	,000
	REGR factor score 3 for analysis 1	1,529	,231	,224	6,611	,000

a. Dependent Variable: summaryoutcome

However, even better sensitivity of testing might be obtained by an additional optimal scaling procedure.

3.2   Optimal Scaling with Spline Smoothing

Optimally Scaled Regression of the principal components variables will, subsequently, be performed. The Optimal Scaling program of SPSS was used for the purpose. It replaces continuous predictor variables with discrete predictor variables. SPSS uses spline smoothing. Splines are cut pieces of linear graphs, that, even with 2 degrees and 2 interior knots, already improves the data fit of the model used.

Command

Analyze....Regression....Optimal   Scaling....Dependent   Variable: enter Summaryoutcome (Define Scale: mark spline ordinal 2.2)....Independent Variables: enter REGR factor score 1, 2, and 3 (all of them Define Scale: mark spline ordinal 2.2)....Discretize: Method Grouping, Number categories 7)....click OK.

The underneath tables are in the output sheets.

Model Summary

Multiple R	R Square	Adjusted R Square	Apparent Prediction Error	Expected Prediction Error	
				Estimate[a]	Std. Error
.841	.708	.698	.292	.320	.039

Dependent Variable: summaryoutcome
Predictors: REGR factor score   1 for analysis 1 REGR factor score   2 for analysis 1 REGR factor score   3 for analysis 1

a. Mean Squared Error (10 fold Cross Validation).

ANOVA

	Sum of Squares	df	Mean Square	F	Sig.
Regression	174.773	8	21.847	71.988	.000
Residual	72.227	238	.303		
Total	247.000	246			

Dependent Variable: summaryoutcome
Predictors: REGR factor score   1 for analysis 1 REGR factor score   2 for analysis 1 REGR factor score   3 for analysis 1

Coefficients

	Standardized Coefficients		df	F	Sig.
	Beta	Bootstrap (1000) Estimate of Std. Error			
REGR factor score 1 for analysis 1	.720	.044	4	269.117	.000
REGR factor score 2 for analysis 1	.263	.045	2	34.075	.000
REGR factor score 3 for analysis 1	.256	.049	2	27.235	.000

Dependent Variable: summaryoutcome

The tables summarize unregularized optimal scaling of principal components analysis data. The unregularized principal components analysis produced slightly less sensitivity than the traditional linear regression does with R-values of 0.847 and 0.841, and F-values of 207.874 and 71.988. And, so optimal scaling was not beneficial here. However, after regularization using ridge, lasso or elastic net regression, the F-statistics may start being better than the unregularized F-statistics. Regularization can be defined as correcting discretized variables for overfitting, otherwise called overdispersion.

3.3   Optimal Scaling with Spline Smoothing Including Regularized Regression Using Either Ridge, Lasso, or Elastic Net Shrinkages

Ridge regression is an important method for shrinking b-values for the purpose of adjusting overdispersion. Lasso regression is slightly different from ridge regression, because it shrinks the smallest b-values to 0. Elastic net regression is again similar to lasso, but has been made suitable for larger numbers of predictors. In order to fully benefit from optimal scaling, a regularization procedure for the purpose of correcting overdispersion is needed. A ridge path model will be first used. SPSS statistical software is used.

Command
Analyze....Regression....Optimal   Scaling....Dependent   Variable: enter summaryoutcome (Define Scale: mark spline ordinal 2.2)....Independent Variables: enter REGR factor score 1, 2, and 3 (all of them Define Scale: mark spline ordinal

2.2)....Discretize: Method Grouping, Number categories 7)....click Regularization
....mark Ridge....click OK.

The output is the underneath graph and table.

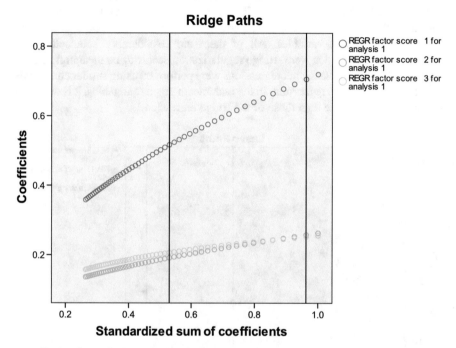

X-axis reference lines at optimal model and at most parsimonious model within 1 Std. Error.

Coefficients

	Standardized Coefficients		df	F	Sig.
	Beta	Bootstrap (1000) Estimate of Std. Error			
REGR factor score 1 for analysis 1	.516	.026	4	382.761	.000
REGR factor score 2 for analysis 1	.191	.030	4	39.695	.000
REGR factor score 3 for analysis 1	.206	.033	3	39.461	.000

Dependent Variable: summaryoutcome

The graph shows the adjusted b-values of the best fit scale model (left vertical line), the b-values are given in the table. The graph shows, how the b-value of different predictors gradually increase, as the shrinking factor λ decreases (from the left to right end of the graph). The right vertical line is the situation, where the spread in the data has increased by one standard error above the best model (left line), and this model has thus deteriorated correspondingly. The table shows the F-statistics of the three component variables. All of them are considerably better, than the unregularized F-statistics were. Ridge regularization seems to be meaningful. Subsequently, Lasso and elastic net regressions were performed using similar commands as the ones with Ridge regression. In the underneath graphs and tables, it is observed that results were worse than those of the Ridge regression.

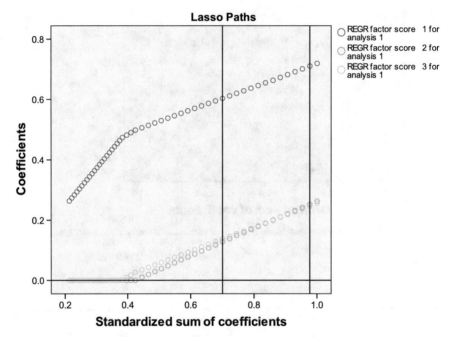

X-axis reference lines at optimal model and at most parsimonious model within 1 Std. Error.

Coefficients

| | Standardized Coefficients | | | | |
	Beta	Bootstrap (1000) Estimate of Std. Error	df	F	Sig.
REGR factor score 1 for analysis 1	.603	.041	4	219.017	.000
REGR factor score 2 for analysis 1	.127	.048	3	7.107	.000
REGR factor score 3 for analysis 1	.136	.050	3	7.578	.000

Dependent Variable: summaryoutcome

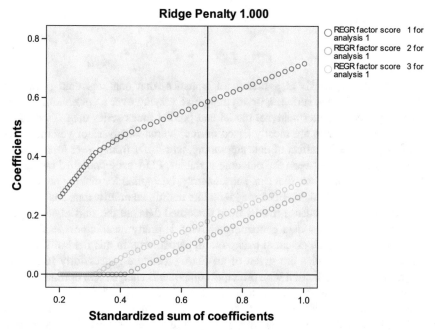

X-axis reference line at selected model (most parsimonious model within 1 Std. Error of optimal model).

Coefficients

	Standardized Coefficients		df	F	Sig.
	Beta	Bootstrap (1000) Estimate of Std. Error			
REGR factor score 1 for analysis 1	.587	.036	4	264.721	.000
REGR factor score 2 for analysis 1	.126	.052	4	5.793	.000
REGR factor score 3 for analysis 1	.180	.053	3	11.467	.000

Dependent Variable: summaryoutcome

4 Conclusion

Functional data analysis is a branch of statistics, that analyzes data providing information of curves, surfaces, or anything else varying over a continuum. Often, it is being used to better analyze, model and predict times series data. Traditional analyses of time series are mostly based on multivariate analyses of variance, that completely ignore the effect of data smoothing, and suffer from power loss due to a positive correlation between the outcome variables. FDA uses smoothed curves for obtaining novel functions, that can, subsequently, be applied for making predictions in general practice. And, so, FDA may look like traditional multivariate analysis, but it is more powerful, because results from smoothed data are the start of something new. FDA, if used as a data generating activity, is pretty meaningless, and sound prior hypotheses are an essential background, particularly, in this field of statistics. In this chapter examples are given of an FDA procedure successfully including: (1) data dimension reduction with principal components analysis, (2) optimal scaling of the reduced data with spline smoothing, (3) ridge, lasso, and elastic net regularization for increased precision, in a single analysis, and leading to a final analysis result with a better precision, than that of the traditional regressions of the same data, particularly if ridge regression has been applied.

Reference

To readers requesting more background, theoretical and mathematical information of computations given, several textbooks complementary to the current production and written by the same authors are available: Statistics applied to clinical studies 5th edition, 2012, Machine learning in medicine a complete overview, 2015, SPSS for starters and 2nd levelers 2nd edition, 2015, Clinical data analysis on a pocket calculator 2nd edition, 2016, Understanding clinical data analysis, 2017, all of them edited by Springer Heidelberg Germany

Chapter 26
Functional Data Analysis (FDA) Advanced

Mortality Hazards of Boys Born in 1960 in 23 Countries

Abstract FDA is an important tool for the analysis of curves, surfaces and volumes, especially when the data points in the curves are highly correlated, and when curves vary between individuals. The idea, to represent the curves using predefined or data-driven basis-functions, provides much flexibility, and overcomes the multicollinearity-problem. This problem can also be addressed by parametric models for the curves, e.g. using differential equations, but realistic models exist for only a very limited number of applications. Alternative statistical tools, such as random-effects or multilevel models, usually, lack the flexibility to deal with highly autocorrelated data with substantial variation in shape between individuals. As an example data from the Human Mortality Database were applied. Averaged mortality rates in West-, East- and three non-European countries for boys born in 1960 were computed. The FDA methodology was able, to compute mortality hazards in East-European versus other countries with the help of acceleration rates, and regression coefficients .

Keywords Functional data analysis · Predefined data functions · Data-driven basis functions · Overcoming the multicollinearity problem · Human mortality database

1 Introduction, History, and Background

Functional data analysis is a novel statistical methodology, implementing (1) smoothing techniques, (2) principal components analysis, and (3) optimal scaling into a single data analysis. In the Chap. 25 of this edition an example is given of a functional data analysis of the effects of gene expression levels on the efficacy drug treatment. In the current chapter the mortality hazards of boys born in 1960 in 23 countries will be analyzed.

© The Author(s), under exclusive license to Springer Nature Switzerland AG 2021
T. J. Cleophas, A. H. Zwinderman, *Regression Analysis in Medical Research*,
https://doi.org/10.1007/978-3-030-61394-5_26

Functional data analysis (FDA) is focused on the analysis of data of curves, surfaces, volumes, anything that varies over some continuum/dimension, or functions in general. Here, we will mainly use the term "curve", but the methods have much wider application, than to curves only. As with classical statistical models, the curves might both be the dependent variable, as well as the independent variable. In the former case, aims of the statistical analysis are, to describe mean, variation and other statistics of a sample of curves, and to analyze the association of the curves with predictor variables. A well known example concerns growth-curves of children, which might be different for boys or girls with various ethnic backgrounds. In case the curves are the independent variable, aspects of the curves are used to predict one or more outcome variables. An example are curves, surfaces or volumes in fMRI (functional magnetic resonance) images, that are predictive of, e.g., the onset of Alzheimer's disease.

General format of a data-set, that might be analyzed with FDA is a sample of N individuals, say patients, each of which is measured repeatedly. For infant growth-curves for instance, it is customary to define a birth-cohort of N newborns, follow them in time, and measure their lengths at various ages. The lengths (or weights) at different ages of a particular infant forms a <u>curve</u> of length (or weight) against time. So, the curve of infant i ($i = 1,...,N$) is formed by the n_i lengths (or weights) $\left(y_{i1}, y_{i2}, \ldots, y_{i,n_i}\right)$ at ages $t_{i1}, t_{i2}, ..., t_{i,n_i}$. With FDA, it is not needed to have balanced data: both the number of measurements (n_i) might vary between infants, as well as the ages ($t_{i1}, t_{i2}, ..., t_{i,n_i}$), at which the lengths/weights were measured, may vary between infants. With fMRI data, typically 3D-MRI images of a sample of N patients are analyzed, meaning that voxel-characteristics in the images are quantified over the x-, y- and z-dimension of the image. A characteristic, that varies in a patient between voxels over one dimension, forms a curve, and, if the characteristic varies over two or three dimensions, then surfaces or volumes are formed. So, in this fMRI case, the data of patient i ($i = 1, .., N$) is a three-dimensional matrix, where $y_{i,jkm}$ is the measurement of patient i in voxel at location (j,k,m) in 3D space. Like with repeated measurements in time, it is not necessary, that all locations (j,k,m) are the same in each patient, nor that there are equally many voxels in all patients.

Many classical statistical tools for single outcome data have been generalized to the analysis of curves, and this includes regression analysis, ANOVA and principal and canonical components analysis. Here we will address only few of these methods, for a general reference the reader is referred to the work of Ramsay and Silverman (2005).

2 Statistical Model

Key ideas of FDA are (1) that the data, i.e. the observed values ($y_{i1}, y_{i2}, ..., y_{i,n_i}$), may be rather noisy, and (2) that the <u>*true*</u> change-pattern of the measurements $y_{i1}, y_{i2}, ...,$ y_{i,n_i} over the relevant continuum has a smooth shape, which might vary between individuals. FDA often uses slopes, and curvatures of the curves, as are reflected in first-, second-, and higher-order derivatives of the curves. By using information in the derivatives FDA -in some sense- approximates purely functional models, as may be defined by differential equations.

The statistical model for smoothing repeated measures of individual i ($i = 1, ..,N$) at time points t_{i1}, t_{i2}, ..., t_{i,n_i} starts with a model, for instance, the following

$$y_i[t_{ij}] = f_i(x_i[t_{ij}]) + e[t_{ij}], \qquad (26.1)$$

where $y_i[t_{ij}]$ is the observed value of individual i at time t_{ij}, $f_i(x_i[t_{ij}])$ being the signal or true value of individual i at time t_{ij}, which may depend on covariate values, $x_i[t_{ij}]$, measured at time t_{ij}, and $\varepsilon[t_{ij}]$ is an independent, and identically distributed noise term, often assumed to be normally distributed with mean zero and unknown variance σ^2. The goal is, to estimate a smooth function $f_i(..)$, that best describes the observations (y_{i1}, y_{i2}, ..., y_{i,n_i}). There are many ways to estimate smooth functions, such as kernel smoothers, and splines, but in FDA, this is often done with basis-functions. This means, that the smooth function $\widehat{f}_i(x_i[t_{ij}])$ is modeled with an additive function of basis functions $\phi_k(..)$

$$\widehat{f}_i(x_i[t_{ij}]) = \sum_{k=1}^{K} c_{ik}\phi_k(x_i[t_{ij}]) \qquad (26.2)$$

where K is the number of basis-functions, and c_{ik} are coefficients to be estimated. The smoother-function $\widehat{f}_i(..)$ is found by minimizing the mean squared distance between the observations y_{i1}, y_{i2}, ..., y_{i,n_i} and their fitted values $\widehat{f}_{i1}(..), \widehat{f}_{i2}(..), ..., \widehat{f}_{i,n_i}(..)$ but penalized by the curvature of $f_i(..)$ as represented by the derivatives of $f_i(..)$:

$$\frac{1}{n_i} \sum_{j=1}^{n_i} (y_{ij} - f_i(x_i[t_{ij}]))^2 + \lambda_i \int_{x_{i1}}^{x_{i,n_i}} (f_i^m(x_i))^2 dx_i \qquad (26.3)$$

where $f_i^m(..)$ is the mth derivative of $f_i(..)$, and λ_i is a penalty parameter which has to be estimated from the data. Typically, m is taken to be 2, which means, that the smoother is of degree 2 m-1 = 3. Generally, at higher order the first term of formula (26.3) is often close to zero, but the smoother-function $f(..)$ is then often highly curved/irregular, and consequently the second term will than have a large value. At lower orders, the smoother-function $f(..)$ will be less close to the observation, thus the first term of (3) will be less small, but the function will be smooth, meaning, that the second term will be smaller. By varying the penalty-term λ the optimal smoothing can be found, that minimizes the sum of the first and second term.

The basis-functions $\phi_k(..)$ can have many different shapes, such as power functions, polynomial functions, sine- and cosine-functions, as in Fourier-transformations, or more general wavelet functions. Most common, however, are (cubic) B-splines, that also are often used in smoothing-splines.

3   An Example

3.1   Step 1 – Smoothing

The underneath figures illustrate the smoothing step for the yearly mortality hazards for a Dutch boy born in 1960 (data from the Human Mortality Database: www. mortality.org, accessed August 8, 2017). The first figure contains the observed hazards from 1960 to 2014, thus for age 0 till age 54 years of this boy. The red curve in this figure represents the smoothed hazards, using B-splines of order 6 with penalization of the fourth derivative with a small penalty-value of $\lambda = 0.01$. The second figure illustrates the residuals between observed, and fitted hazards, and the third and fourth figure give the estimated first, and second derivatives of the hazards-curve, which we denote as velocity and acceleration of the mortality hazards curve.

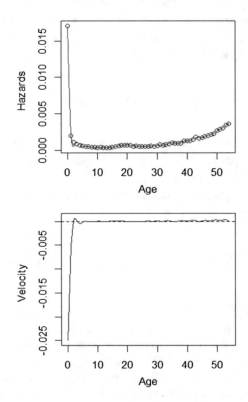

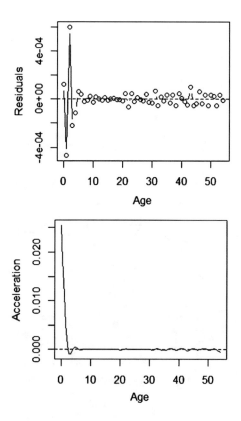

The analysis of the above four figures can be generalized to the hazards of boys born in 1960 in 23 different countries, as is illustrated in the next figures with respectively the observed hazards, the velocity-curves, the fitted hazards-curves, and the acceleration-curves.

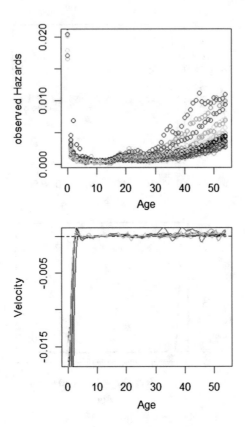

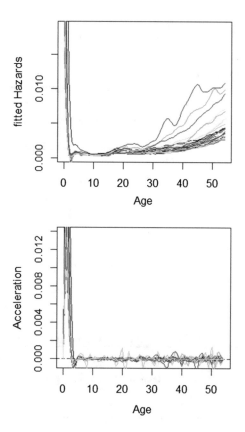

From about age 20–30 onwards, the mortality hazards in some countries are quite higher, than in most countries. But this is a rather smooth cumulative process, because neither the velocity- nor the acceleration-curves differ that much between the countries.

3.2   Step 2 – Functional Principal Components Analysis (FPCA)

Another way to illustrate variation across a sample of functions is by decomposition of the curves, using functional principal components. See figures underneath. This analysis yields eigenvalues and eigenfunctions, sometimes called harmonics, representing the most important sources of variation. For the above dataset of the mortality-hazards between 1960 and 2014 of boys born in 1960 in 23 countries, the relative eigenvalues or percentage explained variation of the first four components were 65.8%, 31.%, 1.4% and 0.6%. The screeplot, and scatterplot, the first two harmonic component scores, and the first two eigenfunctions of the mortality hazards data, are in the underneath two figures. Scores of the 23 countries on the first two-components are given in the underneath third and fourth figure. The first component seems to contrast the curves of especially Portugal (PT) and -perhaps-Spain (ES), while the second component seems to contrast the mortality-curves of East-European countries (Russia, Belarus, Hungary, Poland and Slovenia) versus curves of West-European countries, USA, Australia and Japan.

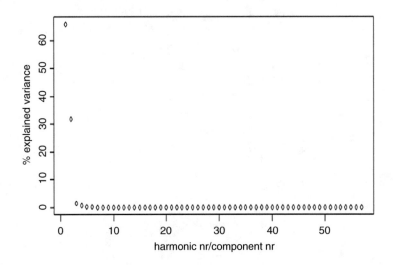

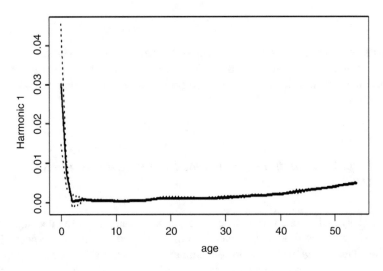

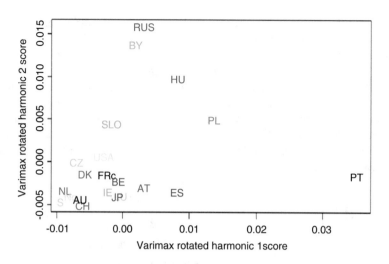

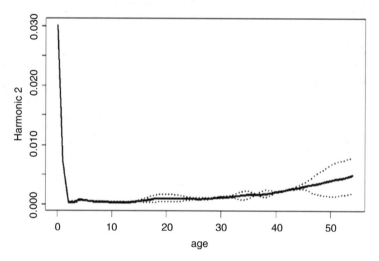

3.3   Step 3 – Regression Analysis

We finally grouped curves of East-European countries (grp 1: RUS, BY, HU, PL, SLO, CZ), West-European countries (group (grp) 2: NO, S, FI, DK, NL, BE, ES, PT, AT, FR, IE, LU, CH) and non-European countries (grp 3: JP, AU, USA), and performed functional regression of the curves on the grouping-variable to test, whether the curves are statistically different. The averaged mortality rates of the three groups of countries are shown in the underneath first figure with the associated acceleration curves in the second figure. The estimated regression coefficients with

95% pointwise confidence intervals for the East-European countries versus the other countries is shown in the third figure. Clearly, the confidence intervals exclude the zero-value for a large part of the age-range, and the mortality is therefore statistically significantly increased in the East-European countries, compared to that of the other countries. The acceleration curves are quite comparable between the three groups of countries, which indicate, that the mortality-curves have comparable shape over the age-range. Averaged mortality rates in West-, East- and the three non-European countries for boys born in 1960 are given. Shown are also the acceleration rates, and the regression coefficient (and 95% confidence intervals) of East-European versus the other countries.

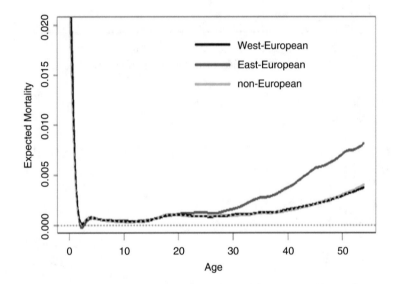

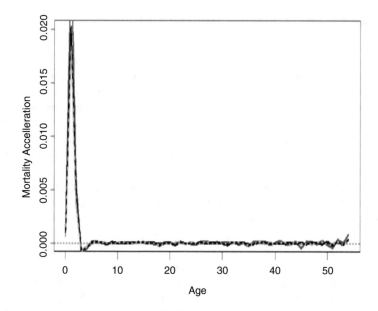

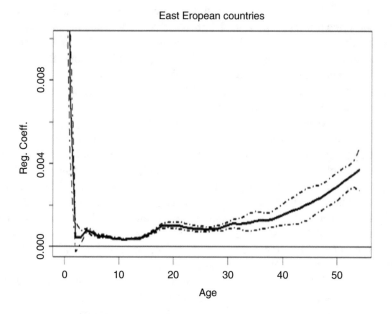

4   Applications in Medical and Health Research

FDA is a common statistical tool for the analysis of curve-data in biomedical research too. A recent paper (Lee et al. 2017) analyzed energy expenditure data of 109 children between ages 5-18 years with room respiration calorimeters. Energy expenditure (EE) was measured minute-by-minute in all children during 24 hours. So the data consisted of 24 hours*60 minutes = 1440 values per child, but in this example only measurements during sleep were analyzed, which reduced the curve to 405 measurements per child. This kind of data is highly variable, noisy with a strong autocorrelation pattern (see Fig. 1 of Lee et al. 2017). Using FDA, the general pattern, averaged over children, was, that during sleep EE first rapidly decreases during, about, 30 minutes, and, thereafter, further decreases, but at a slower rate until before wakening. This general pattern was seen in both obese, and non-obese children, although the obese children had lower mean EE throughout the sleep-period (See Fig. 4 of Lee et al. 2017).

Shen et al. (2017) used functional principal components analysis (FPCA) to analyze repeated systolic and diastolic blood pressure (BP) measurements in 1282 women during pregnancy with normotensive BP antenatally. The BP curve during pregnancy was formed by between 4 and 7 BP measurements per women during pregnancy. FPCA of these data showed, that three or four components were sufficient to describe the variation of the BP measurements, and that the first component explained already about 90% of the total variation. BP trajectories showed, that after an initial drop of -on average- about 8/10 mmHg diastolic/systolic BP in the first 4 weeks of pregnancy, BP steadily increased thereafter to about 4 mmHg more than it did antenatally (see Fig. 2, Shen et al. 2017). The immediate BP-drop, however, was found to depend on antenatal BP.

Dean et al. (2016) used functional principal components analysis and functional partial least squares (FPLS) to relate the occurrence of mucositis or dysphagia in 351 patients treated with radiation therapy for head and neck cancers to dose-volume-histograms (DVH) of the radiation therapy (RT) in organs at risk. The DVH is a curve, relating the varying RT dosage to the number of voxels, receiving those particular doses in an organ of interest. These histograms (0.2 Gy to 2.6 Gy at 0.2 Gy intervals: thus 13 points per patient) were available in all 351 patients, as well as, whether the patients suffered from mucositis or dysphagia at 8 weeks after RT. FPCA, and FPLS improved prediction (i.e. the c-index) of both RT side-effects in comparison to classical penalized logistic regression (see Table 1, Dean et al. 2016). The FPCA and FPLS models were, moreover, less complex in the sense, that only the first principal/PLS components were selected in the prediction models, whereas all 13 Dose-Volume variables of the DVH were selected in the penalized logistic model.

5 Conclusion

FDA is an important tool for the analysis of curves, surfaces and volumes, especially when the data points in the curves are highly correlated, and when curves substantially vary between individuals. The idea to represent the curves using predefined (e.g. splines or wavelets) or data-driven (e.g. FPCA) basis-functions provides much flexibility, and overcomes the multicollinearity-problem, which would be encountered, when the individual points of the curve were analyzed. This latter problem can also be addressed by parametric models for the curves, e.g. using differential equations, but realistic models exist for only a very limited number of applications. Alternative statistical tools, such as random-effects or multilevel models, usually, lack the flexibility to deal with highly autocorrelated data with substantial variation in shape between individuals.

FDA for the analysis of curves is less widely implemented as classical statistical tools for scalar observations. Most commonly used is the fda-library in the R statistical program. Like R itself, the package is freely available from the Comprensive R Archive Network (CRAN: www.r-project.org). The package contains many example datasets, and syntax for training of the ideas of FDA. R provides several other fda-libraries, offering extensions of the tools available in fda.

References

Dean JA, Wong KH, Gay H et al (2016) Functional data analysis applied to modeling of severe acute mucositis and dysphagia resulting from head and neck radiation therapy. Int J Radiation Oncol Biol Phys 96(4):820–831

Lee JS, Zakeri IF, Butte NF (2017) Functional data analysis of sleeping energy expenditure. PLoS One 12(5):e0177286

Ramsay JO, Silverman BW (2005) Functional data analysis, 2nd edn. Springer, New York

Shen M, Tan H, Zhou S et al (2017) Trajectory of blood pressure change during pregnancy and the role of pre-gravid blood pressure: a functional data analysis approach. Sci Rep 7

Chapter 27
Quantile Regression

A Form of Regression Particularly Suitable for Skewed Data like Current Covid-19 Data

Abstract With linear regression, if a normal distribution is not assumed, summaries consistent of quantiles like 0,1, 0,2, 0,3 etc. will be an adequate alternative for computing regression coefficients. They may be more precise predictors of the outcome than the traditional least square regression coefficients. In this edition SPSS statistical software has already often been applied for analyzing various regression models. The quantile regression methodology has been released by SPSS statistical software April 2020. This is fortunate, because readerships familiar with this software are currently much involved in health problems due to Covid-19, with skewed data due to excessive age-dependent morbidities and mortalities. There is much to say about quantile regression, and this chapter provides you with just a primer of quantile regression information. Yet it should be enough to get you started. Also a data example was applied to compare it with traditional linear regressions and robust linear regressions. Not only better precision, but also better insight into the relationships between predictor and outcome variables was obtained.

Keywords Continuous outcome regression · Median · Quantiles · Quantile regression · Linear regression · Robust regression · Generalized linear models

1 Introduction, History, and Background

Quantile regression does model the relationship between a set of predictor (independent) variables and a target (outcome) variable in the form of percentiles (otherwise called quantiles), instead of the traditional means. Most often the median, otherwise the 0,5 quantile, is applied. The median is the midpoint of a frequency distribution. If the frequency distribution is normal (Gaussian, also called parametric), the median will equal the mean value of the frequency distribution. If skewed, it will be computed as the value in the middle. For example,

1. the median of the values 1, 3, 3, 6, 7, 8, 9 will be the value 6,
2. the median of the values 1, 2, 3, 4, 5, 6, 8, 9 will be the value 4,5, it is the midpoint between 4 and 5.

Both, with the examples (1) and (2), the median is slightly different from the mean,

$$\text{mean } (1) = \sum \text{values}/n \text{ values} = 37/7 = 5,29 \text{ instead of } 6,00,$$

$$\text{mean } (2) = \sum \text{values}/n \text{ values} = 38/8 = 4,75 \text{ instead of } 4,5.$$

A quantile literally means a fraction of data, and is virtually identical to a percentile (percentage of the data). The median is the 0,5 quantile, but other quantiles are possible, for example the 0,1, 0,2, 0,3, 0,4 quantile etc. Quantile regression will usually apply the 0,5 quantile, instead of the mean of the data, but a better data-fit will be sometimes obtained with other quantiles.

The idea of estimating a median, instead of a mean regression slope, was proposed already in 1760 by Boskovicz, a Jesuit priest of Dubrovnik (Croatia), building on astrological theories of Newton (Cambridge UK), and he first used the term least absolute criterion, a precedent of the least squares model, as invented by Legendre (Paris France) in 1805. Also Laplace (Paris France) participated in the development of medians and the regression with medians. Median regression computations for larger data sets are quite tedious, as compared to the least squares method, for which reason it has historically generated little popularity among statisticians, until the widespread adoption of computers in the last part of the twentieth century. Another important contributor to quantile regression has been Roger Koenker (economist, University of Illinois 2001).

Quantile regression is a type of regression analysis, that is used, when you want to estimate the conditional median of the target (dependent) variable. Essentially, quantile regression is an extension of linear regression, and it is used, when you make no assumptions about the distribution of the residuals. Quantile regression helps you to obtain a more comprehensive analysis of the relationship between predictor and outcome variables. It can be used in ecology research, healthcare, risk management, and more. Quantile regression is included as a groundbraking feature of the 2020 version of SPSS statistical software, version 26. Particularly, big data models like the recent Covid-19 population data are skewed (Why corona-virus research data are skewed, www.bloomberg.com, May 1, 2020, and innumerable other papers on the internet), and may benefit from assumption-less models like quantile regressions, but have so far not been used for the purpose. This chapter will demonstrate, with the help of small data examples, how quantile regression are of excellent analytic support as compared to traditional and robust linear regression analysis.

2 Traditional Linear and Robust Linear Regression Analysis

In this Chap. a data example will be first analyzed with robust and traditional linear regression. Robust regression is more powerful than traditional linear regression. It has already been addressed in the Chap. 19. Robust regressions, as they are often based on medians rather than means, are, particularly, suitable for detecting and handling outlier data. Also they tend to better fit non-normal data like current Covid-19 data with excessive age-dependent mortality / morbidity risks. The Chap. 19 example of a robust linear regression of a very simple linear regression data model with just a single predictor and outcome variable will be used once again. Numbers of stools on a new laxative as outcome and the numbers stools on an old laxative as predictor were assessed with traditional linear regression analysis using SPSS statistical software. Simple linear regression produced a p-value of 0.049. The data file applied in this Chap. is called robustregressession.sav, and is in extras.springer. com. It consists of a crossover study of two treatments, an old treatment (bisacodyl) used as predictor, and a new treatment (a novel laxative) used as outcome variable. Both predictor and outcome variable are expressed as number of stools per month on treatment.

newtreat	oldtreat	agecats	patientno
24,00	8,00	2,00	1,00
30,00	13,00	2,00	2,00
25,00	15,00	2,00	3,00
35,00	10,00	3,00	4,00
39,00	9,00	3,00	5,00
30,00	10,00	3,00	6,00
27,00	8,00	1,00	7,00
14,00	5,00	1,00	8,00
39,00	13,00	1,00	9,00
42,00	15,00	1,00	10,00
41,00	11,00	1,00	11,00
38,00	11,00	2,00	12,00
39,00	112,00	2,00	13,00
37,00	10,00	3,00	14,00
47,00	18,00	3,00	15,00
30,00	13,00	2,00	16,00
36,00	12,00	2,00	17,00
12,00	4,00	2,00	18,00
26,00	10,00	2,00	19,00
20,00	8,00	1,00	20,00
43,00	16,00	3,00	21,00
31,00	15,00	2,00	22,00
40,00	114,00	2,00	23,00
31,00	7,00	2,00	24,00
36,00	12,00	3,00	25,00
21,00	6,00	2,00	26,00
44,00	19,00	3,00	27,00
11,00	5,00	2,00	28,00
27,00	8,00	2,00	29,00
24,00	9,00	2,00	30,00
40,00	15,00	1,00	31,00
32,00	7,00	2,00	32,00
10,00	6,00	2,00	33,00
37,00	14,00	3,00	34,00
19,00	7,00	2,00	35,00

newtreat = new treatment
oldtreat = old treatment
agecats = age categories
patientno = patient number

Start by opening the data file in your computer with SPSS installed.

Command
Analyze....Regression....Linear....Dependent: enter new laxative....Independent: enter old laxative....click OK.

Model Summary

Model	R	R Square	Adjusted R Square	Std. Error of the Estimate
1	,335[a]	,113	,086	9,53445

a. Predictors: (Constant), old treatment

ANOVA[a]

Modal		Sum of Squares	df	Mean Square	F	Sig.
1	Regression	380, 283	1	380, 283	4, 183	,049[b]
	Residual	2999, 888	33	90,906		
	Total	3380, 171	34			

a. Dependent Variable: new treatment

b. Predictors: (Constant), old treatment

Coefficients[a]

Model		Unstandardized Coefficients		Standardized Coefficients	t	Sig.
		B	Std. Error	Beta		
1	(Constant)	28,521	1,951		14,616	,000
	old treatment	,137	,067	,335	2,045	,049

a. Dependent Variable: new treatment

The above tables are in the output sheets. The old treatment is a borderline significant predictor at p = 0.049 of the new treatment. More statistical power is desirable. Instead of traditional linear regression also a GENLIN (Generalized linear regression-generalized linear regression) procedure can be followed in SPSS using maximum likelihood estimators instead of traditional F- and t-tests. The results are likely to produce a bit better precision.

Command
Generalized Linear Models....Generalized Linear Models....mark: Custom.... Distribution: select Normal....Link function: select identity....Response: Dependent Variable: enter new treatment....Predictors: Factors: enter old treatment....Model: Model: enter oldtreat....Estimation: mark Model-based Estimator....click OK.

The underneath table is in the output.

Parameter Estimates

Parameter	B	Std. Error	95% Wald Confidence Interval		Hypothesis Test		
			Lower	Upper	Wald Chi-Square	df	Sig.
(Intercept)	40,000	4,2650	31,641	48,359	87,958	1	,000
[oldtreat=4,00]	-28,000	6,0317	-39,822	-16,178	21,550	1	,000
[oldtreat=5,00]	-27,500	5,2236	-37,738	-17,262	27,716	1	,000
[oldtreat=6,00]	-24,500	5,2236	-34,738	-14,262	21,999	1	,000
[oldtreat=7,00]	-12,667	4,9248	-22,319	-3,014	6,615	1	,010
[oldtreat=8,00]	-15,500	4,7684	-24,846	-6,154	10,566	1	,001
[oldtreat=9,00]	-8,500	5,2236	-18,738	1,738	2,648	1	,104
[oldtreat=10,00]	-8,000	4,7684	-17,346	1,346	2,815	1	,093
[oldtreat=11,00]	-,500	5,2236	-10,738	9,738	,009	1	,924
[oldtreat=12,00]	-4,000	5,2236	-14,238	6,238	,586	1	,444
[oldtreat=13,00]	-7,000	4,9248	-16,652	2,652	2,020	1	,155
[oldtreat=14,00]	-3,000	6,0317	-14,822	8,822	,247	1	,619
[oldtreat=15,00]	-5,500	4,7684	-14,846	3,846	1,330	1	,249
[oldtreat=16,00]	3,000	6,0317	-8,822	14,822	,247	1	,619
[oldtreat=18,00]	7,000	6,0317	-4,822	18,822	1,347	1	,246
[oldtreat=19,00]	4,000	6,0317	-7,822	15,822	,440	1	,507
[oldtreat=112,00]	-1,000	6,0317	-12,822	10,822	,027	1	,868
[oldtreat=114,00]	0[a]	.	.	.	.	.	.
(Scale)	18,190[b]	4,3484	11,386	29,062			

Dependent Variable: new treatment
Model: (Intercept), oldtreat

a. Set to zero because this parameter is redundant.

b. Maximum likelihood estimate.

Still better precision can be obtained by the use of **robust** standard errors, i.e., Hubert-White estimators for linear models.

Command

Generalized Linear Models....Generalized Linear Models....mark: Custom.... Distribution: select Normal....Link function: select identity.... Response: Dependent Variable: enter new treatment....Predictors: Factors: enter old treatment....Model: Model: enter newtreat....Estimation: mark Robust Estimator....click OK.

The underneath table is in the output.

Parameter Estimates

Parameter	B	Std. Error	95% Wald Confidence Interval		Hypothesis Test		
			Lower	Upper	Wald Chi-Square	df	Sig.
(Intercept)	40,000	.	.	.	.	1	,000
[oldtreat=4,00]	-28,000	.	.	.	.	1	,000
[oldtreat=5,00]	-27,500	1,0607	-29,579	-25,421	672,222	1	,000
[oldtreat=6,00]	-24,500	3,8891	-32,122	-16,878	39,686	1	,000
[oldtreat=7,00]	-12,667	3,4102	-19,351	-5,983	13,796	1	,000
[oldtreat=8,00]	-15,500	1,4361	-18,315	-12,685	116,485	1	,000
[oldtreat=9,00]	-8,500	5,3033	-18,894	1,894	2,569	1	,109
[oldtreat=10,00]	-8,000	2,1506	-12,215	-3,785	13,838	1	,000
[oldtreat=11,00]	-,500	1,0607	-2,579	1,579	,222	1	,637
[oldtreat=12,00]	-4,000	.	.	.	.	1	,000
[oldtreat=13,00]	-7,000	2,4495	-11,801	-2,199	8,167	1	,004
[oldtreat=14,00]	-3,000	.	.	.	.	1	,000
[oldtreat=15,00]	-5,500	3,4369	-12,236	1,236	2,561	1	,110
[oldtreat=16,00]	3,000	.	.	.	.	1	,000
[oldtreat=18,00]	7,000	.	.	.	.	1	,000
[oldtreat=19,00]	4,000	.	.	.	.	1	,000
[oldtreat=112,00]	-1,000	.	.	.	.	1	,000
[oldtreat=114,00]	0[a]	.	.	.	.	.	.
(Scale)	18,190[b]	4,3484	11,386	29,062			

Dependent Variable: new treatment
Model: (Intercept), oldtreat

a. Set to zero because this parameter is redundant.
b. Maximum likelihood estimate.

Out of the stool scores with old treatment, 6 scores produced p-values of <0,05 with the Model-based Estimator, while it produced up to 14 p-values <0,05 with the Robust Estimator. Most of the p-values even as small as <0,0001.

If your results are borderline significant like in the above case, then loglikelihood regression testing and robust regression testing can, obviously, provide you with better statistics, and, thus, better statistical power of testing, than traditional testing can. Similarly better statistics can be obtained with the help of quantile regression, and, in addition, a better insight in the relationships between predictor and outcome variables.

3 Quantile Linear Regression Analysis

Quantiles (fraction of the data) and percentiles (percentage of the data) indicate almost the same. Usually, linear regression assumes, that the y-variable has a normal distribution, and can be summarized by means. The least square computation of the regression coefficient is obtained by the use of means of the x- and the y-values (Chap.1 of this edition).

Y_m = mean of observed Y-values.
X_m = mean of observed X-values.

$$B = \frac{\sum (X\text{-}X_m)(Y\text{-}Y_m)}{\sum (X\text{-}X_m)} = \frac{\sum XY\text{-}n\,X_m Y_m}{\sum X^2\text{-}nX_m^2}$$

If a normal distribution is not assumed, summaries of quantiles, like the 0,10, 0,20, 0,30 quantile etc. will be an adequate alternative for computing regression coefficients. The data file called robustregressession.sav as used in the above section, is used again.

newtreat	oldtreat	agecats	patientno
24,00	8,00	2,00	1,00
30,00	13,00	2,00	2,00
25,00	15,00	2,00	3,00
35,00	10,00	3,00	4,00
39,00	9,00	3,00	5,00
30,00	10,00	3,00	6,00
27,00	8,00	1,00	7,00
14,00	5,00	1,00	8,00
39,00	13,00	1,00	9,00
42,00	15,00	1,00	10,00
41,00	11,00	1,00	11,00
38,00	11,00	2,00	12,00
39,00	112,00	2,00	13,00
37,00	10,00	3,00	14,00
47,00	18,00	3,00	15,00
30,00	13,00	2,00	16,00
36,00	12,00	2,00	17,00
12,00	4,00	2,00	18,00
26,00	10,00	2,00	19,00
20,00	8,00	1,00	20,00
43,00	16,00	3,00	21,00
31,00	15,00	2,00	22,00
40,00	114,00	2,00	23,00
31,00	7,00	2,00	24,00
36,00	12,00	3,00	25,00
21,00	6,00	2,00	26,00
44,00	19,00	3,00	27,00
11,00	5,00	2,00	28,00
27,00	8,00	2,00	29,00
24,00	9,00	2,00	30,00
40,00	15,00	1,00	31,00
32,00	7,00	2,00	32,00
10,00	6,00	2,00	33,00
37,00	14,00	3,00	34,00
19,00	7,00	2,00	35,00

newtreat = new treatment
oldtreat = old treatment
agecats = age categories
patientno = patient number

Start by opening the file in a computer with SPSS statistical software version 26 installed.

Command
Analyze....Regression....Quantile Regression....click Target Variable: enter new treatment....click Covariate(s): enter old treatment....click Criteria....mark Specify single quantiles....value(s): enter 0,1, 0,2, 0,3, 0,4, 0,5, 0,6, 0,7, 0,8, 0,9....various

criteria can be left as already indicated....click Continue...click Display....Print mark parameter estimates....mark Plot or tabulate top 3 effects....in Model Effects move old treat to Prediction Lines....click Continue....click OK.

In the output sheets are the underneath interactive tables and graphs.

Quantile Regression

Model Quality[a,b,c]

	q=0,1	q=0,2	q=0,3	q=0,4	q=0,5	q=0,6	q=0,7	q=0,8	q=0,9
Pseudo R Squared	.138	.116	.098	.080	.077	.044	.033	.011	.018
Mean Absolute Error (MAE)	14.6155	10.2765	8.9266	7.9458	7.4857	8.0712	8.5923	9.5918	12.4643

a. Dependent Variable: new treatment

b. Model: (Intercept), old treatment

c. Method: Simplex algorithm

Parameter Estimates by Different Quantiles[a,b]

Parameter	q=0,1	q=0,2	q=0,3	q=0,4	q=0,5	q=0,6	q=0,7	q=0,8	q=0,9
(Intercept)	12.832	19.981	22.846	26.019	29.636	34.51	Double-click to activate	38.914	36.875
old treatment	.234	.170	.144	.123	.091	.04		.010	.375

a. Dependent Variable: new treatment

b. Model: (Intercept), old treatment

Model Quality[a,b,c]

	q=0,1	q=0,2	q=0,3	q=0,4	q=0,5	q=0,6	q=0,7	q=0,8	q=0,9
Pseudo R Squared	.138	.116	.098	.080	.077	.044	.033	.011	.018
Mean Absolute Error (MAE)	14.6155	10.2765	8.9266	7.9458	7.4857	8.0712	8.5923	9.5918	12.4643

a. Dependent Variable: new treatment

b. Model: (Intercept), old treatment

c. Method: Simplex algorithm

Quantile = 0,1

Parameter Estimates[a,b]

Parameter	Coefficient	Std. Error	t	df	Sig.	95% Confidence Interval	
						Lower Bound	Upper Bound
(Intercept)	12.832	2.9206	4.394	33	.000	6.890	18.774
old treatment	.234	.1002	2.331	33	.026	.030	.438

a. Dependent Variable: new treatment

b. Model: (Intercept), old treatment

Quantile = 0,2

Parameter Estimates[a,b]

Parameter	Coefficient	Std. Error	t	df	Sig.	95% Confidence Interval	
						Lower Bound	Upper Bound
(Intercept)	19.981	3.4406	5.808	33	.000	12.981	26.981
old treatment	.170	.1181	1.438	33	.160	-.070	.410

a. Dependent Variable: new treatment

b. Model: (Intercept), old treatment

Quantile = 0,3

Parameter Estimates[a,b]

Parameter	Coefficient	Std. Error	t	df	Sig.	95% Confidence Interval	
						Lower Bound	Upper Bound
(Intercept)	22.846	3.4147	6.691	33	.000	15.899	29.793
old treatment	.144	.1172	1.231	33	.227	-.094	.383

a. Dependent Variable: new treatment

b. Model: (Intercept), old treatment

Quantile = 0,4

Parameter Estimates[a,b]

Parameter	Coefficient	Std. Error	t	df	Sig.	95% Confidence Interval	
						Lower Bound	Upper Bound
(Intercept)	26.019	3.7158	7.002	33	.000	18.459	33.579
old treatment	.123	.1275	.962	33	.343	-.137	.382

a. Dependent Variable: new treatment

b. Model: (Intercept), old treatment

Quantile = 0,5

Parameter Estimates[a,b]

Parameter	Coefficient	Std. Error	t	df	Sig.	95% Confidence Interval	
						Lower Bound	Upper Bound
(Intercept)	29.636	2.8514	10.394	33	.000	23.835	35.438
old treatment	.091	.0979	.929	33	.360	-.108	.290

a. Dependent Variable: new treatment

b. Model: (Intercept), old treatment

Double-click to activate

Quantile = 0,6

Parameter Estimates[a,b]

Parameter	Coefficient	Std. Error	t	df	Sig.	95% Confidence Interval	
						Lower Bound	Upper Bound
(Intercept)	34.519	2.6744	12.907	33	.000	29.078	39.960
old treatment	.048	.0918	.524	33	.604	-.139	.235

a. Dependent Variable: new treatment

b. Model: (Intercept), old treatment

Quantile = 0,7

Parameter Estimates[a,b]

Parameter	Coefficient	Std. Error	t	df	Sig.	95% Confidence Interval	
						Lower Bound	Upper Bound
(Intercept)	36.580	2.3472	15.584	33	.000	31.805	41.355
old treatment	.030	.0806	.372	33	.712	-.134	.194

a. Dependent Variable: new treatment

b. Model: (Intercept), old treatment

Quantile = 0,8

Parameter Estimates[a,b]

Parameter	Coefficient	Std. Error	t	df	Sig.	95% Confidence Interval	
						Lower Bound	Upper Bound
(Intercept)	38.914	1.6904	23.021	33	.000	35.475	42.353
old treatment	.010	.0580	.164	33	.871	Double-click to activate	.128

a. Dependent Variable: new treatment

b. Model: (Intercept), old treatment

Quantile = 0,9

Parameter Estimates[a,b]

Parameter	Coefficient	Std. Error	t	df	Sig.	95% Confidence Interval	
						Lower Bound	Upper Bound
(Intercept)	36.875	1.9573	18.840	33	.000	32.893	40.857
old treatment	.375	.0672	5.583	33	.000	.238	.512

a. Dependent Variable: new treatment

b. Model: (Intercept), old treatment

The above parameter estimates show the magnitude of the selected quantiles. Out of them the quantiles 0,1 and 0,9 produced p-values of 0,026 and 0,0001.

Plot of the Estimated Parameters

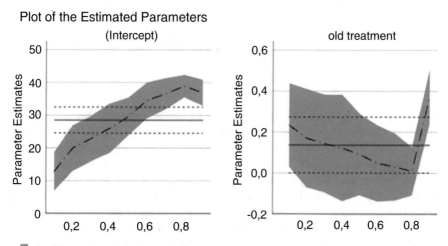

■ Confidence intervals of the parameter estimates

-- Parameter estimates at the different regression quantiles

— Parameter estimates for the ordinary linear regression with the same predictors

···· Confidence intervals bounds for the ordinary linear regression with the same predictors

The above graph shows the patterns of parameter estimates with their 95% confidence intervals. Also the confidence intervals of the traditional linear regression of the same data are given. The graph shows, that, particularly, in the low and very high quantiles, the quantile regression performed better than did the traditional method.

Prediction: old treatment

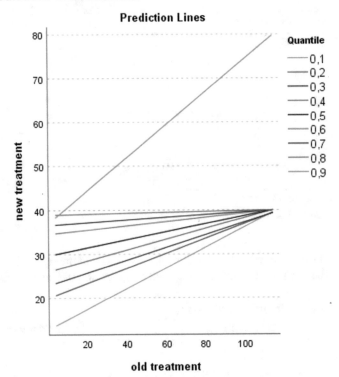

The above graph gives the patterns of the quantile regression lines. It can be observed, that the slopes of the 0,9 quantile, and that of the 0,1 quantile give the steepest slopes. We should add, that in the example of this chapter, both traditional linear regression and robust linear regression, did not provide the amount of information, and of statistical power, that quantile regression did. Also insight in the pattern profiles of quantile regression coefficients and the p-values estimating difference from zero were obtained. Problematic regression are often due to non-normal frequency distributions, and outliers. We should add, that another method for improving precision of problematic regression analyses is the weighted least square methodology as explained in the Chap. 21. It is a kind of poor man's quantile regression, less complex and flexible, and only based on adjustment for heteroscedasticity.

4 Conclusion

There is much more to say about quantile regression, such as how the coefficients, and their standard errors are estimated. Also comparing models and assessing nonlinear quantile regressions are possibilities. This chapter is just a primer but will be enough to get you started. We should add, that, both traditional linear regression and robust linear regression, did not provide the amount of information and of statistical power, that quantile in this small example did, including pattern profile of quantile regression coefficients and the p-values estimating the difference from a regression coefficient of zero. We should add, that another method for improving precision of your regression analysis is the weighted least square methodology as explained in the Chap. 21. It is less complex, and based on adjustment for heteroscedasticity, rather than replacement of the means with quantile data for data summaries. The authors believe, that quantile regression will be an important step forward in the analysis methodology of current Covid-19 data.

Reference

To readers requesting more background, theoretical and mathematical information of computations given, several textbooks complementary to the current production and written by the same authors are available: Statistics applied to clinical studies 5th edition, 2012, Machine learning in medicine a complete overview, 2015, SPSS for starters and 2nd levelers 2nd edition, 2015, Clinical data analysis on a pocket calculator 2nd edition, 2016, Understanding clinical data analysis, 2017, all of them edited by Springer Heidelberg Germany

Index

© The Author(s), under exclusive license to Springer Nature Switzerland AG 2021
T. J. Cleophas, A. H. Zwinderman, *Regression Analysis in Medical Research*,
https://doi.org/10.1007/978-3-030-61394-5

Printed in the United States
by Baker & Taylor Publisher Services